Drones for Transportation Logistics and Disaster Management

Scrivener Publishing
100 Cummings Center, Suite 541J
Beverly, MA 01915-6106

Publishers at Scrivener
Martin Scrivener (martin@scrivenerpublishing.com)
Phillip Carmical (pcarmical@scrivenerpublishing.com)

Drones for Transportation Logistics and Disaster Management

Edited by

A. Prasanth

Rajesh Kumar Dhanaraj

Munish Sabharwal

Vandana Sharma

and

Seifedine Kadry

Scrivener Publishing

WILEY

This edition first published 2026 by John Wiley & Sons, Inc., 111 River Street, Hoboken, NJ 07030, USA and Scrivener Publishing LLC, 100 Cummings Center, Suite 541J, Beverly, MA 01915, USA

For more information about Scrivener publications please visit www.scrivenerpublishing.com.

Wiley Global Headquarters
111 River Street, Hoboken, NJ 07030, USA

For details of our global editorial offices, customer services, and more information about Wiley products visit us at www.wiley.com.

The manufacturer's authorized representative according to the EU General Product Safety Regulation is Wiley-VCH GmbH, Boschstr. 12, 69469 Weinheim, Germany, e-mail: Product_Safety@wiley.com.

Library of Congress Cataloging-in-Publication Data

ISBN 9781394386420

Cover image: Generated with AI using Adobe Firefly
Cover design by Russell Richardson

Set in size of 11pt and Minion Pro by Manila Typesetting Company, Makati, Philippines

Printed in the USA

10 9 8 7 6 5 4 3 2 1

Contents

Preface

In most inaccessible and emergency situations, drones have played an important role. The application of drones is relevant in various disastrous and commercial sectors. But the implementation of drones for extended services in logistics and disastrous situations requires standards, policies, and procedures to be strictly implemented. Also, future drones are projected to have many real-time advanced applications as government regulations evolve. The implementation of current drones in several industrial and service sectors has extended the possibility of handling transportation and logistics. As the future of transportation is based on unmanned vehicles it is indeed important to identify the challenges involved and identify futuristic applications of drones.

The edited book discusses several important aspects and surveys the recent technological advancements of future drones, the challenges faced in current practices, and their advanced applications. Further, it describes the future intelligent and resilient transportation systems backed with the Internet of Vehicle Things, the problem faced by Big Data Analytics and their optimization techniques for inhouse supply-chain management. Also, establishing a framework for handover management on the Internet of Drone-based networks is a daunting task. The role of 5G communications for handling communication in drones, its related security, and legal aspects raises concerns and are addressed here. The perspective of all the stakeholders is discussed using case studies of the existing applications at the global level and with the inclusion of industrial use cases.

The edited book will take you through essential components of Drone systems such as concepts, methodologies, technologies, modeling, designs, statistics, implementations, and maintenance. You'll use real-life technologies such as smart devices, virtual systems, IoT-based technologies, AI-centric solutions, smart sensors, and a Telecom network to train and evaluate simple to complex models for building the Drone systems.

The key aim of the edited book is to provide relevant concepts, theoretical frameworks, emerging technologies, analytics techniques, new models, and the latest empirical research findings in the area of smart cities, Transportation Logistics, Disaster Management, smart devices, smart solutions, etc. It will be written for professionals who want to improve their understanding of the strategic role of trust at different levels of the information and knowledge society that is a trust at the level of global communication, economy, education, and organizations as actors in the Drone systems. By the end of this book, you'll be able to learn and design an end-to-end solution for Transportation Logistics and Disaster Management with the aid of smart drones.

1

Journey to Transportation and Logistics Management Using Drone: Digitization and Technological Evolution

Seema Rani[1], Misbah Anjum[1]*, Shivangi Chawla[2] and Vandana Sharma[3]

[1]Amity Institute of Information Technology, Amity University, Noida, India
[2]Department of Computer Science, Jamia Millia Islamia, Delhi, India
[3]Computer Science Department, Christ University, Bengaluru, India

Abstract

The speedy adoption and amalgamation of drone technology in every sector of life have correlated environmental implications. It has come to reality due to revolutionization on the technology front and the movement for a digitized world by adopting digitization in the course of action. The drones have the capacity to decrease the rate of carbon emissions at significant levels in transportation and logistics management. So, this is becoming the need of an hour to apply the technology for the well-being of an individual. The comprehensive study and assessment in this chapter will address the implications of an environment connected to drone-based transportation and logistics operations. It will comprise the assessment based on academic studies, industry reports, and environmental assessments in terms of carbon emissions, energy consumption, and ecological footprint. The chapter will highlight the concerns and contributions in view of mitigating the environmental implications linked with the deployment of drones in transportation and logistics, enabling stakeholders to develop strategies that foster sustainability in the industry.

***Keywords*:** Drones, environment, transportation, logistics, sustainability, industry

**Corresponding author*: bhatmisbah6@gmail.com
Vandana Sharma: ORCID: 0000-0002-8636-2365

A. Prasanth, Rajesh Kumar Dhanaraj, Munish Sabharwal, Vandana Sharma and Seifedine Kadry (eds.) Drones for Transportation Logistics and Disaster Management, (1–24) © 2026 Scrivener Publishing LLC

1.1 Introduction

Rapid technological improvements have revolutionized several industries, including transportation and logistics management. Globally commerce and economic growth depend critically on the logistics and transportation sectors. Digital technologies are becoming ever more important in the transport and logistics sectors as technology developments continue to change several sectors. As technology breakthroughs continue to transform numerous industries, digital technologies are becoming more important in transportation and logistics. These technologies have the potential to transform how items are moved, tracked, and managed, resulting in greater efficiency and cost-effectiveness. Drones are an emerging technology in the transportation and logistics business. Drones are among the most promising and disruptive technologies in recent years.

Drones, also known as Unmanned Aerial Vehicles (UAVs), have swiftly progressed from being mostly used for recreational purposes to being an important tool in transportation and logistics management [1]. These small, remote-controlled aircraft have the potential to simplify and enhance various aspects of supply chain management, such as keeping track of inventory, warehouse management, and last-mile delivery [2]. The Journey to Transportation and Logistics Management Using Drones examines how drones may be used to reform key supply chain phases. This journey delves into the technological aspects of drone design and operation, as well as the logistical issues required for successfully integrating drones into existing transportation and distribution networks. This study offers light on how drones boost productivity, lower costs, and reduce environmental impact in domains ranging from automated storage and inventory management to last-mile delivery in both urban and rural areas.

Traditional transportation and logistics management has suffered from congestion, high operating costs, and time-consuming processes. Drone technology addresses these issues in a unique way by providing an efficient and versatile form of transportation [3]. Drones have various advantages, including the ability to fly autonomously, travel fast between places, and avoid road network and congestion limits [4]. The advantages of drone technology have created new opportunities in transportation and logistics, defense, agriculture, and public safety. Industry leaders and researchers are interested in this technology because of its potential to transform how things are moved and managed across the supply chain.

Drone transportation and logistics management are intriguing examples of digitalization and technological innovation. The combination of drone technology and internet platforms creates limitless opportunities for information exchange, allowing decision-makers to make more informed decisions, optimize routes, and reduce environmental impact. This convergence of technology creates an enthralling frontier in which age-old difficulties are tackled head-on with ground-breaking solutions, revolutionizing the very nature of transport and logistics. At the center of this technical growth is the critical subject of sustainability.

With the urge to bring down the carbon footprints and embrace eco-friendly in today's time becoming the emerging necessity for realization of drone ideas in reality as guiding light. Their extraordinary capability and operational functionality circumnavigate formidable scenarios and dense locations or cities position them as sustainable substitutes to conventional means of transportation. The amalgamation of drones seamlessly fixes the need of an hour with environment friendly means in the transportation sector. Drone technological developments, legal reforms, and the incorporation of digital tools have all contributed to this transition. The following subsections will brief about development and advancements along with concerns in the context of sustainability.

The future of Transportation Logistics and Disaster Management with Drones is covered in later chapters. "Challenges of Big Data Implementation in Drones-based Logistics" examines the rise of drones to tackle e-commerce last-mile delivery issues. Big Data's function in logistics efficiency, drone hazards, and optimization, as well as research trends, is examined.

Practical Drone Logistics Big Data Methods. Next, the "Frameworks for Handover Management for the Networks of Future Drones" tackles HOM-FDN and aerial communication. In the ever-growing field of drone technology, effective transfers across network nodes optimize resources and ensure continuous, productive drone operations. "Sustainable Practices for Energy Optimization" in UAV Systems discusses HOM-FDN, FDN, and aerial communication. The growing field of drone technology requires efficient network node transfers to maximize resource use and drone productivity. "Convergence of Internet of Vehicle Things and Drones" emphasizes drone-IoVT compatibility across disciplines.

"5G Communication in Drones for Surveillance" covers 5G integration, drone network design, security and privacy, legislative, and ethical issues, and future trends and advances. The "Impact and Assessment of Artificial Intelligence- enabled UAV for Real-Time Data Streaming Application" will discuss UAV technology, applications across industries, data analytics implications, problems and limitations, and ethical and regulatory consequences. In "Blockchain-Based Security and Privacy Solutions for Drones Systems", blockchain technology is used to secure drone communication.

A modular UAV with an amphibious landing and edging sense segment is discussed in "Design and Development of Modular and Multifunctional UAV for Amphibious Landing and Surround Sense Module".

"Implementing Mission Critical Public Safety through Drone Network Communication" examines public safety drone communications. Remote areas, hazardous conditions, and slow reaction times hinder traditional public safety operations. In "Assessing the Impact of Drones on Students' Engagement and Learning Outcomes" the current study design, participant selection, and data collecting for drone-based student learning outcomes assessment will be covered. "Implementation of Delivery Drones in Logistics Business Process" discusses how delivery drones could change logistics. In "Social and Ethical Issues of Using Drones in Logistics and Disaster Management" the community, technology, environment, health, and safety ethical issues are examined.

The "Regulatory Overview and Challenges of Drone Technology in Safety Regulation" examines drone safety regulations and issues. For an understanding of drones in transportation and logistics, the case studies on "Drones-based Application of Warehouse Logistics", "Data-Driven Emergency Response Management for UAV- based Future Transportation", and "Remote Assessments and Aerial Imaging Using UAV for Disaster Management and Precision Agriculture with Immediate Response" are discussed.

1.1.1 Drone Development and Advancements

Drones were mainly utilized for military objectives. However, drones have become smaller, more inexpensive, and easier to control as technology has advanced. Drone design, battery life, propulsion systems, and sensor advancements have allowed drones to transport bigger payloads over longer distances. The incorporation of Global Positioning System (GPS), Light Detection and Ranging, cameras, and other sensors increased navigation, obstacle avoidance, and data collecting [5].

Developing Regulations and Revolutionary Uses

Drones were initially used in logistics for mapping, aerial photography, and surveying, paving the path for their application in more complex activities. Companies began experimenting with drones for last-mile delivery, particularly in distant or difficult-to-reach places, and then expanded their usage in warehouses for asset monitoring, stock tracking, and inventory management. In parallel, governments and aviation organizations developed regulatory frameworks to ensure the safety of drone operations in civilian airspace. These organizations make it simpler to utilize commercial drones by developing certification, licensing, and operating guidelines. Drones' logistical capabilities have been substantially enhanced by legislative advancements, such as the authorization of operations beyond visual lines of sight.

Integration of Technology and Environmental Advantages

Activities involving drones were altered by the introduction of automation technologies such as Artificial Intelligence (AI) and Machine Learning (ML), which enabled autonomous activities such as obstacle recognition, route optimization, and real-time decision-making. Prominent companies such as Amazon, UPS, and DHL have successfully tested drone delivery services for emergency response, medical supplies, and consumer products. Drones have also become an important component of supply chain management systems, easing warehouse operations and enabling real-time freight tracking. Drones helped to minimize carbon emissions by optimizing delivery routes and lowering reliance on traditional ground transportation. These sustainability benefits were amplified by electric drones, demonstrating how critical it is to address environmental concerns in order to ensure the industry's long-term viability.

1.1.2 Sustainability Concern

Drone technology has both beneficial and negative environmental implications, which must be carefully examined to ensure long-term sustainability. One of the most pressing concerns is energy consumption, as drones' carbon footprint is heavily reliant on the energy source used for charging. Drones may contribute to environmental degradation if they are powered by non-renewable sources. Furthermore, improper management of drone batteries can pollute land and water, making their disposal an environmental hazard. Noise pollution, especially in metropolitan areas, is another issue that has an impact on inhabitants' quality of life and disrupts ecosystems. Drone usage may potentially increase airspace congestion, jeopardizing air quality and safety and necessitating effective management and control measures.

Drones have a big positive impact on the environment. They can significantly reduce carbon emissions compared to traditional transportation by optimizing routes and avoiding traffic, particularly in last-mile delivery circumstances. Drones may use advanced algorithms and real-time data to determine the most efficient routes, reducing fuel consumption and environmental impact. The use of renewable energy sources, such as solar-powered charging stations, helps to improve sustainability even further. Furthermore, precision agriculture depends largely on drones to optimize resources and reduce the environmental effect of traditional farming. Furthermore, sensor-equipped drones can detect emissions from industry and transportation, providing authorities with valuable information for enforcing rules and regulating pollution, eventually promoting improved environmental practices.

While the use of drones in transportation and logistics raises valid environmental issues, their advantages to sustainability should not be overlooked. Through strategic planning, innovation, and a commitment to ethical practices, stakeholders may utilize drone technology's transformational potential to build a more sustainable future for the industry. Industry leaders, legislators, and innovators must work together to achieve a balance between efficiency and environmental responsibility in the ever-changing landscape of transportation and logistics management.

In this chapter, we look at how digitization and technical advancements have permitted the incorporation of drones into transportation and logistics management. We will look at the important developments that have opened the path for their use in this industry, as well as the advantages, obstacles, and revolutionary influence they have on global supply chains. The rest of the chapter is structured subsequently.

1.2 Literature Review

Drone technology, digitalization, and the larger panorama of technological growth have all contributed to substantial changes in transportation and logistics management. Drones, also known as UAVs, are becoming increasingly important in improving the efficiency, cost-efficiency, and sustainability of supply chain operations. Several studies have been conducted to analyze the present state of drone technology in transportation and logistics, highlighting its benefits and drawbacks. The interaction of digitization and sustainability has also been examined, particularly with reference to data created in transportation logistics. Furthermore, the potential utility of drones in medical logistics has been acknowledged, particularly considering the COVID-19 pandemic [6].

The performance of retail logistics using drone technology has been analyzed, exposing limitations. Drone integration in logistics is rapidly gaining popularity. Drones are being used for last-mile deliveries, which reduces delivery times and costs [7]. They emphasize the potential of drones to transform the e-commerce business. Drone technology is advancing, with an emphasis on payload capacity, endurance, and autonomy. *J. Anderson and R. Anderson* (2020) researched the development of drone technology and its impact on the logistics business [8]. They discuss how to employ AI and machine learning algorithms to enhance route planning and decision-making.

With extended use of drones in multidisciplinary fields for successful planning in a complicated environment, a lot of regulatory concerns become apparent. This demands a crucial regulatory framework to empower secure and robust drone operations [9]. The agreement and compliance to laws will play a vital role in promising application of drones in transportation. Drone technology aids in offering sustainable logistics by diminishing the carbon emission. Researchers examine the impact on the environment with drones supported logistics operation and revealing how they fit into the larger trend of green logistics [10]. The study [11] examined the function of data analytics and IoT in drone logistics. Further they uncovered how data generated from drones in application boost and progress the planning activity and inventory management. The summarization of researchers' key findings and extension suggested in the domain of application and evolution of drone technology is presented in Table 1.1.

1.3 Fundamental Elements of Drone Technology

The advancement in Drone technology is continually evolving at a tremendous pace. The main factors that have recently gained huge importance among research practitioners are improved battery technology, enhanced sensors, an amalgamation of AI and ML, and smart manufacturing techniques [22]. These are some of the latest factors that are observed in the latest drones and essential elements of drone technology. They are exhibited in Figure 1.1 which is briefly described subsequently.

(a) Batteries: Almost all drones are powered by rechargeable lithium-polymer or lithium-ion batteries. It is one of the most important fundamentals because it helps in determining the drone's flight time and payload capacity. Longer flight time is helpful in logistics operations as it helps drones cover a wider range of areas and is further helpful in making more deliveries in a single go.

(b) Propulsion Systems: Drones have propellers that help in generating thrust and are further helpful in controlling the movement of drones.

Table 1.1 Research insights of various studies.

References	Insights	Practical implications
[12]	The study presented is related to associate existing literature in the context of eco-friendly and continuing logistics. Regrettably, it fails to furnish any details regarding the subject of "Journey to Transportation and Logistics Management using Drone: Digitization and Technological Evolution."	- Provides innovative ideas for managing the harmony of transportation and nature - Presents enlightening perspectives on enhancing the excellence of logistics services
[13]	The document unveils an orderly arrangement of written works that serves as a blueprint for an examination of the point where digitization and sustainability converge in the realm of transportation logistics. This examination will particularly concentrate on the information produced while operating in the field.	- Development of a shared understanding of industrial processes - Examine the relationship between digitalization and sustainability.
[14]	The given publication is headed "Applications and Research Aspects for Drone-Based Models in Logistics: A Classification and Review," but it makes no reference to the specific study "Journey to Transportation and Logistics Management Using Drone: Digitization and Technological Evolution."	- A framework for classifying drone-based logistics models - Determining research gaps in the field

(*Continued*)

Table 1.1 Research insights of various studies. (*Continued*)

References	Insights	Practical implications
[15]	This document presents an exploration of existing writings regarding the employment of unmanned aerial vehicles in the realms of transport and logistical administration, with a particular emphasis on the digitization process and the advancement of technology.	- Identifies the present state of drone technology in healthcare - Proposes strategies to address adoption hurdles
[16]	The exploration of literature delves into the realm of drones in the arena of transportation and the management of logistics, with a particular emphasis on digitization and the ever-evolving world of technology.	- Current study reveals flaws in drone technology. - The links between drone technology and retail logistics are shaky.
[17]	The document unveils an inventive blueprint for implementing a methodical examination of scholarly works pertaining to digitalization advancements in the realm of transportation logistics. However, it should be noted that the paper does not delve into the realms of unmanned aerial vehicles or the process of transportation and logistics management.	- A lack of openness in transportation logistics processes. - A diverse technological landscape and application-individual procedures
[18]	The document fails to make any explicit reference to the subject matter "Embarking on an Expedition towards Drone-Driven Transportation and Logistics Management: Embracing Digitization and the Advancement of Technology."	- Assist stakeholders in grasping existing and future prospects. - Inform decision-making in the deployment of disruptive technologies

(*Continued*)

Table 1.1 Research insights of various studies. (*Continued*)

References	Insights	Practical implications
[8]	The paper provides a literature review on the use of drones in transportation and logistics management, focusing on digitization and technological evolution.	- Identifies the potentials and constraints of drones in SCM and logistics - Establishes a research agenda for future study
[19]	This document unveils a groundbreaking strategy for executing a methodical examination of published works pertaining to the employment of business evaluation techniques in the realm of transport logistics. This approach can be effectively employed in the analysis of the subject of digitalization and the evolution of technology in transportation and logistics management, specifically with regard to the utilization of unmanned aerial vehicles.	- Identifying appropriate business evaluation methodologies - Advancement of the current level of research
[20]	The paper at hand delves into the fascinating realm of drones and their pivotal role in connecting the realms of Industry 4.0 and ITS ecosystems. However, it regrettably abstains from delving into the intricacies of the literature review concerning the captivating topic of "Journey to Transportation and Logistics Management using Drone: Digitization and Technological Evolution"	- Unleashing the birth of an unparalleled digital ITS habitat - Executing a fully digitalized transportation system with UAVs

(*Continued*)

Table 1.1 Research insights of various studies. (*Continued*)

References	Insights	Practical implications
[21]	Drone integration into logistics is fast gaining traction. Drones are being utilized for last-mile deliveries. This study emphasizes the potential for drones to revolutionize the e-commerce industry.	- Drones have the potential to increase last-mile delivery speed, efficiency, accessibility, and customer pleasure. - Drone technology integration into corporate operations is still in the works.

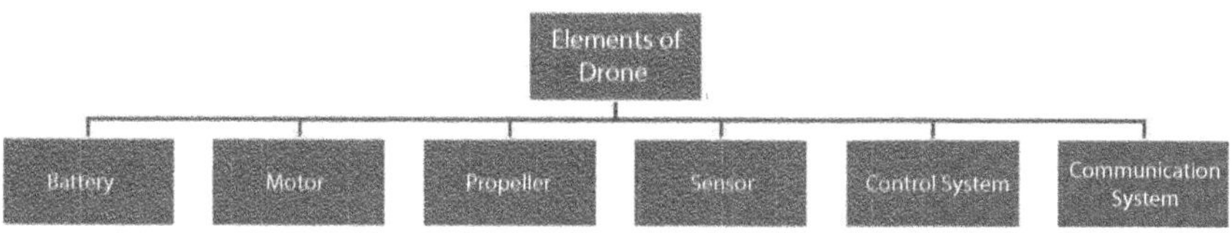

Figure 1.1 Basic elements of drone.

(c) Navigation and Control Systems: These systems include components like GPS, gyroscopes, and barometers. These are helpful in following a proper route which further helps in reaching the target destination in minimal time. In terms of Logistics management, it helps in ensuring that deliveries are made to the correct destination.

(d) Sensors: Drones are powered with many sensors like ultrasonic sensors, Image Sensors, and high-resolution cameras. These sensors altogether help in detecting obstacles, also prevent them from collision, and most importantly helpful in navigating through complex routes. In Logistics, the purpose of sensors can be realized while ensuring safe and efficient deliveries [23].

(e) Communication Systems: Drones require reliable communication systems to receive signals from operators which further helps in transmitting telemetry data. In Logistics, communication systems can be helpful in remote monitoring and control which further helps in enhancing operational efficiency and safety.

(f) Payload carriage: It basically consists of the body part of the drone. In logistics, it can be helpful in carrying goods to their targeted destination. It must be designed to securely transport cargo while preserving the drone's stability.
(g) Data Processing and Analysis: Drones continuously generate data while moving from one place to another, like images of places, and sensory data. Data Analysis tools help in extracting useful information such as checking weather updates, traffic monitoring, and path optimization. In logistics management, these tools help in making accurate decisions which further boost operational efficiency.

These fundamental elements are playing a significant role in the advancement and revolutionizing drone technology. These elements altogether have the potential to make this technology more capable, versatile, and accessible for a wide range of industries starting from agriculture, the entertainment industry, traffic monitoring, logistics management, and many more significant fields.

1.4 Evolution of Drone Technology

The evolution of drone technology is rapidly evolving and the rapid advancements in this buzzing technology resulted in the fifth-generation drone being very small in size, more powerful, and affordable as earlier cameras and sensors were used to be very expensive.

The evolution of drone technology is summarized as follows:

(a) Early UAV: These UAVs were launched in the mid-1800s and it was the first recognized drone which was used by the British Navy for Aerial Target purposes [24]. After that many subsequent models of drones were manufactured. In 1918, Kettering Bug was launched which was the first radio-controlled aircraft. In 1935, DH.82B Queen Bee was launched which was also a radio-controlled drone that was used for aerial gunnery practice and target training by numerous military forces. In 1960, a drone model named MQM-57 Falconer was introduced with the purpose of observing enemy movement. In 1990, a drone model named Gnat 750 was introduced which has the advanced

Table 1.2 Comparative study of modern drones.

Year of manufacture	Name of drone	Payload capacity	Cost	Notable features	Uses
2013	DJI Phantom 1	0.44 pounds	Low	High Resolution Camera which makes it a pioneer in consumer drone.	In Photography and Videography
2003	AeroVironment RQ-11 Raven	1.5 pounds	High	Very easy to handle and portable which is designed for surveillance missions.	Maintenance and Inspection Purpose
2020 [26]	DJI Matrice 300RTK	15.4 pounds	High	Provides Real-time Video footage.	Inspection Purpose
2018	senseFly eBee X	2.2 pounds	Moderate to High	The modular design helps in handling payloads easily.	Inspection Purpose
2009	Aeryon Scout	0.5 pounds	High	Its VTOL capability helps in various military operations.	In Search and Rescue Operations
2008	Draganflyer X6	0.5 pounds	Moderate to High	Compact and Portable design helps in spraying pests on farms.	In Agriculture

capabilities of traffic monitoring. In 2001 RQ-1 Predator was launched which was basically designed to be used by militants for armed uses.

(b) Modern Drones: After the manufacturing of quadcopter drones in the late 2000s, the popularity of drones significantly increased. Earlier it was used only for military services but with more widespread versatility and range of applications, now drones are used for many different purposes like photographic animation, maintenance, delivery, search & rescue, and Agriculture [25].

In 2013, the first consumer drone was designed named DJI Phantom 1. This drone was basically designed for aerial photography and videography. Similarly, for maintenance and Inspection purposes, the first drone that was designed was the AeroVironment RQ-11 Raven. It was initially developed for military use but later it was also adopted for inspection use. In recent years some dedicated drones have been launched that were basically designed for specialized maintenance and inspection purposes, like DJI Matrice 300RTK and the senseFly eBee X. These drones are equipped with dynamic cameras with wide apertures which help in inspection tasks.

Amazon introduced the concept of delivery by drone in 2013, which was known as Amazon Prime Air, which aims to use drones for delivering packages to customers. In the 2000s, a drone named Aeryon Scout was designed which was basically to be used for search and rescue operations. It is a VTOL (Vertical take-off and Landing) drone that is equipped with the latest sensors and high-resolution cameras. These features help it to provide aerial surveillance and collect useful insights during search and rescue missions.

In the early 2000s, a drone named Draganflyer X6 was introduced for agricultural purposes. It was a hexacopter drone i.e., a drone having six propellers, configured with a wide range of sensors and cameras which further makes it suitable for various agricultural activities like crop monitoring, pest control, and precision agriculture. Table 1.2 presents the comparative study of modern drones.

1.5 Use Case in Various Sectors

Drone Technology has recently gained tremendous popularity in various applications, from agriculture to transportation and logistics. Some of the potential use cases of drones depicted in Figure 1.2.

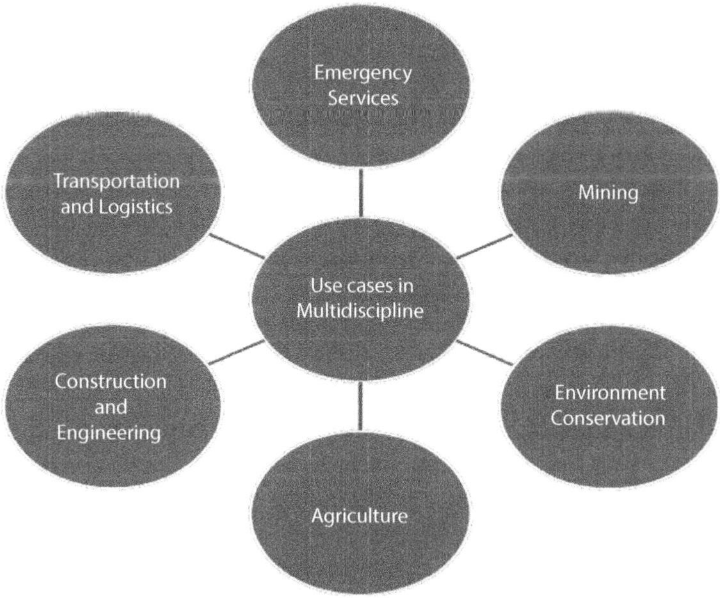

Figure 1.2 Use cases in multidiscipline.

(a) **Agriculture**: Some of the latest drones like XAG V50, DJI Agras T30 are equipped with high-resolution cameras and high payload capacity. These features support monitoring crop health and carrying a large volume of pesticides. The large carrying capacity of pesticides leads to a significant reduction in the time of the sprinkling process. Drones are significantly used for tracking and monitoring livestock in grazing areas.

(b) **Construction and Engineering**: It has been observed that drones are significantly helpful in construction sites as these drones capture high-resolution images which helps project managers in identifying the potential uses. Drones are significantly helpful in improving safety concerns as this technology helps in inspecting minute damage, which further improves efficiency.

(c) **Mining and Resources**: Drones are powered with various geological sensors which significantly help in identifying mineral deposits across remote areas. Drones are also used to assess environmental consequences and are further helpful in monitoring compliance with regulations.

(d) **Environmental Conservation**: Drones are significantly helpful in wildlife preservation. These drones help in tracking how many birds have migrated and are helpful in combating poaching. The forest officials are also using drones in order to monitor and combat illegal activities in protected natural areas.

(e) **Transportation and Logistics**: Companies like Google, and Amazon are continuously exploring and working on this technology for the delivery of medical supplies and other essential amenities. Drones can also be helpful in performing inventory counts and can monitor various stock levels in warehouses.

(f) **Emergency Services**: Some of the latest drones are equipped with thermal cameras which help in finding missing persons in challenging terrains, which helps remarkably in search and rescue. Drones are also helpful during earthquake disasters as they help in identifying which buildings are in danger which further helps the rescue team to give priority to those buildings and further helps in quick evacuation. Table 1.3 exhibits prominent drones used in the above-discussed use cases.

Drone is one of the powerful tools with varied applications ranging from agriculture, Mining, Infrastructure, and much beyond. The drone has the potential to capture high-resolution images, enhance operational efficiency, and significantly improve decision-making processes which altogether helps in increasing accuracy and can significantly reduce time.

1.6 Application in Transportation and Logistics

The drone has the dormant potential to bring a remarkable change in the existing transportation or delivery system. The applications of drones in transportation and logistics will bring huge relief, especially to the people who reside in urban areas with heavy traffic congestion. Drone will be helpful in reaching out to those areas that were difficult to navigate by other modes of transportation. Some of the dormant applications of drones in transportation and logistics are discussed in this section:

(a) **Last-Mile Delivery**: Drone can be used for delivering goods to customer's doorstep, particularly in remote or

Table 1.3 Prominent drones in different sectors.

Year of release	Name of drone	Purpose	Accuracy and precision	Operational efficiency
2022 [27]	XAG V50	The basic purpose of this drone is to optimize crop health.	High	High
2022 [28]	Autel Robotics Evo Max 4T	The primary goal is to provide high-resolution aerial imaging and mapping for site inspection.	Moderate to High	Moderate
2016 [29]	DJI Phantom 4 Pro	The primary purpose is to conduct an aerial examination.	Moderate to High	High
2020 [8]	Parrot Anafi USA	The primary functionalism of these drones is in environmental monitoring through aerial data collection.	High	High
2023 [30]	BonV	The primary objective is to ease the accessibility and efficiency of logistics solutions.	High	High
2023 [31]	M-Eye	These drones are designed to be used in rescue operations to provide security.	High	High

high-traffic congestion areas [32]. Drones also help in delivering the package at a high speed which is almost impossible by any other mode of transportation. There are companies like Amazon, Google, and UPS who are continuously working on this area in order to improve their customer's experience.

(b) **Warehouse Inventory Management**: Drone can be remarkable in inventory management. Drones will help speed up the process of warehouse operations by automating routine tasks like inventory counts, scanning of products, picking, as well as reclamation. In addition, drones can also be used to record product temperature, humidity, and other essential activities.

(c) **Maritime Cargo Delivery**: Drone can be significant in inspecting vessels and cargo, examining and mapping waterways, and monitoring environmental conditions. It will be helpful in reducing the cost and further providing extensive security by eradicating the need for manual inspections [33].

(d) **Disaster Hindrance and Aid**: Drones can be helpful during natural calamities, in carrying medicine, food packets, and other emergency equipment in those areas where conventional modes of transportation are disrupted. Drones having high payload capacity can be helpful in such situations in providing relief materials in those affected areas.

(e) **Port and Harbor Inspections**: Drones are equipped with high-resolution cameras and with their thermal capabilities; they can provide a 360-degree view of the entire port area. This will further help security personnel in tracking as well as monitoring activities in real-time, further helping in identifying potential threats [34].

These are some of the latest applications that are in use, and some are yet to be explored in those areas. This application will enhance efficiency, slacken the existing cost as it would eliminate the labor costs and altogether will improve the supply chain in the transportation and logistics sector. As technology advances, one can observe more advancements in this technology and its use cases; also it would be a small push towards automation.

Challenges: Drone has massive potential in various fields but there are a few shortcomings that can impact their adoption and effectiveness in a few cases. Some of the dormant challenges are:

(a) **Limited Payload Capacity**: Drones are observed to have limited payload capacity. It can be an obstacle where drones need to uplift bulk material at the same point in time.

(b) **Collision Avoidance**: Safety is one of the most important concerns because there are few drones that are equipped with monocular cameras. This camera helps in detecting and avoiding obstacles. There is a need to have such cameras in every drone to prevent collisions. Researchers are significantly using these cameras in order to detect objects in actual time using machine learning algorithms.

(c) **Cost**: Drones are costly and if some collision happens during their time of flight, then operational maintenance is very expensive. This can be one of the barriers to adoption, especially for small-scale businesses.

(d) **Environmental Consequences**: Drones are considered not to be environment friendly because they generate tremendous noise pollution particularly when operating in urban areas which also causes disturbance to surrounding people and wildlife.

(e) **Short Battery Life**: Almost all drones have limited battery life which restricts their range and operative duration. Short battery life results in a reduction of flight time which usually ranges from a minute to a few hours. Secondly, frequent battery changes also mitigate the efficiency of drone operations. This continuous change in battery is also time-consuming which can be a barrier to a mission.

(f) **Malicious Use of Drones**: The continuous evolution of drone technology has made it possible to range in geographical terrain where reachability of human beings is almost unreasonable for image capturing. At the same time, it has been observed that drones are manipulated for malicious purposes for example infiltration, smuggling, or attacks on deprecatory infrastructure. This scenario makes necessary intervention from national security agencies to safeguard illegal use of it.

1.7 Conclusion

The assimilation of drone technology in the domain of transportation and logistics has made a worthy contribution in the modernization of a multitude of industries. In its initial phase, drones were developed for military purposes, have been rapidly remodeled for a variety of devices in multidisciplinary applications. Drones' integral elementary blocks are superior quality batteries, sensors, artificial intelligence and more, have accelerated in transformation and widespread presence. The gradual growth in advancement of drone technology from its nascent unmanned aerial vehicles to modern has expanded the sphere of their capabilities, operations, and applications. The probable and enormous uses of drones in an effective way includes a variety of domain applications from different sectors such as agriculture, healthcare with emergency services, construction companies, mining, environmental preservation and remarkably, transportation and logistics.

Nowadays, altered and evolved drones with potential capabilities have revolutionized the process of delivering in geographically apart warehouses, shipment in marine or cargo, and readiness in disaster relief. These improvements help in retaining the promises to considerably develop operational efficiency, limiting costs, and upgrade in supply chains process. Further, limitations such as constrained battery life, consequential implication to the environment, and unaffordable costs are few hindrance calls for resolution. Restricted payload capacity, avoidance of collisions, exorbitant costs, environmental repercussions, limited battery life, and the potential for nefarious use are a few of the obstacles that necessitate resolution.

In summary, the assimilation of drones into transportation and logistics management epitomized a transformative expedition characterized by technological progress, augmented efficiency, and cost-effectiveness. Although there exist hurdles to surmount, the promise of drones in enhancing various facets of supply chain management is unequivocal. As progressive and continued advancements in technology evolution, it is expected to imagine a paramount position in forming the outlook of different sectors, tendering innovative openings and resolutions to established corners.

References

1. Nuryanti, L., A vehicle routing problem optimization with drone using tabu search algorithm and analytical hierarchy process. *Majalah Ilmiah Pengkajian Industri, 15*, 1, 2021.
2. Garg, V., Niranjan, S., Prybutok, V., Pohlen, T., Gligor, D., Drones in last-mile delivery: A systematic review on Efficiency, Accessibility, and Sustainability. *Transp. Res. Part D Transport Environ., 123*, 103831, 2023.
3. Sabour, M.H., Jafary, P., Nematiyan, S., Applications and classifications of unmanned aerial vehicles: A literature review with focus on multi-rotors. *Aeronaut. J., 127*, 1309, 466–490, 2023.
4. Sajid, M., Mittal, H., Pare, S., Prasad, M., Routing and scheduling optimization for UAV assisted delivery system: A hybrid approach. *Appl. Soft Comput., 126*, 109225, 2022.
5. Limbasiya, T., Teng, K.Z., Chattopadhyay, S., Zhou, J., A systematic survey of attack detection and prevention in connected and autonomous vehicles. *Veh. Commun., 37*, 100515, 2022.

6. Rhiat, A., Chalal, L., Saadane, A., A smart warehouse using robots and drone to optimize inventory management, in: *Proceedings of the Future Technologies Conference (FTC) 2021*, vol. 1, Springer International Publishing, pp. 475–483, 2022.
7. Zennaro, I., Finco, S., Calzavara, M., Persona, A., Implementing E-commerce from logistic perspective: Literature review and methodological framework. *Sustainability*, *14*, 2, 911, 2022.
8. Rejeb, A., Rejeb, K., Simske, S.J., Treiblmaier, H., Drones for supply chain management and logistics: a review and research agenda. *Int. J. Logist. Res. Appl.*, *26*, 6, 708–731, 2023.
9. Cho, G., Unmanned aerial vehicles: Emerging policy and regulatory issues. *J. Law Inf. Sci.*, 22, 2, 201–236, 2013.
10. Min, S. and Kim, K.H., Drone-assisted logistics for sustainable and resilient supply chain management. *Sustainability*, 11, 12, 3263, 2019.
11. Li, A., Zhang, L., Sun, S., IoT and data analytics in drone-based logistics: A comprehensive review. *Comput. Ind. Eng.*, 156, 107206, 2021.
12. Ren, R., Hu, W., Dong, J., Sun, B., Chen, Y., Chen, Z., A systematic literature review of green and sustainable logistics: Bibliometric analysis, research trend and knowledge taxonomy. *Int. J. Environ. Res. Public Health*, *17*, 1, 261, 2020.
13. Riedle, S., Sardarabady, N.J., Kasher, J.D., Sustainability Investigations based on Digitalization Technologies in the Field of Transportation Logistics: A Systematic Literature Review Protocol, in: *IEEE International Conference on Industrial Engineering and Engineering Management (IEEM)*, IEEE, pp. 0944–0949, 2022.
14. Moshref-Javadi, M. and Winkenbach, M., Applications and Research avenues for drone-based models in logistics: A classification and review. *Expert Syst. Appl.*, *177*, 114854, 2021.
15. Law, C.T., Moenig, C., Jeilani, H., Jeilani, M., Young, T., Transforming healthcare logistics and evaluating current use cases of UAVs (drones) as a method of transportation in healthcare to generate recommendations for the NHS to use drone technology at scale: a narrative review. *BMJ Innov.*, *9*, 3, 2023.
16. Muricho, M.W. and Mogaka, C.O., Drone technology and performance of retail logistics. *J. Sustain. Dev. Transport Logist.*, *7*, 1, 73–81, 2022.
17. Kasher, J.D., Riedle, S., Sardarabady, N.J., Digitization technologies in transport logistics: A systematic literature review protocol, in: *IEEE 28th International Conference on Engineering, Technology and Innovation (ICE/ITMC) & 31st International Association For Management of Technology (IAMOT) Joint Conference*, IEEE, pp. 1–13, 2022.
18. Dong, C., Akram, A., Andersson, D., Arnäs, P.O., Stefansson, G., The impact of emerging and disruptive technologies on freight transportation in the digital era: current state and future trends. *Int. J. Logist. Manage.*, *32*, 2, 386–412, 2021.

19. Sardarabady, N.J., Kasher, J.D., Riedle, S., Evaluation Approaches for Measuring Economic Efficiency of Digitization Technologies in Transport Logistics: A Systematic Literature Review Protocol, in: *2022 IEEE International Conference on Industrial Engineering and Engineering Management (IEEM)*, IEEE, pp. 0881–0887, 2022.
20. Beke, E., Bódi, A., Katalin, T.G., Kovács, T., Maros, D., Gáspár, L., The role of drones in linking industry 4.0 and ITS Ecosystems, in: *IEEE 18th International Symposium on Computational Intelligence and Informatics (CINTI)*, IEEE, pp. 000191–000198, 2018.
21. Turgut, M. and Şeker, B., Drone technology in transportation management: A systematic review and framework for future research. *J. Aviat.*, *7*, 2, 251–261, 2023.
22. Abro, G.E.M., Zulkifli, S.A.B., Masood, R.J., Asirvadam, V.S., Laouiti, A., Comprehensive review of UAV detection, security, and communication advancements to prevent threats. *Drones*, *6*, 10, 284, 2022.
23. Fahlstrom, P.G., Gleason, T.J., Sadraey, M.H., *Introduction to UAV systems*, John Wiley & Sons, 2022.
24. Biswas, S., Muttangi, R., Patel, H., Prince, S., Edge ai based autonomous uav for emergency network deployment: A study towards search and rescue missions, in: *International Conference on Wireless Communications Signal Processing and Networking (WiSPNET)*, IEEE, pp. 268–272, 2022.
25. Ribeiro, D., Santos, R., Shibasaki, A., Montenegro, P., Carvalho, H., Calçada, R., Remote inspection of RC structures using unmanned aerial vehicles and heuristic image processing. *Eng. Fail. Anal.*, *117*, 104813, 2020.
26. Rejeb, A., Abdollahi, A., Rejeb, K., Treiblmaier, H., Drones in agriculture: A review and bibliometric analysis. *Comput. Electron. Agric.*, *198*, 107017, 2022.
27. Hatoum, M.B. and Nassereddine, H., The use of drones in the construction industry: applications and implementation, in: *ISARC. Proceedings of the International Symposium on Automation and Robotics in Construction*, vol. 39, IAARC Publications, pp. 542–549, 2022.
28. Lee, S. and Choi, Y., Reviews of unmanned aerial vehicle (drone) technology trends and its applications in the mining industry. *Geosyst. Eng.*, *19*, 4, 197–204, 2016.
29. Brik, B., Ksentini, A., Bouaziz, M., Federated learning for UAVs-enabled wireless networks: Use cases, challenges, and open problems. *IEEE Access*, *8*, 53841–53849, 2020.
30. Zhang, J.Z., Srivastava, P.R., Eachempati, P., Evaluating the effectiveness of drones in emergency situations: a hybrid multi-criteria approach. *Ind. Manage. Data Syst.*, *123*, 1, 302–323, 2023.
31. Eskandaripour, H. and Boldsaikhan, E., Last-mile drone delivery: Past, present, and future. *Drones*, *7*, 2, 77, 2023.
32. Carvalho, G.F., Andrade, F.A., Ramos, G.S., Zachi, A.R., De Barros, A.L., Pinto, M.F., Sliding Mode Controller Applied to Autonomous UAV Operation

in Marine Small Cargo Transport. *IEEE J. Miniaturization Air Space Syst.*, *4*, 4, 345–357, 2023.
33. Chelioti, K.H., Tsaimou, C.N., Tsoukala, V.K., Unmanned Aerial Vehicles: A Survey on Monitoring Advancements for Port Infrastructure Applications. *J. Engin. Proj. Prod. Manage.*, *13*, 3, 2023.
34. Jamil, S., Abbas, M.S., Roy, A.M., Distinguishing malicious drones using vision transformer. *AI*, *3*, 2, 260–273, 2022.

2

Challenges of Big Data Implementation in Drone-Based Logistics

Mani Deepak Choudhry[1]*, M. Sundarrajan[2], S. Jeevanandham[3] and V. Saravanan[4]

[1]Department of Computing Technologies, SRM Institute of Science and Technology, Kattankalathur, Chennai, Tamil Nadu, India
[2]Department of Networking and Communications, SRM Institute of Science and Technology, Kattankalathur, Chennai, Tamil Nadu, India
[3]Department of IT, Sri Ramakrishna Engineering College, Coimbatore, Tamil Nadu, India
[4]Department of Computer Science, College of Engineering and Technology, Dambi Dollo University, Dambi Dollo, Oromia Region, Ethiopia

Abstract

Over the last few years, e-commerce has caused issues in cost-effective last-mile delivery. In response, companies are turning towards autonomous delivery technologies such as drones. Drone delivery platforms are poised to cut shipping costs and transit times by exploiting drones' applicability beyond current use cases. Additionally, the effect of big data on logistics is evident, as large data sets are employed to increase operational efficiency across the supply chain. This chapter summarizes existing works on logistics support by drones, emphasizing research issues and trends currently and in the future. It focuses on the risks and challenges of drones in logistics networks and provides insights into how Big Data can be optimized. Moreover, practical methods of applying Big Data in logistics are presented. The chapter discusses major issues and briefly sketches new research directions for UAV-based package delivery systems.

***Keywords*:** Big Data, drone delivery, logistics management

**Corresponding author*: manideepakjns22@gmail.com

A. Prasanth, Rajesh Kumar Dhanaraj, Munish Sabharwal, Vandana Sharma and Seifedine Kadry (eds.) Drones for Transportation Logistics and Disaster Management, (25–42) © 2026 Scrivener Publishing LLC

2.1 Introduction

Drone logistics, or drone delivery means utilization of Unmanned Aerial Vehicles (UAVs) to move properties and parcels between places. Drone technology can potentially revolutionize a range of logistics stages and end-to-end supply Chains [1]. Green logistics is an important part of contemporary transportation, where different vehicles effectively move goods from one place to another. The vehicles employed for this function comprise ships, boats, railroad trains, trucks, cars, bicycles, airplanes, and even drones. Though every mode of transport has its specific function, trucks tend to dominate urban logistics, especially in last-mile delivery operations [2]. Trucks, which are versatile and can transport loads, have become the workhorse of last-mile delivery in cities [3]. However, the constraints that accompany this common practice must be identified. These constraints create environmentally sound alternatives for last-mile delivery, such as using bicycles. One of the most significant drawbacks of trucks for last-mile delivery is that they cannot reach remote and inaccessible rural areas [4]. Such areas are not likely to have proper road infrastructure for big trucks, and thus it becomes costly and inefficient to reach them. In such cases, other vehicles such as bicycles can play a vital role in bridging the delivery gap. Bicycles are portable and light and can traverse very narrow and intricate roads that trucks can just reach. There is also the immediate necessity of providing for the ecological cost of using trucks, much more so because they run on diesel or gasoline. Such use of fossil fuel does not just foul the air but also creates more carbon emissions globally.

In contrast, bicycles are green, as they are human-driven or electric-motor-aided in specific cases. The green option produces no emissions and does not lead to air pollution, rendering it a preferable sustainable last-mile delivery solution. Traffic congestion is also an issue that impacts truck-based last-mile delivery operations. The winding movement of trucks through urban areas results in delayed parcel delivery, and customer and logistics company inconvenience. The flexibility and ability of bicycles to cut through traffic jams present a solution here. They can pass through crowded roads with ease, and packages are delivered quickly even in the rush hour.

The transportation and logistics industries are now in the process of unprecedented and explosive growth, fueled by a robust wave of innovation. This expansion is led mainly by the booming e-commerce sector, thanks to the mass availability of internet connectivity worldwide. Ongoing innovations in delivery channels have played a decisive role in this shift, leading to a dramatic year-on-year increase in shipments that various companies must dispatch. E-commerce, facilitated by the growing pervasiveness of web connectivity, has transformed the purchase and sale of merchandise [5]. Online shopping has become an easy and popular mode of shopping for consumers across the world, with a huge range of products available at their doorstep. The transportation and logistics sectors have undergone a significant change in response to this demand. Firms have been funding cutting-edge solutions to address the challenges presented by the explosion in package volumes. These solutions cover numerous aspects of logistics, ranging from warehousing and order fulfillment to last-mile delivery [6]. For example, automation and robotics are being integrated into warehouses to maximize the speed and efficiency of order processing.

Along with this, the delivery mechanisms have also been changed because of these innovations. The organizations are searching for new ways of quicker and more efficient package delivery. Unmanned aerial deliveries, self-driving vehicles, and intelligent navigation systems are advancing technologies designed to enhance the performance of logistics operations [7]. The result of these advances is a year-on-year steady rise in the volumes of parcels that companies carry. This volume boost has opportunities as well as challenges. On the positive side, it holds the potential for higher revenues and market share. On the negative side, it demands scalable and sustainable mechanisms to meet rising demand.

Consumer demand in the online marketplace has increased substantially in recent years. Consumers have come to expect speedy delivery and inexpensive shipping of products ordered online. This trend in demand has pushed freight expenses higher each year. Businesses need to absorb the costs of transporting packages to more and more disparate sets of addresses around the city. To overcome this problem, companies tend to consolidate parcels headed for the same geographic location, thereby minimizing transportation expenses as packages go to various destinations.

The actual transportation process usually goes smoothly until the parcels need to be sorted for the last part of their journey to the recipient. After packages arrive at their final destination, transportation becomes very costly, with the most expensive being the "last mile" of the delivery cycle. This is the stage, also known as the "last mile challenge," that is an arduous task for most logistics' companies, especially those dealing with e-commerce organizations. The last mile is, undoubtedly, the most expensive and complex part of the practice of liberation. Since delivery expenses are no longer largely shared, and in numerous instances not shared whatsoever, every package's delivery becomes a one-of-a-kind and unique endeavor. This individualized, one-package strategy greatly increases costs and complexities, affecting both logistics providers and their e-commerce customers. The last mile problem highlights the urgent necessity for innovative solutions and strategies to maximize this final stage of delivery so that customer demands fast, economical service can be fulfilled without jeopardizing the sustainability and profitability of the logistics business [8].

The domain of scientific investigation of autonomous drone delivery systems offers a diverse set of potential avenues and essential research questions of heavy import to the efficacy and fiscal viability of these cutting-edge systems. Answering such questions requires the development of optimization and analytical methods aimed at traversing the complicated complexities inherent in the characteristics and dynamics of these systems. This complexity arises through numerous factors, such as the large number of drone fleets in populous cities, the complex and uncertain trip origin-destination dynamics, and the need for autonomous operations. These dynamic behaviors make modeling and analysis of drone delivery systems very difficult.

This chapter focuses on the determination of key research problems emerging in the area of drone delivery systems. It does this by performing an extensive literature review, examining several related studies and research papers, and pointing out several suggested methods to deal with these issues. As a contribution to the literature, this chapter categorizes recent studies into various thematic streams regarding drone delivery issues and evaluates the feasibility of applying Drone-based shipment services in metropolitan areas. Significant issues explored in this chapter are vehicle routing problems, strategic deployment of recharging sites, complexity of landing maneuvers, determination of the optimal size of the fleet, and assignment of tasks between drones. Additionally, this work presents an evaluation of the performance of various methodologies, identifies gaps and limitations of each method, and gives an indication of future avenues of research for this evolving area.

The review underscores the need for continued exploration and innovation in the domain of autonomous drone delivery systems. By addressing these core issues and continually refining the methods and strategies employed, researchers and industry professionals can add to the successful integration of drone delivery services in urban areas, fulfilling the growing demands for efficient and cost-effective last-mile logistics solutions while simultaneously addressing the unique challenges posed by this dynamic and evolving field.

This chapter aims to enhance comprehension of the future adoption of drones in emergent nations by tackling the following investigative queries:

1. What impediments hinder the application of UAVs in philanthropic logistics in both urbanized and emergent nations?
2. What are the interconnections between influences and their significance?
3. How have these obstacles affected the adoption of UAVs in both advanced and emerging nations?
4. What real-world and investigative insights can be drawn from the study?

The remainder of the chapter constitutes: Section 2.2 provides a literature survey that offers insights into prevalent challenges in the contemporary supply chain industry and reviews prior studies regarding the prospective utilization of drone logistics. Section 2.3 and its subsections elucidate the defined concept of Big Data (BD) in the survey's context, encompassing extensive data measures and characteristics inherent to this field. The chapter then progresses to Section 2.4, where it addresses the open challenges associated with the novel smart transportation methodologies and models. Lastly, Section 2.5 offers a concise summary of conclusions and future trends.

2.2 Related Works

UAVs have experienced a notable surge in their popularity over recent years, being applied in both commercial and private contexts. Particularly, the incorporation of shipment through drones within the logistics division holds the promise of transforming various facets of this sector. Drones have diversified their uses beyond their original military purpose, primarily due to the inherent risks and dangers associated with manned aircraft, as discussed. This chapter seeks to explore the primary barriers to drone adoption in the logistics industry while also providing a rating of the minor barriers. The study undertook an extensive review of both national and international literature, aiming to identify areas where the use of drones has become prevalent.

The use of UAVs, commonly known as drones, has been rapidly and universally adopted since their discovery. One of the key factors driving their use is their self-sufficient capability, enabling drones to complete operations more quickly and with reduced danger compared to traditional methods [9]. Organizations worldwide already incorporate or are in the process of incorporating viable and do-it-yourself drones into operations for numerous tenacities, demonstrating the versatility of these unmanned systems.

1. **Humanitarian Aid [10]:** Drones are now being used to conduct humanitarian missions of aid, delivering crucial support in times of disaster and crises. They have the capability of dispersing medicine, food, and other commodities to far-reaching or inaccessible communities, significantly augmenting relief processes.
2. **Precision Agriculture [11]:** In farming, drones are also playing a crucial role in precision agriculture. They can monitor plants, scan earth quality, and accurately spread nutrients and insecticides. This technology increases crop production and reduces wastage of resources.
3. **Biological Conservation [12]:** Drone technology is used in biological conservation activities, particularly in the protection and monitoring of threatened species and their habitats. They allow researchers to carry out aerial surveys, monitor wildlife, and capture data vital to conservation efforts.
4. **Logistics [13]:** The logistics sector is adopting the use of drones for more efficient and rapid deliveries. Drones can deliver packages and commodities to remote areas or densely populated cities, greatly minimizing delivery costs and time.
5. **Urban Planning [14]:** The usage of drones in urban construction and planning is growing. They offer data collecting and aerial surveys that support land use planning, city administration, and infrastructure design.
6. **Surveillance [15]:** Drones provide an efficient way to monitor large regions, improve situational awareness, and assist law enforcement in a variety of operations in the security and surveillance domain.

The varied set of uses to which drones are being put highlights their vast potential [16]. Drones have moved beyond their original military uses and have become essential assets in numerous industries, providing innovative solutions to multifaceted problems. Their capacity to fly independently and gather information from an aerial vantage point has changed how organizations go about tasks and projects, making them more efficient, less expensive, and safer. With ongoing advancements in technology, the horizon of drone use is expected to widen even more, providing new opportunities for businesses and societies globally.

In the modern age, the accelerated increase in data creation in transportation and mobility has surpassed the capacities of data capture and analysis of the early 2000s. This evolution is caused by drivers such as the emergence of digitalization paradigms, including Internet of Things (IoT) [17–19], declining data storage costs, and the cost-effectiveness of feelers and personal equipment. These innovations have transformed our comprehension of transport and mobility dynamics through varied means of aggregation, communicating, stowing, consolidating, regaining, and dispensing transportation information, bringing a new dawn of insight into these complexities.

In transportation and mobility, the vastness, heterogeneity concerning sources, types, and formats, and the built-in variability of data, with high rates of change that make them difficult to interpret, have necessitated the rising part of data-intensive activities. These operations, for example, data integration [20], discovery [21], interrogation, and probing for High-capacity Synchronous Systems [22], form basic building blocks while designing and putting in place Intelligent Transportation Systems (ITS). Alas, today's ITS deployment environment is fraught with shortages of monitoring information and analysis functionalities [23]. Thus, the problems at hand emphasize the necessity of efficient and pragmatic solutions. Besides, in the face of a surge in real-time data streams, a strategic emphasis on advanced compilation, integration, and progressive knowledge acquisition techniques has the potential to unveil unparalleled perception pertinent to policy enhancements for shipping assistance, tasks, and procedures.

The literature reviewed underscores the significant interest in utilizing drone technology for various supply chain operations, yet it reveals a notable reliance on certain assumptions that necessitate closer scrutiny before actual drone implementation can occur. For example, the placement of charging stations is often driven by demand points, overlooking critical considerations related to station location and infrastructure development. Ongoing research and patent exploration are focused on potential charging station sites, including charging towers, building terraces, and lamp posts, but these options require thorough analysis for their viability. A conspicuous research gap exists concerning how drone delivery services, particularly with a heightened focus on charging aspects, can be integrated into developing economies, coexisting with existing trucking systems, warehouses, and distribution facilities. Furthermore, there is a dearth of literature about the development of practical models for strategically locating charging stations on city maps for drone delivery services, which would be financially viable for logistics companies to invest in. In addition, limited attention has been dedicated to exploring the managerial implications of implementing drone delivery in developing economies, especially considering the substantial initial investment required.

2.3 Big Data in Transit

The supply chain industry is experiencing a transformative shift with the emergence of Big Data, enabling the resolution of longstanding data-driven challenges and opening doors to novel opportunities, applications, and services. These data streams emanate from various sources, including Global Positioning System (GPS) navigators and mobile phones, providing an invaluable opportunity to analyze the utilization of transportation infrastructure by customers with an unprecedented level of granularity in both temporal and spatial dimensions. This influx of data presents a challenge and an opportunity for the transportation sector, making it increasingly essential to connect the power of Big Data to enhance traffic management and infrastructure utilization while delivering improved services to the public.

In the transportation sector, Big Data encompasses the amalgamation of gathering substantial volumes of data with high velocity, heterogeneity, and constant evolution within the domain, coupled with the utilization of advanced techniques and models for efficient storage, retrieval, management, processing, and analysis of this amassed information. This definition expands on Gartner's [24] characterization of Big Data as information assets with high scale, rapid flow, and diverse formats, requiring innovative processing techniques to improve decision-making, uncover insights, and optimize processes. The huge volume of data is envisioned as a comprehensive set of mechanisms that empower the effective handling (storage, processing, and accessibility) of all data essential for devising innovative strategies to offer safer, environmentally friendlier, and more efficient transportation, while also enabling users to personalize and customize their transportation experiences. In essence, Big Data has the potential to harness information and address transportation challenges on an unprecedented scale. This technological revolution has unveiled unprecedented prospects for novel shipping frameworks, aids, and use cases that had remained uncharted territory due to the lack of data collection and processing capabilities of this magnitude.

The integration of the BD archetype in the province of transit is driven by two essential features. The primary factor is strongly linked to the rise of urban expansion and the development of Intelligent Cities, creating a rich environment for the emergence of novel BD sources. The next key driver stems from the arrival and ongoing advancement of groundbreaking technological methods, rapid mobile networks, cloud computing, and the IoT. These technologies have not only facilitated the collection of vast datasets but have also ushered in innovative approaches for the processing and utilization of Big Data, making it increasingly feasible to address complex challenges and leverage data-driven insights in the realms of transportation and mobility.

In the context of Smart Cities [25], intelligent mobility systems need to be developed and deployed due to the coexistence and growth of diversity in modes of transport and mobility trends within urban environments. Addressing mobility challenges at the social level requires an understanding of how people travel around and between cities and of how to make transportation operations sustainable, safe, usable, and continuous. The challenges have initiated research into examining sustainable transport methods like bicycles and electric transport, with particular emphasis on ensuring existing infrastructure and other mobility services are preserved as they grow and develop. The long-term objective is to create efficient, sustainable, and responsive transport systems for Smart Cities' evolving needs.

The label "disruptive technologies" in the area of Smart Transportation and Mobility applications and services [26] identifies a group of basic elements making up the Information and Communication Technology (ICT) model. These technologies can be classified according to their function in the data and services lifecycle, including data gathering, infrastructure provision, data processing, data fetching, and security with consideration of the main paradigms, protocols, and methods involved with each technology. Within the area of data gathering [27], the IoT paradigm has been well-featured in recent years, led by developments in machine-to-machine communications [28]. Regarding infrastructure provisioning, Cloud Computing [29] has taken the forefront. Open Data stands out as a significant paradigm explored for data retrieval [30], highlighting the importance of open and accessible data sources.

Figure 2.1 schematically depicts the convergence of three pivotal paradigms that have increasingly become foundational technological elements.

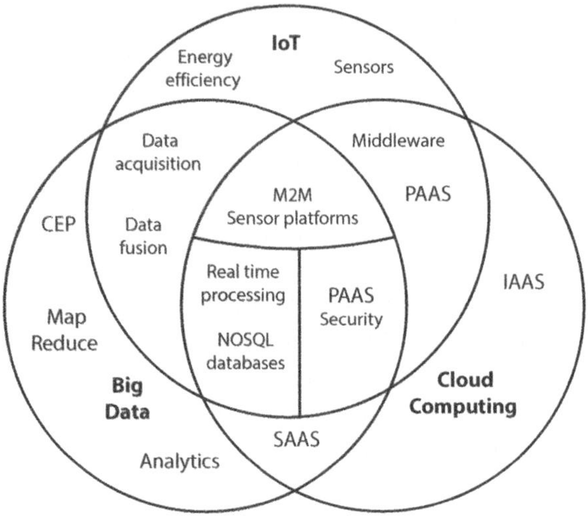

Figure 2.1 A schematic diagram showcasing the convergence of the IoT, cloud computing, and Big Data paradigms.

Three fundamental characteristics, as outlined in the Model 3Vs introduced by Laney [31], are typically associated with Big Data: volume, velocity, and variety. While various facets, such as precision, representation, and prominence, have been explored by numerous researchers, this chapter emphasizes the key factors, as they play a crucial role in shaping how data is collected, analyzed, managed, stored, shared, and processed.

Big Data is a defining concept that encompasses datasets so vast they overwhelm traditional methods of gathering, organization, meting out, and investigation. These datasets have grown to impressive petabyte sizes, marking a transformative shift. In the transport and mobility sectors, geolocated data from social networks is rapidly gaining prominence. The concept of velocity highlights the incredible speed at which new data is generated and shared, enabling real-time analysis without any need for prior storage - such as in effective real-time traffic control systems. The diverse data acquired from various channels enhances integration and creates significant opportunities for individuals, governments, and businesses.

Advanced Data Analysis, standing at the apex of large data-centric frameworks, merges techniques like pattern discovery, adaptive learning, and strategic optimization. It has developed strong momentum as a revolutionizing technological innovation, especially in the transportation industry [32–34]. Compared to traditional methods, it offers a clear competitive edge in two areas. First, studies indicate that simple analytical models run on big data sets tend to produce more accurate and meaningful results compared to complex algorithms working on limited data sets. Second, it increases model complexity, enabling specialists to derive previously unknown insights. In the field of mobility and transit systems, this sophisticated analytical method provides two main advantages: fast processing and improved operational competence. The ongoing flow of real-time transport data, along with its forecasting ability for assessing parts in smart transport infrastructures, indicates both the urgent challenges and pioneering potential within this field.

2.4 Factors Affecting UAV Implementation in Logistics

2.4.1 Legal Factors

The effective acceptance of new technology in an industry frequently depends upon the creation of definite and positive legal frameworks. In the scenario of UAVs, or what is more colloquially referred to as drones, government control over their usage can differ very widely from state to state. However, some legal considerations cannot be ignored when an attempt is being made to create a UAV system. Aspects like limited flight authorities and lack of sufficient insurance coverage can stall an organization's maiden attempts at drone deployment, normally as a result of overly restrictive drone laws that are difficult to comply with. Legal barriers, including trespassing laws, may also prevent UAVs from effectively responding to emergencies by restricting the areas of operation. Equally, limiting line-of-sight visual restrictions, where the operators must be within specific closeness while piloting a drone, dominate even in most advanced nations, impeding UAV operation outside their operator's immediate sight. In addition, the lack of specific spectrum ranges for UAVs is an ongoing obstacle that remains unsolved in most nations. These legal issues may result in expensive litigation and liabilities when it comes to allocating responsibility in case of drone accidents. These legal aspects must be tackled and harmonized for the successful integration of UAV technology into many sectors.

2.4.2 Financial Factors

The deployment of new technology, like UAVs, is dependent on having sufficient financial support, and in this regard, the financial hurdles are especially steep. Even with the many economic benefits that come with UAVs in delivery and monitoring, there are significant hurdles to their integration into current systems. One major problem with the use of drones in humanitarian logistics is the expense of commercially viable solutions. *Doole et al.* [35] show that a typical delivery drone used in the fast-food industry can be priced at around $4,800 per unit, almost twice the cost of the next most cost-efficient substitute, an e-bike. The absence of specialization in the humanitarian industry leads to the inability to find sensors with specific needs, which increases initial and ongoing maintenance costs even further. Other factors adding to the volatility of maintenance expenses are communication losses, adverse climatic circumstances, and demolished setup. These huge preliminary charges pose a significant challenge to humanitarian organizations willing to invest in UAV technology.

2.4.3 Knowledge and Behavioral Factors

The knowledge and behavioral barriers classified within the domain include obstacles resulting from the nature and behavior of the populations across the regions in question. During the last stage of the contemporary humanitarian relief supply chain, there must be smooth coordination between drones and citizens. This will require the citizens to accept the technology, a point Aydin [36] emphasizes. For example, low public knowledge of UAV technologies, an important obstacle to the introduction of new systems, can undermine public support for deploying drones in humanitarian logistics, as noted by *Yoo et al.* [37]. Another essential consideration is the citizens' environmental perception, which is their capacity to evaluate and decide based on their environment. At the time of rescue operations, people tend to be more anxious, which might impair their interaction with UAVs around them, making the general effectiveness of drones in humanitarian operations less effective. The skill level of UAV operators also matters significantly, as poorly skilled operators will cause mission failures. A clearer picture of the above factors and their consequences in both developed and developing nations can be observed in Table 2.1.

2.4.4 Privacy and Safety Factors

The use of delivery drones is also accompanied by some serious challenges, primarily concerning privacy intrusions, safety risks, and ethical dilemmas. Drones, when exchanging information with ground stations over open channels, are susceptible to different security weaknesses, as emphasized [38]. They also bear the risk of privacy intrusion since they have consumer information and may be equipped with cameras. The information

Table 2.1 The adoption of UAVs in humanitarian logistics is shaped by both informational and behavioral factors, varying across different regions.

S. no.	Factor	Impact in established nations	Impact in emerging nations
1	Lack of Awareness Regarding UAV Technologies	In technologically advanced nations, citizens are generally more aware of UAVs due to their presence in mainstream media and their depiction in films and television programs.	Successful adoption of a new technology also depends a great deal upon public awareness and acceptance. As soon as the public is educated and receptive to a new technology, it prepares the ground for easy adoption and infusion of such a technology in the different functions of society and business.
2	Lack of Environmental Perception Amongst Citizens	Although people in developed nations might be aware of drones and their applications, particularly in rescue missions, it should be noted that the panic and anxiety that usually follow disasters can make it difficult for people to appreciate and respond well to the aid provided by drones.	During crises, individuals tend to be in a state of panic and distress. This increased emotional state can greatly compromise their judgment and decision-making skills, and it becomes difficult for them to notice and react to aid provided by UAVs.

(Continued)

Table 2.1 The adoption of UAVs in humanitarian logistics is shaped by both informational and behavioral factors, varying across different regions. (*Continued*)

S. no.	Factor	Impact in established nations	Impact in emerging nations
3	Vandalism Threats During Missions	In developed nations, vandalism in the context of UAV operations can be triggered by different actors. These may involve residents of the area, especially those whose lands the UAVs traverse over, who could become involved in vandalism activities based on privacy, noise, or other issues related to drone flight.	In developing countries, the vandalization of drones tends to be caused by a lack of public awareness and comprehension. If individuals are not well known about the technology of UAVs or their application, then they end up using destructive means, thinking that the drones are dangerous or violating their privacy.

can be intercepted for theft or fraud utilizing hacking, as noted. Safety is a vital consideration in that drones carrying parcels might incidentally find their way into risky areas, potentially harming people. Ground collision drone crashes are commonly blamed on aspects like low battery levels and loss of partial rotors or batteries [39]. Drones are also likely to crash into each other as a result of lost communication, power issues, or hardware and software failures. In addition, the fear exists that drones may be used for criminal intentions such as terrorism or smuggling. Such privacy and safety concerns can adversely affect the public views of drone-delivery technology, and this effect can be further articulated among marginalized communities. To rectify such fears, it is important to highlight the sophisticated technologies used in the design of drones to remove concerns about security threats. Reduction of risks and issues can be brought about through creating no-fly zones for the drones, incorporating high-level encryption to avert cyber-attacks, limiting the usage of cameras, enforcing rigorous control on access for recording during flights, establishing a minimum altitude condition, and monitoring drone hovering activities. A zoning method can be applied methodically to enable the mass deployment of drones [40].

2.5 Conclusion

This chapter addresses various concepts and gives important insights concerning the subject. The accelerated growth and development of Unmanned Aerial Systems across various fields, from business to civilian uses are highlighted. This growth has been driven by technology, cost, and knowledge of the part these systems play in enhancing society. The chapter explored the use of drone delivery in the logistics firm that would transform product movement and delivery. Drones started out being specifically for use within the military but are increasingly taking their efficient cost-cutting, environmentally friendly characteristics into logistical activity. The key challenges of applying drones to logistics were comprehensively addressed. These impediments range from regulatory and safety to technical restrictions and public sentiment. Overcoming these obstacles is necessary to see the full potential for drone delivery for logistics. The potential for the future of this chapter is wide and diverse, with opportunities for research, innovation, and cross-field study. While Unmanned Aerial Systems continues to transform industries and societies, remaining attuned to new trends and innovations will be important to bringing their full capabilities into play and overcoming the challenges of their integration.

References

1. Thibbotuwawa, A., Bocewicz, G., Nielsen, P., Banaszak, Z., Unmanned aerial vehicle routing problems: A literature review. *Appl. Sci.*, 10, 1–20, 2020.
2. Rejeb, A., Rejeb, K., Simske, S., Treiblmaier, H., Humanitarian drones: A review and research agenda. *Internet Things*, 16, 1–20, 2021.
3. Bilinska-Reformat, K. and Dewalska-Opitek, A., E-commerce as the predominant business model of fast fashion retailers in the era of global COVID-19 pandemics. *Procedia Comput. Sci.*, 192, 2479–2490, 2021.
4. Raj, A., Mukherjee, A.A., De Sousa Jabbour, A.B.L., Srivastava, S.K., Supply chain management during and post-COVID-19 pandemic: Mitigation strategies and practical lessons learned. *J. Bus. Res.*, 142, 1125–1139, 2022.

5. Sorooshian, S., Sharifabad, S.K., Parsaee, M., Afshari, A.R., Toward a modern last-mile delivery: Consequences and obstacles of intelligent technology. *Appl. Syst. Innov.*, 5, 1–16, 2022.
6. Merkert, R., Bliemer, M.C., Fayyaz, M., Consumer preferences for innovative and traditional last-mile parcel delivery. *Int. J. Phys. Distrib. Logist. Manage.*, 52, 261–284, 2022.
7. Mohsan, S.A.H., Khan, M.A., Noor, F., Ullah, I., Alsharif, M.H., Towards the unmanned aerial vehicles (UAVs): A comprehensive review. *Drones*, 6, 1–27, 2022.
8. Aurambout, J.J., Gkoumas, K., Ciuffo, B., Last mile delivery by drones: An estimation of viable market potential and access to citizens across European cities. *Eur. Transport Res. Rev.*, 11, 1–27, 2019.
9. Kunz N, N. and Reiner, G., A meta-analysis of humanitarian logistics research. *J. Humanitarian Logist. Supply Chain Manage.*, 2, 116–147, 2012.
10. Tanzi, T., Apvrille, L., Dugelay, J., Roudier, Y., UAVs for humanitarian missions: Autonomy and reliability, in: *IEEE Global Humanitarian Technology Conference (GHTC 2014)*, pp. 271–278, 2014.
11. Tokekar, P., Hook, J.V., Mulla, D., Isler, V., Sensor Planning for a Symbiotic UAV and UGV System for Precision Agriculture, in: *IEEE/RSJ International Conference on Intelligent Robots and Systems (IROS)*, pp. 5321–5326, 2013.
12. Gonzalez, L.F., Montes, G.A., Puig, E., Johnson, S., Mengersen, K., Gaston KJ, K.J., Unmanned aerial vehicles (UAVs) and artificial intelligence revolutionizing wildlife monitoring and conservation. *Sensors*, 16, 1–18, 2016.
13. Raj A, A. and Sah, B., Analyzing critical success factors for the implementation of drones in the logistics sector using grey-DEMATEL based approach. *Comput. Ind. Eng.*, 138, 1–45, 2019.
14. Feng, Q., Quanlong, Liu, J., Gong, J., UAV remote sensing for urban vegetation mapping using random forest and texture analysis. *Remote Sens.*, 7, 1074–1094, 2015.
15. Semsch, E., Jakob, M., Pavlicek, D., Pechoucek, M., Autonomous UAV surveillance in complex urban environments, in: *2009 IEEE/WIC/ACM International Joint Conference on Web Intelligence and Intelligent Agent Technology*, vol. 2, IEEE, Milan, Italy, pp. 82–85, 2009.
16. Cummings, A.R., McKee, A., Kulkarni, K., Markandey, N., The rise of UAVs. *Photogramm. Eng. Remote Sens.*, 83, 317–325, 2017.
17. Aruchamy, P., Balraj, L., Sowndarya, K., An energy-aware link fault detection and recovery scheme for QoS enhancement in Internet of Things-enabled wireless sensor network. *Comput. Electr. Eng.*, 123, 1–29, 2025.
18. Ali, M.A., Dhanaraj, R.K., Nayyar, A., A high performance-oriented AI-enabled IoT-based pest detection system using sound analytics in large agricultural fields. *Microprocess. Microsys.*, 103, 1–19, 2023.
19. Balraj, L., Prasanth, A., Devi Sowndarya, K.K., A Novel Adaptive Unclonable Utility Model for Secure Communication in Internet of Things Environment,

in: *Proceedings of 14th International Conference on Advances in Computing, Control, and Telecommunication Technologies*, pp. 1442–1447, 2023.

20. Khattak, A.J., Integrating Big Data in Metropolitan Regions to understand driving volatility and implications for intelligent transportation systems, in: *Information Technology and Intelligent Transportation Systems: Volume 1, Proceedings of the 2015 International Conference on Information Technology and Intelligent Transportation Systems ITITS 2015, held December 12-13, 2015, Xi'an China*, Springer International Publishing, pp. 3–4.
21. Andrienko, G., Andrienko, N., Chen, W., Maciejewski, R., Zhao, Y., Visual analytics of mobility and transportation: State of the art and further research directions. *IEEE Trans. Intell. Transp. Syst.*, 18, 2232–2249, 2017.
22. Amini, S., Gerostathopoulos, I., Prehofer, C., Big data analytics architecture for real-time traffic control, in: *2017 5th IEEE International Conference on Models and Technologies for Intelligent Transportation Systems (MT-ITS)*, IEEE, Naples, Italy, pp. 7–15, 2017.
23. Suh, W., Henclewood, D., Guin, A., Guensler, R., Hunter, M., Fujimoto, R., Dynamic data driven transportation systems. *Multimedia Tools Appl.*, 76, 25253–25269, 2017.
24. Beyer, M.A. and Laney, D., *The importance of 'big data': a definition*, pp. 2014–2018, Gartner, Stamford, CT, 2012.
25. Rohini, A. and Sridevi, S., Self-powered wireless solution towards the smart city: Use cases and its applications. *Self-Powered Sens.*, 237–250, 2025.
26. Manyika, J., Chui, M., Bughin, J., Dobbs, R., Bisson, P., Marrs, A., *Disruptive technologies: Advances that will transform life, business, and the global economy*, vol. 180, pp. 17–21, McKinsey Global Institute, San Francisco, CA, 2013.
27. Atzori, L., Iera, A., Morabito, G., The internet of things: A survey. *Comput. Netw.*, 54, 2787–2805, 2010.
28. Holler, J., Tsiatsis, V., Mulligan, C., Karnouskos, S., Avesand, S., Boyle, D., *From machine-to-machine to the Internet of Things: introduction to a new age of intelligence*, Academic Press, London, 2014.
29. Bhardwaj, S., Jain, L., Jain, S., Cloud computing: A study of infrastructure as a service (IAAS). *Int. J. Eng. Inf. Technol.*, 2, 60–63, 2010.
30. Janssen, M., Charalabidis, Y., Zuiderwijk, A., Benefits, adoption barriers and myths of open data and open government. *Inf. Syst. Manage.*, 29, 258–268, 2012.
31. Laney, D., 3D data management: Controlling data volume, velocity and variety. *META Group Res. Note*, 6, 1, 2001.
32. Gunturi, V.M. and Shekhar, S., *Big spatio-temporal network data analytics for smart cities: Research needs, Seeing Cities Through Big Data: Research, Methods and Applications in Urban Informatics*, pp. 127–140, Springer, Switzerland, 2017.
33. French, S.P., Barchers, C., Zhang, W., How should urban planners be trained to handle Big Data?, in: *Seeing cities through Big data: Research, methods and*

applications in urban informatics, P.(V). Thakuriah, N. Tilahun, M. Zellner (Eds.), pp. 209–217, Springer, Switzerland, 2017.

34. Wittmann, J., Electrification and digitalization as disruptive trends: new perspectives for the automotive industry?, in: *Phantom ex machina: Digital disruption's role in business model transformation*, A. Khare, B. Stewart, R. Schatz (Eds.), pp. 137–159, Springer, Switzerland, 2017.
35. Doole, M., Ellerbroek, J., Hoekstra, J., Estimation of traffic density from drone-based delivery in very low level urban airspace. *J. Air Transport Manage.*, 88, 2020.
36. Aydin, B., Public acceptance of drones: Knowledge, attitudes, and practice. *Technol. Soc.*, 59, 2019.
37. Yoo, W., Yu, E., Jung, J., Drone delivery: Factors affecting the public's attitude and intention to adopt. *Telematics Inf.*, *35*, 1687–1700, 2018.
38. Kwon, D., Son, S., Park, Y., Kim, H., Park, Y., Lee, S., Jeon, Y., Design of secure handover authentication scheme for urban air mobility environments. *IEEE Access*, 10, 42529–42541, 2022.
39. Han, P., Yang, X., Zhao, Y., Guan, X., Wang, S., Quantitative ground risk assessment for urban logistical unmanned aerial vehicle (UAV) based on bayesian network. *Sustainability*, 14, 2022.
40. Pedersen, C.B., Rosenkrands, K., Sung, I., Nielsen, P., Systemic performance analysis on zoning for unmanned aerial vehicle-based service delivery. *Drones*, 6, 2022.

3

Frameworks for Handover Management for the Networks of Future Drones

S. L. Jany Shabu[1*], Xiao-Zhi Gao[2], J. Refonaa[1] and D. Poornima[1]

[1]*Department of Computer Science and Engineering, Sathyabama Institute of Science and Technology, Chennai, India*
[2]*School of Computing, Faculty of Science, Forestry and Technology, University of Eastern Finland, Joensuu, Kuopio, Finland*

Abstract

In the present scenario, Drones are prominently referred to as Unmanned Aerial Vehicles (UAVs) that have recently emerged as ground-breaking technology to revolutionize a wide range of businesses and applications. The requirement for reliable Handover Management for Future Drone Network (HOM-FDN) is critical as drones become increasingly commonplace in applications including surveillance, delivery, and infrastructure inspection. This explores the realm of FDN and aerial communication, providing a thorough examination of HOM-FDN. Further, the limitations of HOM in hypothetical drone networks have been discussed in the forthcoming chapters. In order to do complicated tasks, these networks use a fleet of drones, which are often controlled by a central control or a decentralized system. Besides, the survey of current and emerging frameworks for managing handoffs between drones in future networks has been discussed in detail. It helps to provide an in-depth investigation of handover management's significance, investigates the difficulties and limitations encountered, and showcases novel strategies for dealing with the complexities of FDN.

Keywords: Drone, handover management, energy efficiency, future drone networks, nodes

**Corresponding author*: janyshabu.cse@sathyabama.ac.in

A. Prasanth, Rajesh Kumar Dhanaraj, Munish Sabharwal, Vandana Sharma and Seifedine Kadry (eds.) Drones for Transportation Logistics and Disaster Management, (43–66) © 2026 Scrivener Publishing LLC

3.1 An Overview

The use of drones is rapidly expanding into all aspects of modern society [1]. Drones are being considered by several sectors as a means to enhance existing services or launch brand-new ones [2]. For these unconventional purposes, a reliable wireless link with restricted control capacity is required for UAVs to fly beyond the pilot's line of sight [3]. A high throughput and reliable connection is needed for enterprises that want to broadcast HD footage from a drone over a wireless connection [4]. There will likely be a dramatic increase in the number of UEs that access mobile services in the not-too-distant future [5]. A terrestrial BS's ability to provide a consistent level of service to its users is degraded as the number of User Equipment (UE) connected to it grows [6]. The BS will likely be down, preventing UEs from connecting to respective mobile networks and services. Capacity injection techniques, like a mobile BS, are the current standard of care for these sorts of problems [7].

The service provider sets up mobile base stations in a dense region, which raises the median quality of service for mobile users [8]. An UxNB is a radio access node that is installed on a UAV and provides coverage to the UE on board [9]. The term "UxNB" describes a specific kind of radio access node that may be mounted on UAVs, namely NodeB. The 3G UMTS (Universal Mobile Telecommunications System) network standard for base stations is "NodeB". Here, "UxNB" probably means a version of this idea tailored for UAVs, with the "x" denoting the deployment based on UAVs. When it comes to terrestrial base stations for 5G modern radio (NR), the UxNB is the successor to the current NodeB (gNB) [10]. The academic world is keen on UxNB as a means of expanding mobile network coverage. The UxNB has a wide range of potential applications, including crises and densely populated locations. UxNB may be deployed in any area, regardless of topography [11]. One of the primary goals of 6G research is to find ways to provide continuous service to a wide variety of UEs. Depending on the context, the estimated mobile network traffic may change [12]. The capacity of mobile networks must be increased to accommodate high-density events like football games, concerts, and the aftermath of natural disasters [13].

The natural next step is to use the existing cellular network to provide service to UAVs engaged in beyond-line-of-sight operation [14]. In addition to providing ubiquitous coverage, cellular networks should be able to meet the needs of high-throughput capabilities, connections, and event management and oversight [15]. First, through stochastic or semi-deterministic simulations, the authors show that LOS propagation circumstances are the norm for aerial users when communicating with numerous Base Stations (BS) [16]. While this is beneficial for the serving BS's signal, it has been shown to be detrimental for interference from neighboring BSs. They show that UAV, in contrast to ground users, typically operate in disruption limited zones. It means that UAV typically operate in disruption limited zones" implies that drones, or UAVs, work in regions where interference from other users or weak signals is more likely to influence their communications than strong signals themselves. Experiments conducted in the field have verified the findings of other publications. Even though the majority of the first group's testing was carried out in a semi-urban, rural agglomeration, they concluded that the connectivity at a variety of altitudes should be adequate to provide command-and-control operation. The second demonstrated, *via* higher-altitude observations, that the number of interfering BSs increases considerably as the UAV flies higher [17].

Positive implications for the evolution of ubiquitous connectivity were found in this investigation especially in underserved areas and for use in emergencies. As communications technology evolves solar panels get more efficient, lightweight composite materials become available, and autonomous avionics and antennas become more affordable, it has been proposed as a feasible element of future aerial networks. As costs increase over time and as more efficient technologies and materials become available, the use of systems in subsequent networks will become more economically viable. It is expected that in the coming years, as better materials are developed and as key technological advances are made, new enablers will gradually surface. As a result of these studies, it is being seriously explored as a potential technology for next-generation wireless communication systems. Even though the choice of energy source has been considered a vital issue in research, photovoltaic power combined with the storage of energy has been seen as the key means of generating power for platforms because of their large surfaces, which are perfect for solar panel films. Because of its low-delay qualities in comparison to upcoming satellite networks, it can deliver wireless connectivity directly to the customers of terrestrial networks.

As wireless communication architectures undergo continuous change and new use cases emerge the benefits of implementing such a system become more alluring [18]. Machine learning systems that can make smart decisions for a fleet of cargo drones or flying automobiles are here. Floating large BSs, intelligent signals supplements, and mobile data centers are just a few examples. It predicts a future with a very large constellation, which is referred to as a mega-constellation as this will provide countless users/

devices in rural and urban areas with access to high-capacity networks, estimation offloading, and data analytics tools. UAVs like drones offer various advantages to communicating *via* wireless networks. Drones have great potential, yet they need to fulfill several requirements to achieve their goals [19].

In the future, networked drones may be able to provide several smart solutions that have been proposed. Connected drones will see widespread deployment in the forthcoming generation of mobile networks, where they will serve in a variety of roles including delivery systems, monitoring systems, and military networks. As a result, the very idea of surveillance, as well as the procedures and mentalities involved in gathering information, will be drastically altered by the advent of linked drones. Drones will be used by governments, corporations, and ordinary people.

Future cellular networks that incorporate drone deployment will improve communication dependability and stability and make better use of available system bandwidths. As a bonus, they'll offer solid coverage, even in high-traffic locations. Since it might be challenging to set up a network in a hurry during times of crisis, drones can fill in where terrestrial devices fall short. Future wireless communication networks will benefit greatly from the introduction of drones, yet a significant challenge lies in the Handover (HO) control of these devices. When it comes to providing users with consistent and uninterrupted communication services regardless of their location, HO is a crucial component of wireless networks. In a perfect world, the HO mechanism would allow the UE's communication to seamlessly transition from one cell to another as the UE moved from one location to another.

The following are the chapter's most significant contributions:

1. This study provides a comprehensive analysis of Handover Management for Future Drone Networks (HOM-FDN) and its relevance to the field of FDN and aerial communication.
2. HOM is especially important in this case since it allows drones to stay in contact and under control as they traverse a wide variety of terrains. It highlights the need for efficient transfers between network nodes in the rapidly evolving field of drone technology, where this is essential for making the most of available resources and maintaining stable, productive drone operations.
3. Due to the problem's influence on drones' flight times and functionality, the topic of handovers' energy efficiency is being studied in greater depth.

The following is the outline for this chapter. In the first section, introduce the concept of drones and explain why handover management is crucial. Section 3.2 gives some context for the HO idea, with an emphasis on drone HO and reviews of associated works. Applications of drones are discussed in Section 3.3, while possible solutions to problems are outlined in Section 3.4, and performance measures are discussed in Section 3.5. The investigation finishes with a chapter outlining potential paths for drone communication in the future.

3.2 Literature Review

An overview of the associated works on mobile network handover is provided, followed by specifics on handover in Long-Term Evolution (LTE). LTE is still widely utilized despite being the future standard for mobile networks. One of the novel approaches for 5G is ultra-densification. As a result, BSs can now service fewer users and their coverage areas are shrinking. Therefore, due to the impact of densification in 5G networks, handover occurs more frequently.

Aydin et al. [20] incorporated in heavily packed areas, such as stadiums, the usage of drone-mounted (DM) BS (UxNB) is a viable option. As a result of its rapid deployment, UxNB has emerged as a promising technology with potential applications in capacity injection. However, this new technology comes with its own set of security concerns. Capacity infusion through UxNB may raise security concerns due to mutual authentication, the establishment of a communication channel between terrestrial BS and UxNB, and rapid handover operations. This new protocol recommends that users switch from terrestrial to UxNB using a coordinated effort. The suggested technique addresses the problem of UxNB authentication by the BS of Earth. In addition, the system does not require an authorization exchange stage for each UE before the handover.

Colpaert et al. [21] determined the purpose of this study is to help close that information gap and throw light on the issues that currently exist. Case studies of a realistic 5G deployment using mm-wave will be used to evaluate the efficacy of 3D beamforming and handovers in UAV networks. This article examines how well a beam-formed mm-Wave link performs when used by a UAV flying over a populated area. The need for beamforming and the frequent handovers necessitated by the small size of the cells provide a significant problem for such networks. When considering beamforming, however, the size of the antenna array becomes significant, and the effects of azimuth and elevation must be calculated, in particular about the handover rate and outage capacity. A high throughput super reliable wireless link will be required for future drone applications involving fast-moving UAVs. Due to the small beams used in Mm Wave communication, it is thought that this technology would prove useful for UAV communication.

Kurt et al. [22] detailed communication services rely heavily on High Altitude Platform Stations (HAPS), which are network nodes that function in the stratosphere at an altitude of roughly 20 km. The HAPS has become an integral aspect of the future generation of wireless networks to developments in areas such as independent avionics, matrix antennas, photovoltaic effectiveness, and power from battery densities, as well as booming industry ecologies. Based on a thorough and up-to-date literature assessment, presented a future vision and framework for HAPS networks in this article and described in detail how HAPS systems can be used to uniquely benefit urban areas, and emphasize their untapped potential. Recent and exciting developments in the HAPS power and review of payload systems. Cost-effective deployment of HAPS systems is offered by the incorporation of the new Reconfigurable Smart Surface (RSS) technology into the content of the transmitted signal.

Angjo et al. [23] insisted that the present research summarizes the handover management for future mobile networks with an emphasis on connected drones (FMN). As wireless customers or infrastructure in the skies, drones will play a significant role in mobile technology's next evolutionary step. Drones present several opportunities, particularly in the realm of handover management, both for telecommunications-related and unrelated mobile networks. Unlike terrestrial networks, drones operate in a three-dimensional (3D) space, adding another layer of complexity to mobility issues. This publication gives a summary of how researchers are currently tackling the problems that define drones, with a particular emphasis on the handover procedure. A broad understanding of drone integration in heterogeneous networks is provided, and potential issues and their solutions are discussed in detail. In addition, this review provides some suggestions for future studies on connected drones in heterogeneous networks.

Shayea et al. [24] elaborated on multiple studies on the topic of handover management for mobile communication networks (HOM-MCN) using connected drones are summarized in this publication. However, the drone's exposure to antenna side lobes and the free-space transmission behavior at high altitudes are two factors that can affect the radio environment. These distinctions may render conventional mobility models and approaches ineffective for uses using linked drones. Therefore, due to low battery life, packet loss, heavy traffic, and/or rapid flight, drone connections may face severe difficulties. This research helps fill a gap in the existing literature by offering a more narrowly tailored assessment of publications on drone networks, managing mobility for drones, and related topics. Problems with mobility management are emphasized as one of the most significant obstacles to using linked drones. Intelligent handover strategies that make use of deep learning, machine learning, and autonomous resilient procedures are shown to considerably reduce handover problems and related issues when compared to more conventional methods, as discussed in this chapter.

Several well-known approaches to model construction, including the DM, BF & HO, HAPS, FMN, and HOM-MCN, could benefit from minor adjustments. Handover Management for FDN, a model proposed here, however, solves the remaining problems in this area of technology. In the quickly evolving field of drone technology, this underscores the importance of fast transfers across network nodes to make the most of the resources at hand and sustain stable, successful drone operations.

3.3 Handover Management for Future Drone Networks (HOM-FDN)

Future drone applications utilizing rapidly moving UAVs would necessitate a high throughput super reliable wireless network. Due to the small beams used in communication, it is thought that this technology would prove useful for UAV communication. The need for beamforming and the frequent handovers necessitated by the small size of the cells provide a significant problem for such networks. When thinking about beamforming, however, the size of the antenna array becomes important, and the effects of azimuth and elevation must be taken into account particularly as it relates to the handover rate and outage capacity, two aspects of UAV communication that are often overlooked. Numerous strategies have been offered to address these interference concerns. Although it has been recommended that carriers make concessions to support aerial users—for example, by adjusting the BS antenna's tilt, yet can't be sure that this will work. Directed antennas and beamforming, according to several works, will allow the BS to segment users in three dimensions, allowing for more effective ground and air service.

This work helps close the gap in understanding how antenna topologies and beamforming affect handover issues. This work is an intermediate step in developing superior beamforming networks for UAV communications. Different topologies and sizes of antenna arrays have different impacts on changeover behavior and outage cost, which are analyzed. At some point, the benefits of adding additional antenna elements will be outweighed by the problems with signal tracking when using a beam narrower than a pencil. Beam finding will be difficult at first access; however certain work-arounds can be used, such as using broader beams at first to find the user.

Group handover and capacity injection applications are depicted in Figure 3.1. A football game is a good example of an event when a large number of people will be congregating in one location (the stadium) for a short amount of time (say, 90 minutes) steadily increasing. At any given time throughout the game, it is prohibited to have more than one aerial BS for all UEs. As shown in Figure 3.1, UxNB will be more helpful in expanding the BS's capacity. In such a dense deployment situation, capacity

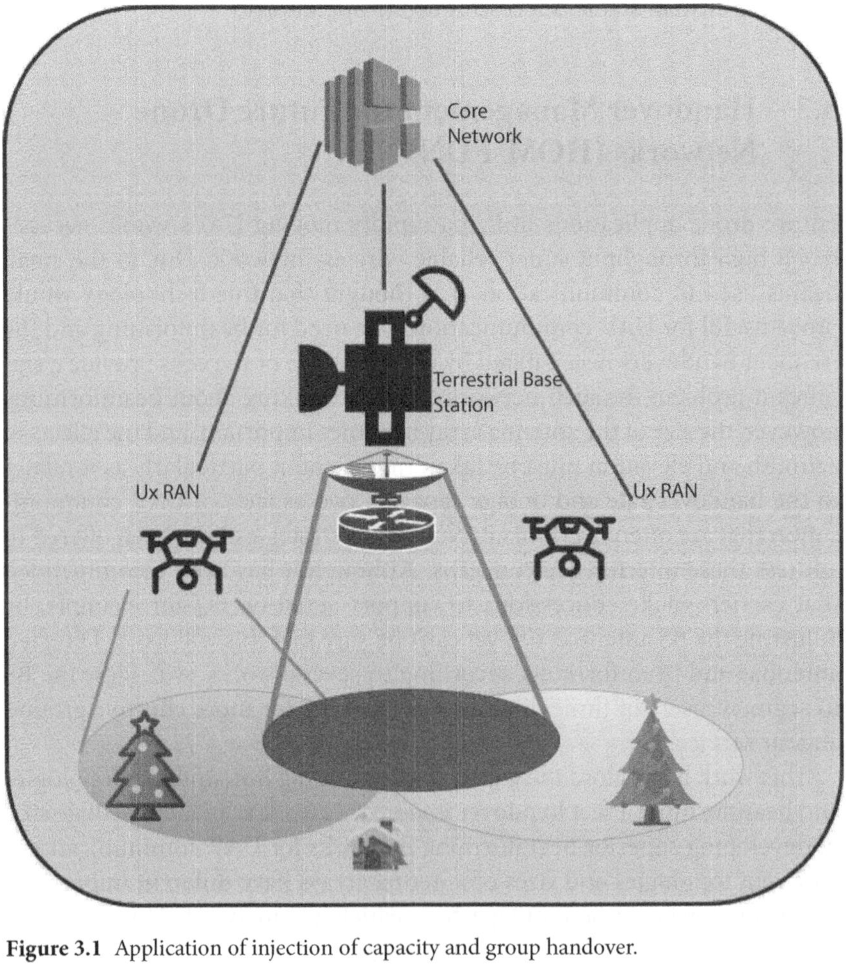

Figure 3.1 Application of injection of capacity and group handover.

injection is the first issue, followed by handover between the overlay cells. Overlay cells occur when many BSs are used in the same area. For the sake of efficiency, the stadium's terrestrial BS that services all UEs will hand off some of the UEs to the UxNB. Many transitions will occur in the meantime. All of these handoffs need to happen in order, one after the other. However, the UxNBs have some constraints, including weight and battery life, which limit their flight time to no more than an hour. A new UxNB will replace the previous one at least once during the course of a football game, which lasts at least 90 minutes. More handoffs will need to be executed as a result of this.

Based on the findings of this research, changing UEs from aerial to UxNB is suggested. Collectively, rather than alone. In densely populated areas, where numerous users have the simultaneous need to connect to a different BS, group handovers are common. A bus full of people is a great setting for observing a group handoff in action. These UEs will see a decline in signal strength from the s-BS at the same time they're getting an increased signal from another BS. Since the UE is responsible for selecting the BS it will get service from, a large number of UEs will simultaneously select the new BS. The new BS's bandwidth will decrease as a result of this. A difficulty in group handoff will arise now. If all UEs have to be transferred individually to the requirements, it will disrupt service and waste energy. The issue with the stadium example is the same. Many UE will seek handover once UxNBs are operational, as they will get better signals from UxNBs. An effective group handoff solution is needed in this situation.

$$UE = min\left(\varnothing \times \sum_{j=1}^{n} BS(x_j^{1/2})\right) \times q(x_j) + HO \tag{3.1}$$

Since UE is the original User equipment function, it follows from (1) which is the open square of the size, which means that further drones can be extracted without impacting the already extracted ones, provided that the user extractor drone maintains its continuity. Therefore, this given equation is responsible for the User-specified content.

In LTE, two distinct handover scenarios depend on whether or not a change in the mobility management entity (MME) already exists as demonstrated in Figure 3.2. This section provides a detailed explanation of inter-BS intra-MME. Without modifying its MME, the UE breaks up its connection with the BS serving (s-BS) and begins communicating with the BS as the target BS (t-BS). This scenario was selected because it

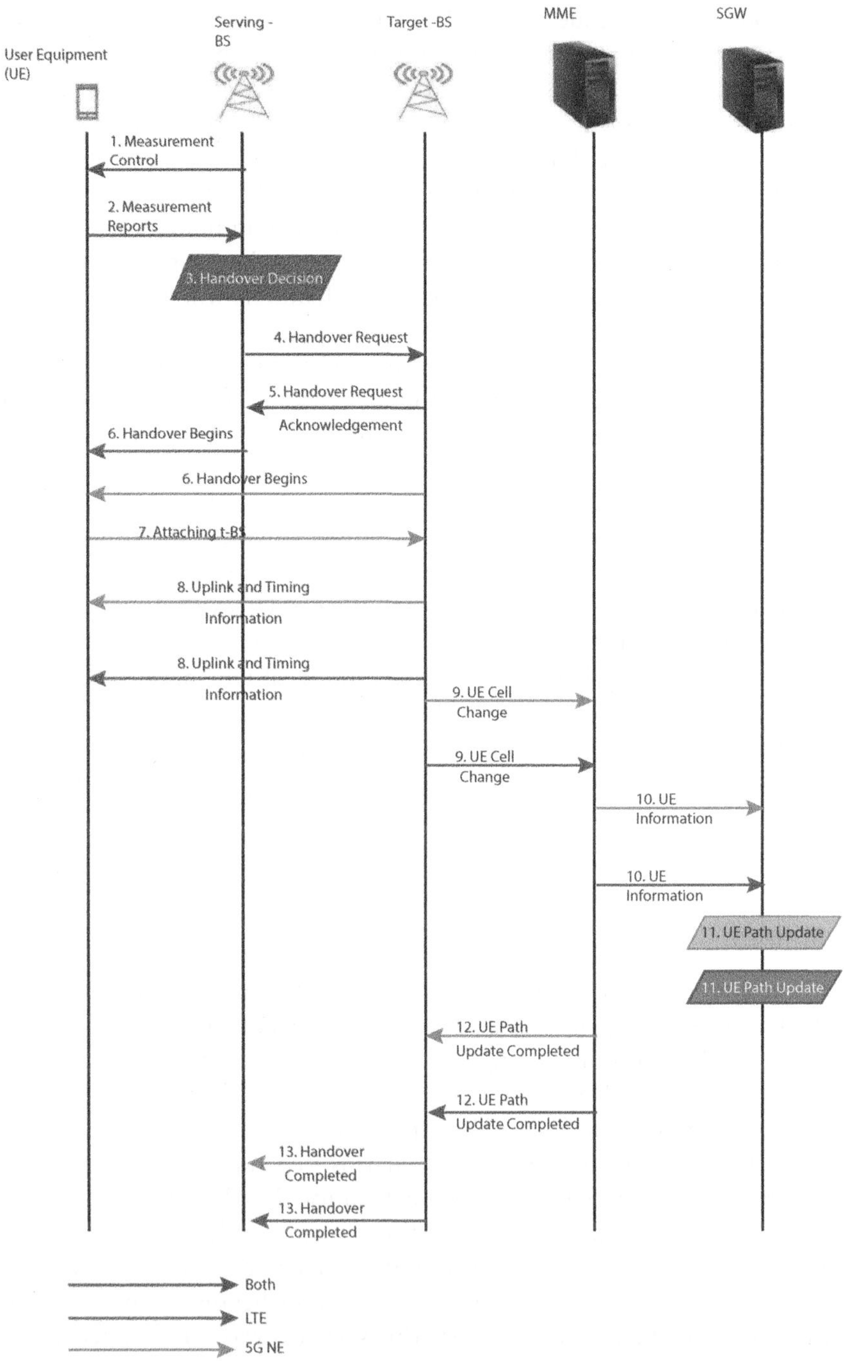

Figure 3.2 LTE and 5G handoff system.

requires fewer interactions than others. The following list shows the handoff procedures.

- The s-BS is responsible for configuring the method used to measure UEs.
- The UE provides the s-BS with a measurement report (MR).
- The research states that the s-BS decides whether or not to perform the handover.
- The s-BS initiates communication with the t-BS by sending a handover request.
- According to its capabilities, the t-BS will relay an acknowledgment to the s-BS.
- The t-BS sends the necessary information to the UE to get it ready for the handover.
- The UE successfully attaches to the target cell.
- Uplink scheduling and allocation data is transmitted to the UE by the t-BS.
- The t-BS alerts the MME when a cell change occurs in the UE.
- The MME relays UE information to the serving gateway SGW.
- SGW revises UE's route.
- The MME alerts the t-BS to perform a path update.
- The handoff is finalized after the t-BS notifies the s-BS.

The handover procedure in 5G networks is quite similar to that of LTE, with a few small differences. The MME here is represented by the Access and (MMF), whereas the Service Gateway Function (SGW) is represented by the User Plane Function (UPF).

$$BS_t = H_R \times H_A = \sum_{j=0}^{n} H_{R_{mj}} H_{R_A} - BS_S \tag{3.2}$$

The above Base station activity is computed using the aforementioned Equation (3.2), where BS_t is the target activity of the Base station, H_R is the Handover request, H_A is the Handover acknowledgment, and the student's attention is computed and BS_S is the serving base station.

Figure 3.3 illustrates a potential network architecture for a swarm of interconnected drones. Drones that are connected to the internet will soon become a significant technology because of the many uses they will find in the future. However, a reliable connection while they are on the go is

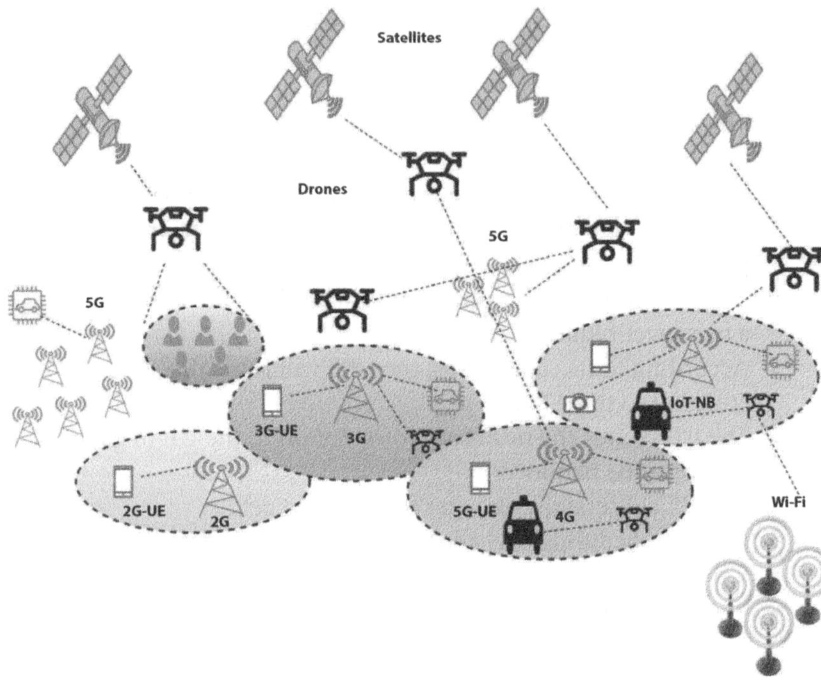

Figure 3.3 Drones in the networks of the future.

the most pressing problem that must be adequately handled. The study's primary objective is to learn about drones in general and HO management specifically. This part therefore presents the overarching ideas of networked drones, the HO method, and the HO management. Drones have seen a meteoric rise in popularity over the past few years due to their adaptability and the variety of problems they can solve. It is feasible for both businesses and consumers to use drones in their operations. If used as BSs, drones can link mobile users wirelessly. They are marketable beyond the realm of communications as well. A mobile user will benefit from reduced latency, increased energy efficiency, and more reliable connection if drones are deployed as BSs. Several quick solutions appropriate for various contexts will be made possible. The linked drone is the future of drone research and precision delivery of services.

$$\sum_{n=1}^{g\_n} D = \sum_{n=1}^{g\_n} \frac{G_0 + g_1 * 1G + g_2 * 2G + \ldots\ldots\ldots\ldots + g_n * nG}{S} \tag{3.3}$$

The Equation (3.3) provides a summation of all drones in the air represented by D and can be used to determine the overall performance. G_0 is the generation of systems, g_2 be the coefficient of the same for each generation, S be the Satellite of the corresponding area of the system.

Drones can be connected to other networks such that they can function as BSs, relay nodes (RNs), or UEs in the air. In the latter case, for instance, data collected by a drone can be sent to a BS *via* an uplink connection. On the other hand, the downlink channel can be used for remote drone operation and monitoring. Drones will collect data for service connections and can be operated to either reach hard-to-reach locations or to manage high connectivity densities. It examines a network in which drones act as relays, transmitting data between mobile devices and the BS. The research aims to find a way to improve cellular coverage for people on the ground. Drones are already being used for a variety of purposes, low-altitude monitoring, disaster relief, logistical application, and communication support are a few of the many possible uses for this technology and these uses are typically enabled by the addition of communication equipment or specialized sensors.

$$UE_i = \{\theta_i - D_i, \quad if\ g_i \in G_s\ for\ S_i \in RN \quad 0 \qquad otherwise \tag{3.4}$$

In Equation (3.4), UE_i stands for the verifier's usefulness of the equipment. The outcome of the auction is determined by $\theta_i - D_i$. The verifier for some satellites is denoted by S_i.

$$HO_{threshold}(m) = (HO_{\hat{n}}(H) + HO_{\hat{n}}(V)) \tag{3.5}$$

Equation (3.5) represents HO handover for threshold management which is the combination of horizontal handover $HO_{\hat{n}}(H)$ and $HO_{\hat{n}}(V)$ represents the vertical handover systems respectively.

$$UE_i(n) = \{1, 0, \widehat{HO}(H) > \hat{n}(n) \forall j \in \{n - l + 1, n - 1\}\ in\ other\ cases + \text{RLF} \tag{3.6}$$

$$BS_{i_{\hat{n}}} n = \{1, 0, \widehat{HO}(V) < \hat{n}(n) \forall j \in \{n - l + 1, n - 1\}\ in\ other\ cases - \text{RLF} \tag{3.7}$$

Equations (3.6) and (3.7), $UE_i(n)$ stands for the verifier's user equipment. $BS_{i_{\hat{n}}} n$ be the base station of the equipment. A subset's verifier, represented by i in the auction notation n, is chosen at random. Using the preceding Equations (3.1), (3.2), and (3.3), the maximum adaptive angle rate of the proposed technique may be calculated.

Effective resource management reduces pollution by achieving the same result while using fewer resources. Although energy resources play a part in accomplishing these goals, boosting energy efficiency has the potential to yield the biggest reward in the shortest amount of time.

Concept of Hand-Over (HO)

When it comes to mobile connections in wireless networks, HO is a crucial strategy for supporting mobility. When enabled, this method allows the UE to maintain its communication link even as it travels to different BSs. The HO method was created to handle mobile connections while UE is mobile, ensuring continuous communication and greater reliability regardless of where users may be. The throughput of the UE is improved, RLF is decreased, and interruption time is shortened to the HO method. Therefore, a higher quality serving network can be achieved by improved HO management. This is why this method receives much interest from telecommunications companies and academic institutions.

In most cases, there are a few different procedures that make up the HO process in a wireless communication system. In the LTE network, for instance, there are three stages to this procedure. In LTE's first phase, the UE transmits the eNodeB (eNB) a Measurement Report (MR) periodically. The MR is used to determine which eNB will take over the UE's connection. The next step is to send the HO request. After the planning phase, the execution phase begins with a communication to the UE about the eNB switch. After that, the UE will exit the serving eNB and try to connect to the desired eNB. The final step of this process is commonly referred to as "completion." The UE's source eNB's allotted resources are freed up, and the system layers are informed to redirect the UE's data traffic.

It describes how effective handover (HO) management in wireless networks sustains UE connectivity during mobility, enhancing throughput and reliability while minimizing Radio Link Failures (RLF) and interruptions. HO's phased approach, particularly in LTE, allows seamless transition between Base Stations (BS), ensuring efficient resource utilization and optimal network performance, which is crucial for minimizing pollution and achieving energy efficiency.

Figure 3.4 depicts the HO between two cell towers as well as the HO between a cell tower and a wide area network (WAN) access point.

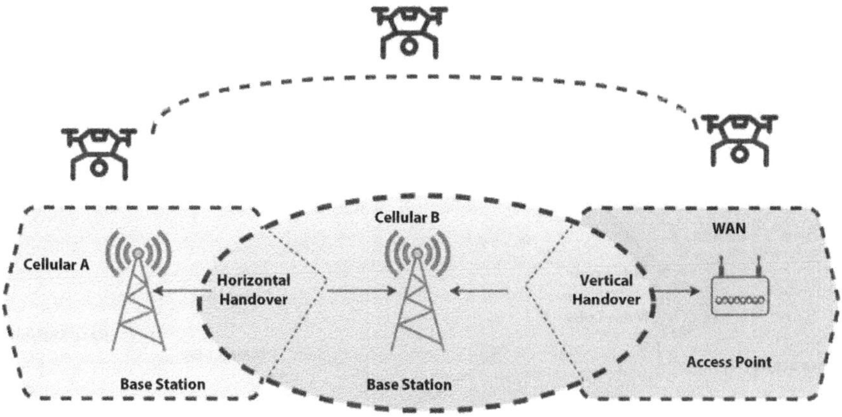

Figure 3.4 Depiction of drones with both horizontal and vertical HO.

Horizontal HO is named because its underlying technology remains constant, while vertical HO gets its name from its orientation. These are the two primary HO scenarios, however, numerous sub-scenarios exist inside each. Equalizing the workload over multiple cells is another area of concern. Some UE may be forced to connect to another BS to receive optimal service and minimal latency if the current BS becomes overloaded. Battery life on devices can be preserved through controlled transmission power to HO's ability to redirect sessions to nearby BSs with lower transmission powers. As a result, the widespread use of networked drones will lead to an increase in HO scenarios.

The primary difficulty encountered by mobile UAVs is exposed in Figure 3.5. Drone-based communication suffers because the antennas' side lobes are oriented upwards while serving airborne users yet downwards when serving ground-based ones. As a result of their low price and high efficiency drones are the most prominent answer to various problems in the future of communication. Since a drone travels in the air, its movement is three-dimensional which presents new difficulties not faced by ground-based systems.

$$BS_{\hat{A}}(m) - BS_B(m) = \frac{\frac{1}{2L}\sum_{j=0}^{L-1} \left| ML\left(\hat{n}(m-j)\right) - SL\left(\hat{n}(m-1-j)\right) \right|}{UE1 + UE2} \tag{3.8}$$

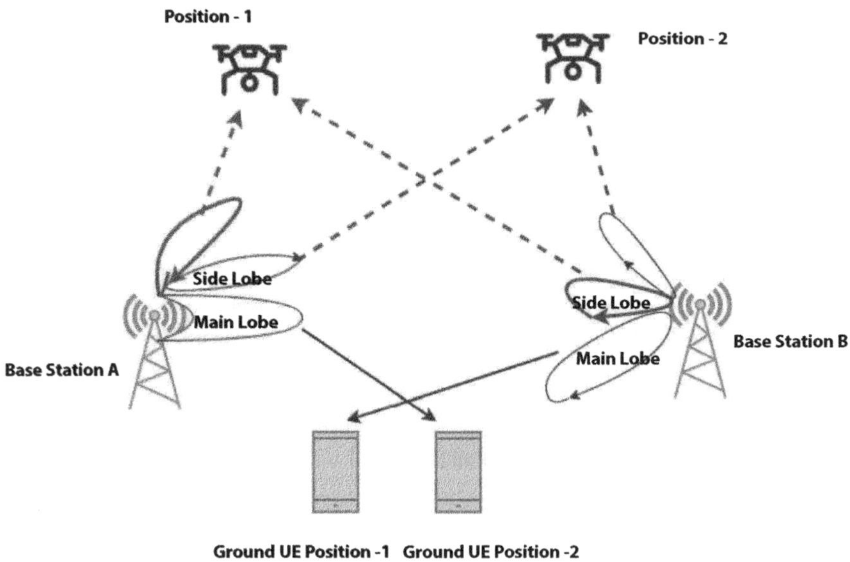

Figure 3.5 UE on the ground and HO of drones in the air.

Since $BS_{\widehat{A}}(m) - BS_B(m)$ is the initial Base Station function A and B respectively, it follows from (8) that is ML main lobe, SL be the side lobe, $UE1$ be the user equipment, $UE2$ be the user equipment.

LoS pathways cause drones to fly higher than they otherwise would, exposing them to more interference from other UAVs. In a nutshell, the HO of these gadgets is affected by many factors, such as elevation, coverage area, population density, etc. Several research has zeroed in on drone mobility management as a means of mitigating these issues and achieving continuous communication. However, a significant difficulty for future networks will be the mobility management of linked drones.

Algorithm 3.1 Emerging UxNB Authentication for Capacity Injection.

Begin

Step 1: Key pairs xi, f(xi)P, and UxNB should be set as private and public, respectively.
Step 2: Send f(xi)P and xi to the ground-based BS.
Step 3: if all pairs (xi, f(xi)P) are correct, then
Step 4: Forward f(x)
Step 5: else
Step 6: UxNB Is Invalid

End

An example of the developing UxNB authentication for capacity injection is demonstrated in Algorithm 3.1. Set the private and public key f(xi) P, xi to the UxNB. Transmit the given pairs to the terrestrial BS. If the pairs are to be functioned then send the corresponding signal f'(x) or else the condition is not valid, then end the corresponding functions.

HOM-FDN is a suggested paradigm and set of management tactics for developing and enacting drone strategies that contribute to environmental economic growth. This research employs the handover notion to provide a secure and reliable prediction and interaction system for AI. Everything has been thoroughly examined and analyzed for accuracy, efficiency, and safety. The algorithm's results are shown to be superior to those of competing approaches.

3.4 Results and Discussion

The simulation research examines the effectiveness, precision, and efficiency of the proposed HOM-FDN. The results from each section are demonstrated as follows:

3.4.1 Accuracy Analysis

The testing technique for the accuracy analysis ratio is illustrated in Figure 3.6. Accuracy is measured along the Y-axis, while the X-axis indicates the number of samples taken. When evaluating and forecasting input data signals, the Accuracy of HOM-FDN shows accuracy performance vs. the sample factor.

3.4.2 Efficiency Analysis

Figure 3.7 depicts one way to calculate the Efficiency analysis ratio. The graph's X-axis represents the total number of samples taken, while the Y-axis displays the Efficiency analysis's findings. Effectiveness of HOM-FDN. It is essential to compare performance against the sample variance factor when assessing and forecasting incoming data signals.

3.4.3 Performance Analysis

One method of calculating the Performance analysis ratio is shown in Figure 3.8. The X-axis indicates the number of samples, while the Y-axis shows how well they performed. When comparing the HOM-FDN's

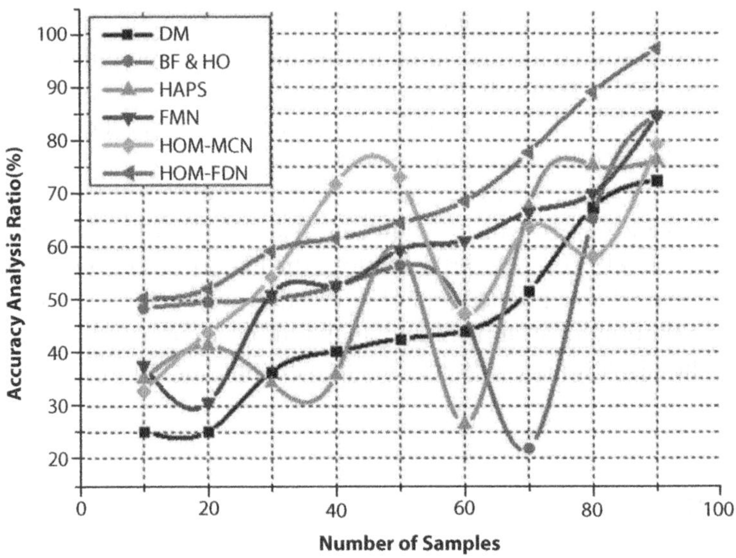

Figure 3.6 Accuracy analysis.

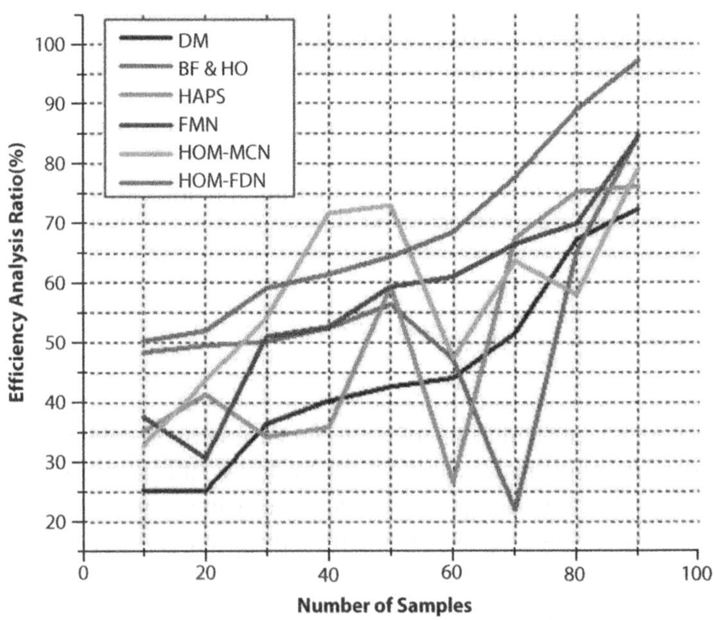

Figure 3.7 Efficiency analysis.

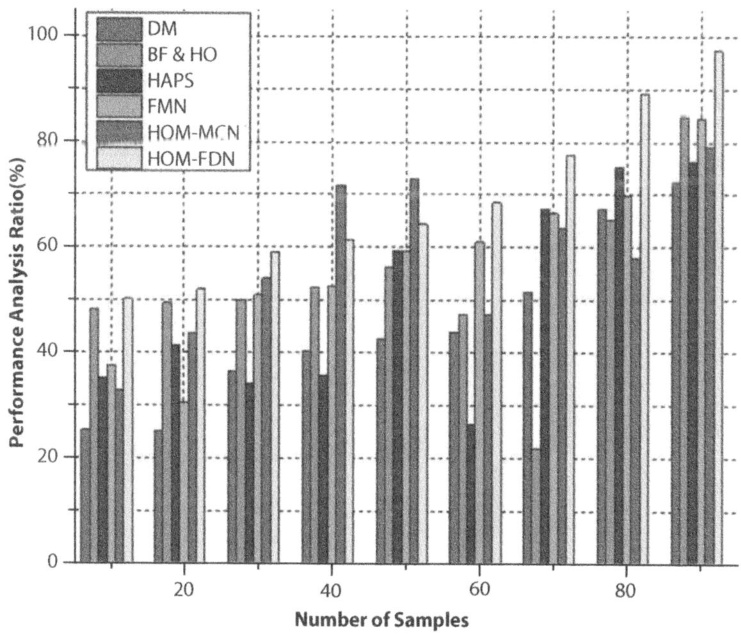

Figure 3.8 Performance analysis.

performance against the temporal variation component of incoming input signals, the efficacy of the HOM-FDN is crucial. The key objectives of this performance analysis are to show the importance of capacity injection for QoS by comparing handover times and the number of command package transfers during group handover.

3.4.4 Safety Analysis

Figure 3.9 exhibits a potential approach for calculating the safety analysis ratio. The HOM-FDN Computational Performance relies heavily on the accuracy of the safety vs. the number of sample variation components to assess and anticipate incoming input signals.

3.4.5 Prediction Analysis

One possible method for determining the prediction analysis ratio is depicted in Figure 3.10. To evaluate and foresee incoming input signals, the HOM-FDN computational prediction largely depends on the accuracy of the forecast vs. the number of sample variation components.

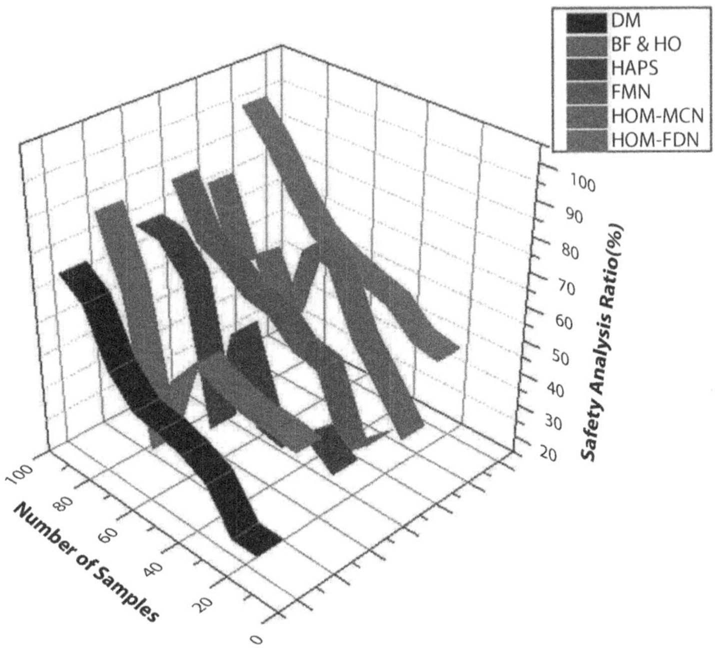

Figure 3.9 Safety analysis.

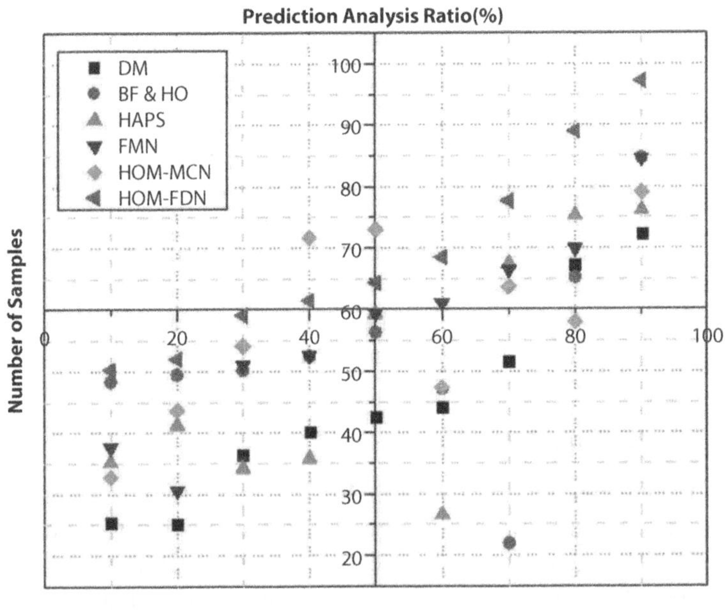

Figure 3.10 Prediction analysis.

The chapter's simulations corroborated the claim that implementing HOM-FDN handover management for the future drone network would enhance prediction precision, performance, safety, and usability. The models are compared in a variety of ways, including by looking at the assumptions made by each model (such as DM, BF & HO, HAPS, FMN, HOM-MCN, and HOM-FDN).

3.5 Conclusion

This chapter outlined the planned situation and described solutions for safe capacity injection and group handoff. For an emergent UxNB to be trusted, it must first be verified as genuine by the nearest terrestrial BS. The key motivations of this chapter are to reduce power consumption by decreasing the amount of control packet broadcasts during the handover and the need to do so in high-density locations. The proposed system allowed the t-BS to perform a private function authentication on a set of UEs without disclosing any security parameter. Private keys of UEs are concealed within the elliptic curve groups because of the discrete logarithm problem. Due to the absence of control packet transfer between the s-BS and the t-BS, the handoff occurs instantly.

The authentication approach can rectify the performance issue present in the Xn link between terrestrial and UxNB. Mobile network security issues need to be actively tackled with the use of artificial intelligence. If the network intends to serve mobile aerial users, recommend using somewhat vertical rectangular antenna arrays. Due to the usual xy-planar motion of UAVs, these arrangements are the most forgiving of horizontal beam misalignment errors. Our research revealed that UAVs flying at higher altitudes experience fewer handovers than those flying at lower altitudes since the relative mobility is lower, yet this comes at the tradeoff of diminished signal strength. In the static case, however, the UAV experiences significantly more side lobes at higher elevations.

In this chapter, the importance of end-to-end security in the 6G network was magnified due to the increasing number of autonomous user devices (UEs). At this point, there may be significant gains from using AI-powered security systems. Due to the proliferation of peer-to-peer connections, it is necessary to disperse authentication ideas instead of centralizing them. Integrating blockchain technology into authentication systems is a good idea which is focused on in this chapter because of its simplicity and its ability to support users located in different locations. The need for a trustworthy middleman was removed by blockchain-based solutions.

Worldwide, 5G mobile networks have now become standard, and work on 6G has already started. It is believed that 6G networks would encounter more intricate security challenges.

References

1. Tanveer, J., Haider, A., Ali, R., Kim, A., Reinforcement Learning-Based Optimization for Drone Mobility in 5G and Beyond Ultra-Dense Networks. *Computers Mater. Continua*, 6, 3807–3823, 2021.
2. Hajiakhondi, Z., Mohammadi, A., Abouei, J., Deep reinforcement learning for trustworthy and time-varying connection scheduling in a coupled UAV-based femtocaching architecture. *IEEE Access*, 9, 32263–32281, 2021.
3. Yang, H., Alphones, A., Xiong, Z., Niyato, D., Zhao, J., Wu, K., Artificial-intelligence-enabled intelligent 6G networks. *IEEE Netw.*, 34, 272–280, 2020.
4. Alshaibani, W.T., Shayea, I., Caglar, R., Din, J., Daradkeh, Y.I., Mobility Management of Unmanned Aerial Vehicles in Ultra-Dense Heterogeneous Networks. *Sensors*, 22, 6013, 2022.
5. Chen, Y., Lin, X., Khan, T.A., Mozaffari, M., A deep reinforcement learning approach to efficient drone mobility support. *arXiv preprint arXiv*:2005.052291, 2020.
6. Hozayen, M., Darwish, T., Kurt, G.K., Yanikomeroglu, H., A Graph-Based Customizable Handover Framework for LEO Satellite Networks, in: *2022 IEEE Globecom Workshops (GC Wkshps)*, pp. 868–873, 2022.
7. Jang, Y., Raza, S.M., Kim, M., Choo, H., Proactive handover decision for UAVs with deep reinforcement learning. *Sensors*, 22, 1200, 2022.
8. Alsamhi, S.H., Almalki, F.A., Ma, O., Ansari, M.S., Lee, B., Predictive estimation of optimal signal strength from drones over IoT frameworks in smart cities. *IEEE Trans. Mobile Comput.*, 22, 1, 402–416, 2023.
9. Al-Rubaye, S., Tsourdos, A., Namuduri, K., Advanced air mobility operation and infrastructure for sustainable connected evtol vehicle. *Drones*, 7, 319, 2023.
10. Khan, S.A., Shayea, I., Ergen, M., Mohamad, H., Handover management over dual connectivity in 5G technology with future ultra-dense mobile heterogeneous networks: A review. *Eng. Sci. Technol. Int. J.*, 35, 101172, 2022.
11. Sorbelli, F.B., Chatterjee, P., Corò, F., Palazzetti, L., Pinotti, C.M., ANovel Multi-Layer Framework for BVLoS Drone Operation: A Preliminary Study. *arXiv preprint arXiv*:2301.05835, 2023.
12. Goudarzi, S., Anisi, M.H., Ciuonzo, D., Soleymani, S.A., Pescapé, A., Employing unmanned aerial vehicles for improving handoff using cooperative game theory. *IEEE Trans. Aerosp. Electron. Syst.*, 57, 776–794, 2020.

13. Mahamod, U., Mohamad, H., Shayea, I., Othman, M., Asuhaimi, F.A., Handover parameter for self-optimisation in 6G mobile networks: A survey. *Alexandria Eng. J.*, 78, 104–119, 2023.
14. Prajith, P., Yokesh, V., Aruchamy, P., Artificial Intelligence-Enabled Fully Echoed Q-Routing and Adaptive Directional Medium Access Control Protocol for Flying Ad-Hoc Networks. *Int. J. Commun. Syst.*, 38, 1–32, 2025.
15. Angjo, J., Shayea, I., Ergen, M., Mohamad, H., Alhammadi, A., Daradkeh, Y., II, Handover Management of Drones in Future Mobile Networks: 6G Technologies. *IEEE Access*, 9, 1–22, 2021.
16. Grasso, C., Raftopoulos, R., Schembra, G., Serrano, S., H-HOME: A learning framework of federated FANETs to provide edge computing to future delay-constrained IoT systems. *Comput. Netw.*, 219, 109449, 2022.
17. Shrestha, R., Bajracharya, R., Kim, S., 6G enabled unmanned aerial vehicle traffic management: A perspective. *IEEE Access*, 9, 91119–91136, 2021.
18. Yogeshwari, S. and Prasanth, R., Open research challenges in wireless technologies: A case study, in: Al-Turjman, F., Devi, Y.M., Eudes, A. (Eds.), *Intelligent Wireless Communications and Internet of Things: Advances and Challenges*. CRC Press, 267–281, 2024.
19. Mozaffari, M., Lin, X., Hayes, S., Toward 6G with connected sky: UAVs and beyond. *IEEE Commun. Mag.*, 59, 74–80, 2021.
20. Aydin, Y., Kurt, G.K., Ozdemir, E., Yanikomeroglu, H., Group handover for drone base stations. *IEEE Internet Things J.*, 8, 13876–13887, 2021.
21. Colpaert, A., Vinogradov, E., Pollin, S., 3D beamforming and handover analysis for UAV networks, in: *2020 IEEE Globecom Workshops (GC Wkshps*, pp. 1–6, 2020.
22. Kurt, G.K., Khoshkholgh, M.G., Alfattani, S., Ibrahim, A., Darwish, T.S., Alam, M.S., Yongacoglu, A., A vision and framework for the high altitude platform station (HAPS) networks of the future. *IEEE Commun. Surv. Tutor.*, 23, 729–779, 2021.
23. Angjo, J., Shayea, I., Ergen, M., Mohamad, H., Alhammadi, A., Daradkeh, Y.I., Handover management of drones in future mobile networks: 6G technologies. *IEEE Access*, 9, 12803–12823, 2021.
24. Shayea, I., Dushi, P., Banafaa, M., Rashid, R.A., Ali, S., Sarijari, M.A., Mohamad, H., Handover Management for Drones in Future Mobile Networks—A Survey. *Sensors*, 22, 6424, 2022.

4

Convergence of Internet of Vehicle Things and Drones: An Interoperability Perspective

V. Yokesh[1]*, N. Sathish[2], S. Lavanya[3] and Pham Chien Thang[4]

[1]Department of Electronics and Communication Engineering, Sathyabama Institute of Science and Technology, Chennai, India
[2]Department of Electronics and Communication Engineering, Sri Venkateswara College of Engineering, Sriperumbudur, India
[3]Department of Computer Science and Engineering, Muthayammal Engineering College, Namakkal, Tamil Nadu, India
[4]Faculty of Journalism and Communication, TNU-University of Sciences, Thai Nguyen, Vietnam

Abstract

The emerging integration of drones and the Internet of Things (IoT) in the vehicle industry, called Internet of Vehicle Things (IoVT), is remarkable. This chapter discusses the cross-section of the IoVT and drones, focusing on the critical issue of interoperability between these diverse technologies. With cars becoming more connected and self-sufficient, drones play a vital role in IoVT, advancing the ecosystem with novel services and applications. This paper highlights the need for integration among information vehicles and drones regarding their communication interfaces, data exchange formats, and other protocols. At their current place in drone technology and IoVT integration, it is mostly about integration. It investigates setting the interoperability requirements for effective communication between drones and vehicles for the joint solution of issues like traffic control, surveillance, and emergency situations.

Keywords: Internet of Things (IoT), Internet of Vehicle Things (IoVT), interoperability, drones, traffic control, surveillance, emergency response

**Corresponding author*: yokesh.inba@gmail.com

A. Prasanth, Rajesh Kumar Dhanaraj, Munish Sabharwal, Vandana Sharma and Seifedine Kadry (eds.) Drones for Transportation Logistics and Disaster Management, (67–94) © 2026 Scrivener Publishing LLC

4.1 Introduction

4.1.1 Overview of IoT in Vehicles and Drones

The concept of smart technology in vehicles and drones is that instead of using manual control, IoT is embedded [1]. IoT in vehicles includes using sensors, actuators, and connectivity solutions for real-time data gathering and analysis. The data can be used for many applications including predictive maintenance, driver assistance, and fleet management. In drones, IoT facilitates embedding sensors for navigation, data collection, and communication, enhancing their capabilities in agriculture, surveillance, and logistics.

For vehicles, IoT technologies are used to monitor engine health, optimize fuel efficiency, and provide connectivity for in-car entertainment as well as navigation systems. On the contrary, it allows handling drones remotely in addition to performing their duties such as mapping, surveillance, data collection, or transmitting data, and swarming or collaborative efforts.

In particular, the primary objective is to build a connected platform in which vehicles and aircraft not only communicate to each other and the operators, but also collectively develop a smart and interactive plant. This connectivity enables positive effects of its operation on the performance, safety, and innovation, as well as on the future state of cars and UAVs naturally and organically coexisting in our daily life without any contribution from the human operator.

4.1.2 Significance of Convergence

In the junction of IoT in automobiles and drones, interoperability is absolutely crucial [2]. Interoperability is the capacity of these technologies to cooperate and interact without regard to hardware, software, or communication protocols. This has numerous important consequences.

Efficiency and Collaboration	Interoperability ensures seamless data exchange between drones and autos, facilitating good cooperation between them. To transmit important information or enhance traffic flow, a drone might, for instance, engage with moving vehicles.

Data Standardization	By methods of interoperability, encouragement of common data standards helps to enable uniform communication between many devices. By improving interoperability, this standardization helps different manufacturers' drones and vehicles to interact better.
Safety	Ensuring safety when autos and drones share the same territory depends on interoperability. Real-time communication of these devices can help to choose the optimal paths, increase general safety measures, and prevent collisions.
Innovation and Integration	Interoperability encourages innovation since it makes multiple technologies able to cooperate and improve one another feasible. For example, drone data can be readily included into vehicle systems to support improved decision-making—such as route optimization grounded on aircraft observations.
Scalability	As the ecosystem of drones and IoT-enabled cars develops, interoperability is growingly important. It ensures that modern technologies can be readily included into present systems, therefore promoting scalability and adaptation to a fast changing technical scene.

4.1.3 Significance of Interoperability

With regard to complex and networked systems such as the IoVT and drones, interoperability is especially crucial in many different technical spheres [3]. These are the primary justifications for the need of interoperability in this context:

Seamless Integration of Technologies	It allows for the seamless integration of diverse technologies, enabling connected vehicles and drones to work together cohesively. Through seamless data, command, and communication flow across various components, this integration improves the system's overall efficacy and efficiency.

Enhanced Collaboration and Coordination	Interoperability fosters collaboration and coordination between connected vehicles and drones. Collaborative efforts are crucial for addressing complex challenges, optimizing traffic management, and creating integrated transportation systems that are responsive to real-time data from both ground and aerial perspectives.
Data Exchange for Informed Decision-Making	Interoperability helps data to be exchanged between drones and linked cars. By giving a complete and real-time knowledge of the surroundings, traffic conditions, and any hazards, this data interchange enables informed decision-making and helps to guarantee safer and more effective operations.
Efficient Resource Utilization	Interoperability optimizes the utilization of resources, including communication bandwidth, computing power, and energy. By streamlining communication protocols and data exchange formats, interoperability ensures that resources are used efficiently, avoiding redundancy and reducing the overall system overhead.
Scalability and Future-Proofing	Interoperability promotes scalability and future-proofing of the system. As the number of connected vehicles and drones increases, interoperable systems can easily accommodate growth without requiring significant modifications. It also allows for the incorporation of emerging technologies and standards.
Flexibility in System Design	Interoperability provides flexibility in designing and deploying systems that include both connected vehicles and drones. This flexibility is crucial for adapting to evolving technologies, accommodating diverse use cases, and integrating components from different manufacturers or service providers without facing compatibility issues.
Cost-Effectiveness and Return on Investment	Interoperability reduces development and implementation costs. When systems are interoperable, organizations can leverage existing infrastructure and technologies, avoiding the need for extensive customizations or proprietary solutions. This results in a more cost-effective implementation and a faster return on investment.

User Experience and Accessibility	Interoperability contributes to a better user experience and accessibility. Users, whether they are operators, drivers, or administrators, benefit from systems that seamlessly work together. Interoperable solutions enhance accessibility, usability, and user satisfaction.
Regulatory Compliance	In order to ensure the safety and legality of operations, industry standards and regulations must be followed, and interoperable systems are more likely to comply with established rules and regulations. Interoperability aids in meeting regulatory requirements and standards.
Innovation and Ecosystem Growth	Interoperability fosters innovation and supports the growth of a collaborative ecosystem. By breaking down silos and encouraging collaboration between different stakeholders, interoperability creates an environment conducive to innovation. It enables the development of new applications, services, and business models that leverage the combined capabilities of connected vehicles and drones.

4.2 Communication Standards for Integration

The IoVT and drone integration depend heavily on communication protocols [4]. By establishing standard protocols, it is ensured that drones and connected vehicles may share data, coordinate operations, and interact successfully. These are the major communication protocols that are pertinent to the IoVT and drone integration.

4.2.1 Vehicular Communication Standards

- ***Dedicated Short-Range Communication (DSRC):*** The DSRC Service contributes to the public's safety by facilitating communications between vehicles and infrastructure. By alerting drivers of a potentially dangerous situation or incident in time for them to take preventative or evasive action, it can save lives. Non-public safety organizations may also utilize the band for their own commercial or private DSRC operations.
- ***Cellular Vehicle-to-Everything (C-V2X):*** C-V2X [5] is a platform for linked mobility that enables cars to communicate with other cars, cyclists, pedestrians, road infrastructure, and

mobile networks, among other things. C-V2X makes use of two complementary transmission modalities in order to provide connectivity within the larger transportation ecosystem.

Firstly, automobiles can connect with other road users (cyclists, pedestrians, or other vehicles) thanks to direct communications (PC5), which operates independently of cellular networks. Second, vehicles may now access real-time information around local traffic and road circumstances thanks to network communications (Uu), which makes use of traditional mobile networks. Through vehicle connectivity and the creation of intelligent transport systems that work together to minimize traffic and pollution, C-V2X has the potential to transform the way we see traffic data to improve travel and boost safety on the roads.

- ***ISO 15118 (Road Vehicles–Vehicle-to-Grid Communication Interface):*** The increasing number of EVs that are connected to the electrical grid allows for precise matching of the grid's capacity with the energy demand thanks to the smart charging mechanism incorporated into ISO 15118 as shown in Figure 4.1. Bidirectional energy transmission is also made possible by ISO 15118, allowing EV energy to be fed back into the grid as needed, thus realizing vehicle-to-grid applications. More grid-friendly, safe, and convenient EV charging is made possible by ISO 15118.

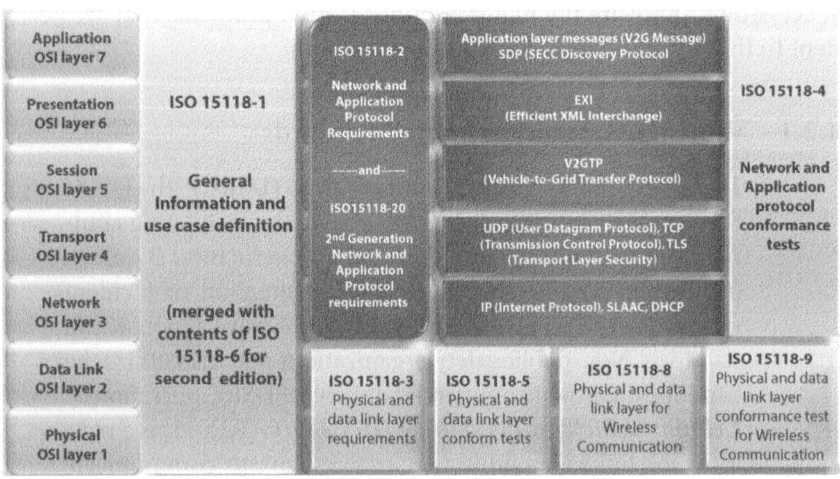

Figure 4.1 Seven layers of ISO/OSI and their connection w.r.t eight parts of ISO 15118.

4.2.2 Drone Communication Standards

- ***MAV link (Micro Air Vehicle Communication Protocol):*** The technique for message structure and application layer serialization is defined by the MAV Link protocol [6]. After that, these signals are sent to the lowest layers of the network—the transport layer and physical layer, for example. The benefit of the lightweight construction of the MAV Link protocol provisions various transport layers and medium types. It can be sent *via* low bandwidth serial telemetry channels that operate at sub-GHz frequencies, such as 433 MHz, 868 MHz, or 915 MHz, Wi-Fi, Ethernet (i.e., TCP/IP Networks), or both.
- ***Dronecode:*** Dronecode is an open-source platform that includes communication standards for drones. Provides a unified framework for communication, mission planning, and control, supporting interoperability among different drone hardware and software.

ASTM F38 UAV (Unmanned Aircraft Systems): Seven new standards covering all significant aspects of small unmanned aircraft systems operations, including design, building, operation, and maintenance requirements, have been adopted by the ASTM International Committee F38 on Unmanned Aircraft Systems. These standards are depicted in Figure 4.2. Now certified by F38, the following seven ASTM standards are intended for any unmanned aerial system (sUAS) that is allowed to operate over

F2908	Specifications for Aircraft Flight Manual (AFM) for a small Unmanned Aircraft System (sUAS)
F2909	Practice for Maintenance and Continued Airworthiness of Small Unmanned Aircraft Systems (sUAS)
F2910	Specification for Design and Construction of a Small Unmanned Aircraft System (sUAS)
F2911	Practice for Production Acceptance of Small Unmanned Aircraft System (sUAS)
F3002	Specification for Design of the Command-and-Control System for Small Unmanned Aircraft Systems (sUAS)
F3003	Specification for Quality Assurance of a Small Unmanned Aircraft System (sUAS)
F3005	Specification for Batteries for Use in Small Unmanned Aircraft Systems (sUAS)

Figure 4.2 Seven standards of ASTM International Committee F38 on UAS.

a specific region and inside airspace as decided by a country's regulatory aviation body.

4.2.3 Cross-Domain Communication Standards

IEEE 802.11p (Wireless Access in Vehicular Environments): [7–9] IEEE 802.11p is a standard for wireless communication in vehicular environments, often used in conjunction with DSRC. Facilitates communicate among vehicles and drones, supporting applications such as traffic coordination, collision avoidance, and cooperative sensing.

- ***MQTT (Message Queuing Telemetry Transport):*** The MQTT protocol is a machine-to-machine (M2M) Internet of Things networking protocol. It is a very lightweight publish-subscribe message transport protocol. When bandwidth is low, this protocol may be helpful for connecting to a remote location. These characteristics allow it to be used in a wide range of situations, including continuous contexts such as those involving communication between machines and the Internet of Things. This publish and subscribe mechanism allows us, as clients, to send and receive messages. It makes it simple for numerous devices to communicate with one another. It is a simple communication protocol designed for low-bandwidth devices, making it perfect for Internet of Things applications.
- ***5G (Fifth Generation) Networks:*** 5G networks will need to accommodate a wide range of demanding latency, throughput, capacity, and availability requirements to enable fresh possibilities. The exciting potential of 5G, an integrated connectivity fabric, lies in its unified design, which allows it to support a wide range of applications with very different reliability and throughput requirements by slicing the network. Operators can enable distinct architectural flavors and deployments for different business models, use cases, or service groups with this network-slicing idea as shown in Figure 4.3. On a shared network infrastructure, they can execute every network instance concurrently.

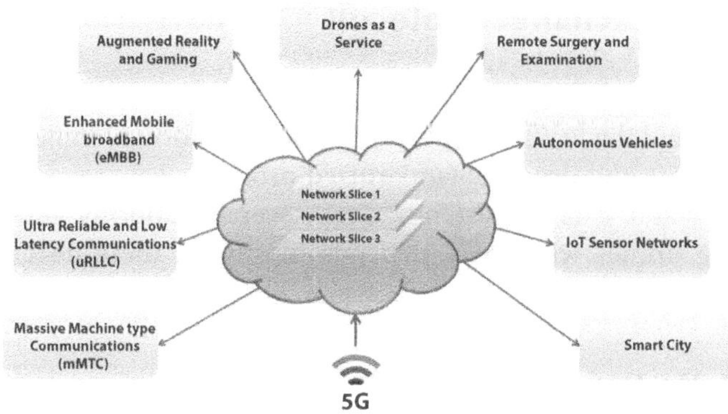

Figure 4.3 5G Networks with network slicing and its architectural flavors.

4.2.4 Interoperability Considerations

- ***Open Geospatial Consortium (OGC) Standards:*** The foundation of FAIR (Findable, Accessible, Interoperable, and Reusable) geographic information is the OGC Standards. Thousands of companies worldwide have been using OGC Standards for the past thirty years to guarantee interoperability and optimize the value of their geospatial data. OGC Standards provide a safe base for geospatial innovation, established and supported consensually by governments and companies all around.
- ***ISO/IEC 20078 (Intelligent Transport Systems – Framework for Cooperative ITS):*** Cooperative Intelligent Transport Systems (ITS) communication finds a structure in ISO/IEC 20078. Encouraging interoperability in mixed-use situations, supports cooperative communication between infrastructure, cars, and other organizations.

Following these communication guidelines will let IoVT and drone integration move more naturally, thereby facilitating effective coordination, data exchange, and cooperation between connected cars and unmanned aerial systems. These criteria enhance the whole efficiency and operation of the combined aviation and transportation systems as well as contribute to build more coherent and compatible surroundings.

4.3 Data Exchange Protocols

Data interchange mechanisms are required in the framework of drone integration and the IoVT to enable the seamless information flow between connected cars and drones. Standardized protocols help to send, receive, and understand data consistently over several platforms and systems. Integrating drones with IoVT depends mostly on these data exchange protocols:

4.3.1 IoVT Data Exchange Protocols

- ***Vehicle Information Access (VIA) Protocol:*** One can repeatedly transmit vehicle-related data including real-time information on position, speed, and status. Permits vital vehicle data to be accessed and shared by IoVT applications for fleet monitoring, traffic control, and cooperative driving.
- ***OpenXC (Open Extensible Car):*** Vehicle data can be shared in a common format with OpenXC, an open-source data exchange platform. Facilitates the sharing of information for research, development, and third-party applications, including sensor readings, vehicle diagnostic data, and other data.
- ***GENIVI (Geneva In-Vehicle Infotainment):*** In order to promote interoperability, GENIVI guidelines establish protocols for in-car infotainment systems. Makes it easier for connected cars to share multimedia and entertainment data, which helps to provide a consistent in-car experience.

4.3.2 Drone Data Exchange Protocols

- ***Drone API:*** Drones and ground control stations can communicate with each other using the open Drone API standard. Promotes interoperability across various drone systems by enabling the sharing of mission data, telemetry, and control orders.
- ***ROS (Robot Operating System):*** Despite being created originally for robots, ROS is frequently utilized in drone applications to facilitate component-to-component communication. Facilitates the drone ecosystem's sharing of sensor data, control orders, and other information.

- ***DDS (Data Distribution Service):*** Drone applications frequently use DDS, a middleware protocol for real-time data dissemination. Allows for the low-latency transfer of time-sensitive data, such as sensor readings and control commands.

4.3.3 Cross-Domain Data Exchange Protocols

- ***MQTT (Message Queuing Telemetry Transport):*** Drones and IoVT are two examples of IoT applications that can benefit from the efficient and lightweight communications technology MQTT. Facilitates publish-subscribe real-time data sharing, allowing connected cars, drones, and backend systems to communicate effectively. Furthermore, MQTT specifies three service quality categories, ranging from lowest to greatest, based on reliability:
- ***CoAP (Constrained Application Protocol):*** A web transfer protocol called CoAP should be utilized with constrained networks that have low availability and low bandwidth. It supports the REST model and has a client/server architecture akin to HTTP. Servers provide resources offered *via* URLs, and clients can submit requirements of the GET, POST, PUT, and DELETE forms. Since the CoAP communication channels are UDP-based and 1:1, delivery cannot be assured. CoAP is designed to function in extremely crowded networks, when nodes are infrequently down and lack intelligence.
- ***JSON (JavaScript Object Notation):*** For data interchange between systems, JSON is a popular lightweight data interchange format. Gives linked cars, drones, and other components a way to exchange structured data in a format that is easy to understand and legible by humans. The levels of MQTT are shown in the Figure 4.4.

Level 0	There is no guarantee of the message delivery
Level 1	The delivery is guaranteed, but it is possible to receive duplicate messages
Level 2	The delivery is guaranteed and there will be no duplicates

Figure 4.4 MQTT levels.

4.3.4 Interoperability Considerations

- ***ODP (One Data Protocol):*** ODP seeks to promote interoperability by standardizing data representation and communication across a range of IoT applications. Facilitates uniform data interchange formats, which lowers complexity and guarantees system and device compatibility.
- ***OCF (Open Connectivity Foundation):*** OCF creates standards for interoperable and safe Internet of Things connectivity. Guarantees consistent data transfer and communication between different Internet of Things devices, such as drones and connected cars.
- ***Schema.org:*** Schema.org provides a shared vocabulary for organized data on the web and sets guidelines for semantics and data formats to enable information exchange among several platforms, systems, and fields.

By using specific data exchange techniques, the IoVT and drone integration may guarantee consistent and compatible data sharing, thereby accomplishing standardized communication [10]. This helps the ecosystem to be more coherent and efficient so that linked cars and drones may cooperate without any difficulties.

4.4 Integration with Edge Computing

A key component of the convergence between the IoVT and drones [11–13] is integration with edge computing. Edge computing is data processing close to the source of generation instead of depending just on centralized cloud servers. Time-sensitive projects, real-time decision-making, and lightening of network bandwidth demand all depend on this approach in special value. These are important factors for combining edge computing with IoT and drones.

4.4.1 IoVT Integration with Edge Computing

- ***Edge Data Processing in Connected Vehicles:*** Using edge computing to locally process data within connected cars. Reduces dependency on centralized cloud resources, improves real-time decision-making for onboard systems, and lowers latency for important apps.

- ***Edge-based Analytics for Traffic Management:*** Real-time analytics at road segments and traffic crossings using edge computing. Provides faster traffic congestion monitoring, traffic signal timing optimization, and localized decision-making for traffic flow control.
- ***Edge-Assisted Autonomous Driving:*** Including edge computing to help functions of autonomous driving. Faster processing of sensor data, local vehicle decision-making enabled by it, and instantaneous responses to changing road circumstances improve safety.
- ***Local Edge Storage for Vehicle Data:*** Edge-based storing and processing of previous vehicle data without depending just on centralized cloud storage allows effective recovery of historical data for analytics, predictive maintenance, and compliance reporting.

4.4.2 Drone Integration with Edge Computing

- ***On-Board Edge Computing for Drones:*** Empowering drones with edge computing capacity for board data processing lowers the need for constant contact with far-off computers, improves real-time sensor data processing, and lets responses to environmental circumstances be faster.
- ***Edge Analytics for Drone Imaging:*** Edge processing of drone-captured pictures and video streams for instant analysis. Provides real-time item identification, terrain mapping, and situational awareness absent of the need to broadcast large volumes of raw data to a central location.
- ***Edge-Assisted Drone Navigation:*** Edge computing for real-time drone flying obstacle avoidance and navigation. Provides flight safety, lessens dependency on constant connection with a central control system, and lets drones make quick decisions depending on local facts.

4.5 Security and Privacy Measures

Data availability, integrity, and secrecy must be guaranteed if connected cars and drones are to operate safely and effectively [14]. The convergence of drone technology and the IoVT calls for careful security and privacy protection.

4.5.1 Security Measures

- ***End-to-End Encryption:*** Establishing strong encryption techniques to guard messages among networked autos, backend systems, and drone communications guarantees that only authorized agencies may access and decode the data, so safeguarding private information against illegal access or modification [15].
- ***Secure Authentication and Authorization:*** Using robust authentication protocols to validate linked autos, drones, and user, user, authenticity. Establishing permission limits to offer the appropriate degree of access. Reduces the likelihood of undesired access, prevents bad actors from gaining system access, and ensures that companies may only make use of their necessary resources.
- ***Secure Boot and Firmware Updates:*** Linked autos and drones can boot with guaranteed only certified and validated firmware, implementing safe firmware updating policies. Guarantees the integrity of the program and reduces the potential of unlawful access by preventing evil modification of device firmware.
- ***Intrusion Detection and Prevention Systems (IDPS):*** Implementing systems that, by constant monitoring network traffic, can identify and terminate detrimental activity, rapidly finds and reduces probable security hazards, therefore improving the general security posture of the integrated system.
- ***Secure Data Storage and Processing:*** For linked autos and drone systems' processed and stored data, using access restrictions and encryption, prevents undesired access by safeguarding important data both in use and in rest.
- ***Tamper-Resist Hardware:*** Choose devices with security features built in to stop actual physical manipulation. By preventing illicit physical access and manipulation, it enhances the general security of drones and linked autos.
- ***Secure Over-the-Air (OTA) Updates:*** By using secure techniques for remote firmware and software upgrades, one guarantees that updates are tamper-proof, validated, and legitimate, therefore preventing malicious actors from using update procedures.

4.5.2 Privacy Measures

- ***Data Anonymization:*** Personal identifiable information (PII) is removed from data prior to transmission or stor age and forbids the linking of some people with user data therefore lowering the likelihood of unwanted tracking or profiling.
- ***Consent Management:*** Establishing protocols to monitor and guarantee user consent for the assembly and usage of private information. Guarantees of consumers' control over the use of their data, therefore promoting transparency and respecting personal privacy preferences.
- ***Privacy by Design:*** including linked infrastructure, connected autos, and drone architecture and design with respect for privacy; guarantees that privacy is immediately considered, therefore reducing the likelihood of privacy violations and facilitating compliance with privacy regulations.
- ***User-Controlled Data Sharing:*** User-owned data sharing allows users to choose with whom and what information is distributed. Provides consumers with the knowledge required to choose whether or not to divulge their data, therefore fostering a more private society.
- ***Data Minimization:*** It is the method of compiling and evaluating the least necessary data for a specific objective. Reducing the amount of private data kept in the system helps to improve privacy by minimizing the probable effects of a data breach.
- ***Transparent Privacy rules:*** Providing consumers with explicit privacy rule knowledge including details on data collecting, storage, and use. Shows a commitment to privacy compliance, builds user confidence, and helps people to make informed decisions.
- ***Secure Communication Channels:*** guaranteeing encrypted and safe communication among networked cars, backend systems, and drone operators; stops illegal listening to sensitive data while it is in route.

By implementing both of these security and privacy protections, IoVT and drones together can better safeguard data, respect user privacy, and ensure the general security and dependability of the connected ecosystem.

4.6 IoVT Regulations

Within the automobile sector, the IoVT functions according to a set of standards that provide consistency, security, and interoperability across different connected car technologies.

4.6.1 Automotive Industry Standards

These standards are developed in collaboration with numerous industry consortia and standards bodies. These are some essential IoVT standards for the automotive sector under Table 4.1:

Table 4.1 Industrial automotive standard.

Standards	Name	Description
ISO 20078-1:2018	***Road Vehicles: Interaction between the vehicle and external devices for diagnostics linked to emissions – Part 1: Overview and use case description***	Standardizes communication for emissions-related diagnostics in road vehicles. Enables consistent diagnostics and emission control across vehicles.
ISO 14229	***Unified Diagnostic Services (UDS)***	Outlines the requirements for diagnostic services that are independent of data links, allowing for standardization of diagnostic communication. Facilitates the diagnosis and maintenance of vehicles through a standardized protocol.
ISO 20078-2:2018	***Road Vehicles – Interaction among vehicle and external equipment for emissions-related diagnostics – Part 2: Guidance on terms, meanings, shortenings and abbreviations***	Guides terms, abbreviations, and shortenings related to emissions-related diagnostics communication. Ensures clarity and consistency in communication standards.

(*Continued*)

Table 4.1 Industrial automotive standard. (*Continued*)

Standards	Name	Description
ISO 20078-3:2018	*Road Vehicles: Interaction between the vehicle and external devices for diagnostics linked to emissions – Part 3: Electrical circuits and diagnostic connectors, security access, vehicle network connection, and performance requirements*	Addresses diagnostic connector and related electrical circuits, security access, communication with vehicle networks, and performance requirements. Defines standards for physical connections and security access in diagnostics.
ISO 11898	*Controller Area Network (CAN)*	Specifies the high-level capabilities for Controller Area Network (CAN) communication. A widely used standard for in-vehicle connectivity, therefore enabling communication between several electronic control systems.
ISO/SAE 21434	*Road vehicles – Cybersecurity Engineering*	Addresses the cybersecurity of road vehicles, including the protection of safety-relevant systems from cybersecurity threats. Essential for ensuring the security of connected vehicles and preventing cyber-attacks.
SAE J3061	*Cybersecurity Guidebook for Cyber-Physical Vehicle Systems*	Delivers strategies for developing cybersecurity processes and strategies for cyber-physical vehicle systems. Assists in establishing cybersecurity best practices in the automotive industry.

(*Continued*)

Table 4.1 Industrial automotive standard. (*Continued*)

Standards	Name	Description
W3C	***Vehicle Information Service Specification (VISS)***	A standard for providing access to vehicle information through web services, enabling interoperability in the IoVT. Facilitates the development of web-based applications and services that interact with vehicles.

4.6.2 Compliance with Road Safety Regulations

Implementing IoVT technologies depends critically on following traffic rules [16]. By including IoVT, additional components meant to ensure the safe running of vehicles on roads and highways are incorporated that have to follow road safety rules. In the framework of IoVT, these are the important elements to consider while following traffic safety regulations:

4.6.3 Drone Regulations

Drone airspace management refers to a collection of guidelines and rules meant to guarantee the effective and safe integration of unmanned aerial vehicles (UAVs) into the airspace [17]. Different companies and regulatory agencies help these standards to grow. Key guidelines and rules pertaining to drone airspace management follow here:

4.7 Cross-Industry Collaboration

Cross-industry collaboration refers to the cooperative efforts between organizations and entities from different sectors to address challenges, leverage expertise, and create innovative solutions [18]. Such collaborations ease the exchange of knowledge, resources, and technologies, leading to the advancement of integrated and holistic solutions that span multiple industries. List of major factors that account for safety regulations and standards of drone regulations are tabulated in Tables 4.2 and 4.3 respectively.

Table 4.2 List of major factors that account for safety regulations.

Factors	Description
Vehicle-to-Everything (V2X) Communication Standards	Make sure that the regulations governing safe and dependable communication between automobiles, infrastructure, and other entities are followed by the V2X communication standards.
Driver Assistance Systems	Follow safety rules when using Advanced Driver Assistance Systems (ADAS). Without sacrificing the driver's focus and composure, these technologies ought to improve driving safety.
Data Privacy and Security	Respect data privacy laws to safeguard the private information that IoVT systems gather and transfer. Put strong cybersecurity safeguards in place to stop illegal access and data breaches.
Vehicle Diagnostics and Emission Standards	Make that IoVT systems abide by laws about emissions and vehicle diagnostics. This is especially crucial for upholding environmental regulations.
Autonomous Driving Regulations	Honor the laws pertaining to autonomous cars for IoVT systems engaged in independent driving. This provides rules on security, testing, and handling other drivers.
Liability and Insurance Regulations	Discuss insurance and liability issues related to IoVT technologies. It is crucial to abide by laws about insurance coverage for connected and autonomous vehicles.
Traffic Management and Control	IoVT solutions must be compliant with laws about traffic control and management. This involves cooperating with intelligent traffic management systems and adhering to traffic signal procedures.
Emergency Services Integration	Make sure emergency services are integrated. Conforms to rules for prompt and efficient action in the event of an emergency or accident.
Human-machine interface (HMI)	To avoid distractions and guarantee secure interactions, abide by user interface requirements if IoVT adds new aspects to the Human-Machine Interface.

Table 4.3 Drone regulation standards.

Standards	Description
ASTM F3411 - Standard Specification for Remote Identification (RID) and tracking of unmanned Aircraft Systems (UAS)	Create guidelines for unmanned aircraft systems (UAS) tracking and remote identification. It supports UTM by enabling the identification of drones in the airspace.
ICAO (International Civil Aviation Organization) Annex 2 - Rules of the Air	Defines the rules of the air, including provisions related to the conduct of aircraft, signaling, and right of way. Relevant for drones operating in controlled and uncontrolled airspace.
ICAO Annex 8 - Airworthiness of Aircraft	Provides standards for the airworthiness of aircraft, including those related to drones. Defines the necessities for the certification of UAVs.
FAA (Federal Aviation Administration) part 107 - Small Unmanned Aircraft Systems (sUAS) Regulations	Describes the rules governing the use of tiny unmanned aerial vehicles (drones) in the US. Covers pilot certification, operational limitations, and safety requirements.
EASA (European Union Aviation Safety Agency) Basic Regulation and U-Space Regulation	The Basic Regulation and U-Space Regulation outline the regulatory framework for the benign integration of drones into European airspace.
ISO 21384-3: 2019 - Unmanned aircraft systems - Part 3: Operational Procedures	Provides standards for operational procedures for unmanned aircraft systems. Addresses aspects such as pilot qualifications, operations manuals, and emergency procedures.

(*Continued*)

Table 4.3 Drone regulation standards. (*Continued*)

Standards	Description
NASA UTM (Unmanned Aircraft Systems Traffic Management)	Focuses on developing technologies and concepts for safely managing low-altitude airspace for drones. Contributes to the development of UTM standards.
ICAO Circular 328 – Unmanned Aircraft Systems (UAS)	Provides guidance on regulatory considerations for unmanned aircraft systems. Addresses certification, operations, and airworthiness.
ANSI (American National Standards Institute)/ASSURE SUAS Standard – ANSI/ASSURE 1:2020	Developed by the Association for System Safety of UAS through Research Excellence (ASSURE), this standard provides testing and evaluation criteria for small unmanned aircraft systems (sUAS).

4.7.1 Synergies between IoVT and Drones

Data Fusion and Enhanced Insights	Connected vehicles generate a wealth of data related to vehicle performance, location, and environmental conditions. Drones provide aerial perspectives and additional spatial data. Combining IoVT and drone data allows for comprehensive insights, aiding in decision-making processes, such as traffic management, environmental monitoring, and infrastructure inspection.
Real-time Monitoring and Surveillance	Connected vehicles contribute to real-time monitoring through integrated sensors and communication capabilities. Drones offer aerial surveillance capabilities, enabling real-time monitoring of large areas. The combination of vehicle and drone data enhances situational awareness, benefiting applications like traffic monitoring, security surveillance, and emergency response.
Efficient Traffic Management	Connected vehicles provide data on traffic conditions, contributing to dynamic traffic management. Drones can survey traffic patterns from the air, offering additional insights for optimizing traffic flow, identifying congestion points, and supporting efficient transportation systems.

Last-Mile Delivery and Logistics	Connected vehicles in logistics contribute to route optimization and real-time tracking of shipments. Drones can complement last-mile delivery services by reaching destinations that are challenging for traditional vehicles. The integration allows for a more flexible and efficient logistics network.
Search and Rescue Operations	Connected cars give search and rescue operations location data and communication tools. Rapid coverage of vast areas by drones allows search efforts to have overhead views and aids in the locating of missing people or damage assessment of disaster-torn areas.
Infrastructure Inspection and Maintenance	Roads and infrastructure conditions are monitored and inspected in part by vehicles. Aerial infrastructure inspections using drones—which provide visual data for maintenance planning—of bridges, electricity lines, and pipelines help to The mix raises infrastructure management's efficiency.
Emergency Response and Disaster Management	Connected cars enable real-time accident and event data for emergency agencies. Through aerial views of impacted areas, damage assessment, and support of search and rescue operations, drones help with disaster response. The combined strengths enhance general emergency management.
Environmental Monitoring	Environmental sensors on some vehicles help to monitor pollution and air quality. For environmental data collecting, drones can cover vast distances covering tracking of changes in ecosystems, monitoring of wildlife, and evaluation of deforestation. Collectively, they present a whole picture of the state of the surroundings.
Collaborative Autonomous Systems	Autonomous systems and vehicle-to-everything (V2X) communication are developing in part from connected cars. Drones can run both independently and cooperatively alongside connected cars. Combining their strengths results in more intelligent transportation systems and coordinated autonomous driving.

Collaborative Surveying and mapping	vehicles help to map data using GPS and other sensors. For surveillance and mapping, drones can provide 3D models and high-resolution maps. The cooperation supports projects in land surveying, building, and urban planning.

IoVT and drones together build a strong ecosystem whereby ground-based and aerial data collecting and analysis cooperate to solve different problems in different sectors. More strong, flexible, and efficient systems resulting from these synergies can change our approach to environmental monitoring, transportation, logistics, and emergency response.

4.7.2 Industry Partnerships

Companies from other industries or within the same industry work together in industry partnerships to accomplish shared objectives, pool resources, and capitalize on one another's advantages. These relationships might be strategic alliances, joint ventures, research and development partnerships, and more. Here are several examples of industry partnerships:

- Technology and Automotive
- Healthcare and Technology
- Financial Services and Technology
- Telecommunication and Entertainment
- Retail and E-Commerce
- Automotive and Energy
- Technology and Aerospace
- Healthcare and Pharmaceutical
- Agriculture and Technology
- Automotive and Technology
- Education and Technology
- Telecommunication and Internet of Things
- Energy and Environmental Sustainability

4.8 Interoperability Challenges

Interoperability challenges refer to the difficulties and obstacles faced when integrating and ensuring seamless communication between different systems, technologies, or entities. These challenges can arise in various domains, including information technology, healthcare, transportation, and more.

4.8.1 Technical Challenges

Technical issues include integrating and guaranteeing smooth communication between several technical systems, components, or platforms. These difficulties can affect several sectors including different technologies. Common technical difficulties with interoperability include these:

- ***Diverse Data Formats and Structures:*** Incompatible data formats and structures make it difficult for systems to interpret and process data correctly during exchanges.
- ***Heterogeneous Protocols and Standards:*** Systems may use different communication protocols and standards, leading to issues in establishing a common ground for interoperability.
- ***API Incompatibility:*** Applications and systems often expose functionalities through APIs (Application Programming Interfaces). Incompatibility in API designs can hinder integration efforts.
- ***Lock of a Unified Data Model:*** Different systems may lack a standardized data model, making it challenging to represent and interpret data consistently across platforms.
- ***Semantic Misalignment:*** Even when data formats are compatible, differences in the meaning or semantics of data can lead to misinterpretation and errors.
- ***Inconsistent Data Quality:*** Ensuring consistent data quality, including accuracy, completeness, and timeliness, is crucial for successful interoperability.
- ***Encryption and Security Challenges:*** Establishing secure communication between systems while managing differences in encryption methods and security protocols poses challenges.
- ***Versioning Issues:*** Changes in software versions or updates can introduce compatibility issues, especially when systems are not designed to handle versioning seamlessly.
- ***Scalability Challenges:*** Ensuring that interoperability solutions can scale effectively to handle increasing data volumes, user loads, and system complexity.

Dealing with these technical difficulties calls for rigorous planning, strict adherence to standards, and application of best practices in system design, development, and integration. Overcoming technological obstacles and promoting effective interoperability can be achieved *via* industry-wide cooperation as well as by means of standardized procedures.

4.8.2 Regulatory Challenges

In the framework of interoperability, regulatory obstacles are those resulting from the regulatory environment controlling the interaction and exchange of data, services, or technologies between several institutions or sectors. Different legal considerations, compliance rules, and regulatory systems can all lead to these difficulties. Common regulatory issues related to interoperability include these:

- ***Data Privacy and Security Regulations:*** Stringent data secrecy and security protocols, such as General Data Protection Regulation, can complicate the sharing and exchange of data, especially when dealing with sensitive information.
- ***Cross-Border Compliance:*** Operating across different jurisdictions with diverse regulatory requirements can pose challenges in ensuring compliance with varying laws governing data protection, security, and digital transactions.
- ***Healthcare Compliance:*** In healthcare, regulatory necessities likewise HIPAA (Health Insurance Portability and Accountability Act) in the U.S. add complexity to the interoperability of health information systems.
- ***Financial Regulations:*** Financial industries are subject to strict regulations, and interoperability efforts must comply with laws such as Sarbanes-Oxley Act, PCI DSS (Payment Card Industry Data Security Standard), and others.
- ***Telecommunication Regulations:*** Telecommunications interoperability faces challenges related to compliance with regulations set by regulatory bodies governing communication services, spectrum allocation, and competition.
- ***Standardization Compliance:*** Compliance with industry standards and protocols may be challenging, especially when different regions or sectors follow different standards.
- ***Intellectual Property Rights:*** The interoperability of systems may involve the exchange of technologies or proprietary

information, raising concerns about intellectual property rights and licensing agreements.

- ***Consumer Protection Laws:*** Ensuring compliance with consumer protection laws when exchanging data or providing services that involve consumer information.
- ***Competition and Antitrust Laws:*** Regulatory considerations related to competition and antitrust laws may impact collaborations and partnerships, especially when multiple entities are involved.

4.9 Application and Future of IoVT and Drones

Almost any application that uses a vehicle can use an integrated vehicle. Monitoring air pollution [19], mapping, environmental monitoring, and landscape archaeology [20], as well as agricultural uses [21], are a few of the aerial applications. All they do is make data access through cloud storage easier. Drones are currently primarily utilized for airborne surveillance. However, data from subsurface sources can also be gathered using them. These are the drones of the future.

Aerial: Any use case for a standard drone can benefit from further integration with the internet. Users can save and retrieve data on the internet from anyplace in the world if they have a device associated with the internet. For example, it can be essential to keep an eye on air pollution levels in the modern world due to the ongoing climate transformation. Thus, air pollution stages in a certain location can be observed by drones, and the data gathered can be viewed online for access by people worldwide. This can also be used to offer and receive suggestions for fixing the problem.

Mining: Drones with Internet of Things capabilities can be deployed underground to search a mine for metals. It can also be equipped with temperature and pressure sensors, which would provide information on the conditions at a specific depth below the surface of the earth. Can use any tiny rotor drones or aerial drones for this purpose.

Aquatic Life: Underwater drones are capable of reaching depths that humans cannot and monitoring aquatic life. Before installation, they can provide crucial information on conditions like the pressure at a specific depth, which can be used to calculate the pressure that will exert pressure on submerged pipelines.

4.10 Conclusion

Drone technology and the IoVT were combining to provide a revolutionary change in intelligent transportation systems that allowed for improved safety, adaptive route management, and real-time data exchange. Smooth communication between IoVT and UAVs guarantees effective traffic monitoring, better environmental surveillance, and optimal logistics. However, resolving issues with latency, data security, and communication protocols is necessary to achieve strong interoperability. Utilizing cutting-edge technologies like edge computing, 5G, and AI-driven analytics can greatly improve these integrated systems' scalability and dependability. Furthermore, to guarantee smooth communication between IoVT and drones in a variety of operational contexts, standardized frameworks and regulatory norms must be established. The merging of IoVT and drone systems has the potential to revolutionize urban management, logistics, and mobility as they develop further, creating future ecosystems that are safer, smarter, and more sustainable.

References

1. Abdelmaboud, A., The Internet of Drones: Requirements, Taxonomy. *Recent Adv. Chall. Res. Trends Sens.*, 21, 17, 2021.
2. McEnroe, P., Wang, S., Liyanage, M., A Survey on the Convergence of Edge Computing and AI for UAVs: Opportunities and Challenges. *IEEE Internet Things J.*, 9, 15435–15459, 2022.
3. Mohindru, V., Singh, Y., Bhatt, R., Gupta, A.K., Unmanned Aerial Vehicles for Internet of Things (IoT): Concepts, Techniques, and Applications, 105–128, 2021.
4. Devakirubai, N., Anitha, V., Sumathi, D., Optimizing Vehicle-to-Vehicle (V2V) Communication Efficiency with KNN-based Dynamic Time slot Allocation, in: *Proceedings of IEEE International Conference on Social and Sustainable Innovations in Technology and Engineering*, pp. 1–6, 2024.
5. Chen, S., Hu, J., Zhao, L., Zhao, R., Fang, J., Shi, Y., Xu, H., Overview. In: Cellular Vehicle-to-Everything (C-V2X), in: *Wireless Networks*, pp. 329–357, 2021.
6. Koubâa, A., Allouch, A., Alajlan, M., Javed, Y., Belghith, A., Khalgui, M., Micro Air Vehicle Link (MAV link) in a Nutshell: A Survey. *IEEE Access*, 7, 87658–87680, 2019.
7. Qiao, G., Zhuang, Y., Ye, T., Qiao, Y., BCDAIoD: An Efficient Blockchain-Based Cross-Domain Authentication Scheme for Internet of Drones. *Drones*, 7, 302, 2023.

8. Xia, T., He, J., Liu, H., Cross-Domain Authentication Technology of UAV based on alliance chain. *Mobile Inf. Syst.*, 2022, 657–665, 2022.
9. Jiang, D. and Delgrossi, L., IEEE 802.11p: Towards an International Standard for Wireless Access in Vehicular Environments, in: *IEEE Vehicular Technology Conference*, Marina Bay, pp. 2036–2040, 2008.
10. Prajith, P., Yokesh, V., Prasanth, A., Sathish, N., Artificial Intelligence-Enabled Fully Echoed Q-Routing and Adaptive Directional Medium Access Control Protocol for Flying Ad-Hoc Networks. *Int. J. Commun. Syst.*, 38, 1–18, 2025.
11. Bemposta Rosende, S., Ghisler, S., Fernández-Andrés, J., Sánchez-Soriano, J., Implementation of an Edge-Computing Vision System on Reduced-Board Computers Embedded in UAVs for Intelligent Traffic Management. *Drones*, 7, 682, 2023.
12. Yazid, Y., Ez-Zazi, I., Guerrero-González, A., El Oualkadi, A., Arioua, M., UAV-Enabled Mobile Edge-Computing for IoT Based on AI: A Comprehensive Review. *Drones*, 5, 148, 2021.
13. Nida, F., Paresh, S., Manik, G., Integration of multi access edge computing with unmanned aerial vehicles: Current techniques, open issues and research directions. *Phys. Commun.*, 52, 101641, 2022.
14. Ajakwe, S.O., Kim, D.S., Lee, J.M., Drone Transportation System: Systematic Review of Security Dynamics for Smart Mobility. *IEEE Internet Things J.*, 10, 14462–14482, 2023.
15. Yahuza, M., Yamani Idna Idris, M., Bin Ahmedy, I., Wahid Adbul Wahab, A., Internet of Drones Security and Privacy Issues: Taxonomy and Open Challenges. *IEEE Access*, 9, 57243–57270, 2021.
16. Lata, A., Manju, D., Kumar, V,.E., Standards and Regulatory Compliances for IoT Security. *Int. J. Serv. Sci. Manage. Eng. Technol.*, 12, 133–147, 2021.
17. Alamouri, A., Lampert, A., Gerke, M., Impact of Drone Regulations on Drone Use in Geospatial Applications and Research: Focus on Visual Range Conditions, Geofencing and Privacy Considerations. *PFG*, 91, 381–389, 2023.
18. Alvarez, E., Corpetti, T., Houet, T., UAV & satellite synergies for optical remote sensing applications: A literature review. *Sci. Remote Sens.*, 3, 100019, 2021.
19. Manjunath, M.N., Abbas, A., Gopinath, U., Athith, N., IOT based air pollution monitoring system using drones. *Int. Res. J. Eng. Technol.*, 7, 384–388, 2020.
20. Casagrande, G., Sik, A., Szabó, G. (Eds.), *Small flying drones: Applications for geographic observation*. Springer, Cham, 2018.
21. Lee, S., Roh, B., Kim, S., The Convergence System of an IoT Standard Service Platform and an Autonomous Drone, in: *IEEE International Conference on Consumer Electronics (ICCE)*, pp. 1–3, 2022.

5

5G Communication in Drones for Surveillance in Future Transportation Activities

N. Sathish[1*], V. Yokesh[2], T. Kuntavai[3] and Mariya Ouaissa[4]

[1]*Department of Electronics and Communication Engineering, Sri Venkateswara College of Engineering, Sriperumbudur, India*
[2]*Department of Electronics and Communication Engineering, Sathyabama Institute of Science and Technology, Chennai, India*
[3]*Department of Electrical and Electronics Engineering, Adhiparasakthi Engineering College, Melmaruvathur, Tamil Nadu, India*
[4]*Cadi Ayyad University, Marrakech, Morocco*

Abstract

Drones have gained vast importance in the surveillance sector due to their easy adaptability in any field, low cost, and low manpower needs. In the fastest-growing environment, the demand for transportation services is also tremendously increasing based on global needs. Drone usage is increasing across a broad spectrum of transportation-related applications. As laws and regulations change, drone integration is helping existing transportation networks operate more safely, effectively, and efficiently overall. The key areas where drones are integrated into transportation services are Surveillance and Monitoring, Traffic Management, Search and Rescue Operations, Package Delivery, Infrastructure Inspection, Surveying and Mapping, Public Safety and Emergency Response, Environmental Monitoring, and so on. The 5G Communication plays a vital role in integrating the various transportation services and the drones. The capabilities, dependability, and efficiency of drone operations are improved by 5G's breakthroughs, making it a game-changer in drone communication.

This chapter focuses on various aspects of 5G communication in reshaping surveillance in future transportation services like integration of 5G and drones,

**Corresponding author*: n.satyish@gmail.com

A. Prasanth, Rajesh Kumar Dhanaraj, Munish Sabharwal, Vandana Sharma and Seifedine Kadry (eds.) Drones for Transportation Logistics and Disaster Management, (95–114) © 2026 Scrivener Publishing LLC

Network Architecture for Drone Surveillance, Security and Privacy Considerations, Regulatory and Ethical Considerations, and Future Trends and Innovations.

***Keywords*:** 5G communication, drones, transportation services, integration of services, surveillance, privacy and security

5.1 Introduction

Technological developments that promise to completely change how we monitor, control, and guarantee the security of our developing transportation networks are driving a dramatic shift in the transportation sector. The modern revolution emerges through the drone-coordinated use of fifth-generation (5G) communication systems for rapid, precise transportation surveillance. These integrated drone-5G systems allow us to reform transportation understanding and relationships because we occupy a transitional position at the intersection of connectivity and mobility systems.

The future of transportation surveillance systems will experience basic transformation due to this paradigm-changing integration, which delivers detailed operational abilities [1]. The technological advancement surpasses ordinary current operational improvements. The 5G technology delivers complete communication method updates, constituting a substantial step toward the evolving communication industry. Future generations can expect unprecedented data transfer speeds along with minimal latency and superior connectivity capabilities through 5G technology.

5.2 Overview of 5G Communication

The newest and most sophisticated mobile network standard, 5G communication technology, promises major advancements over 4G LTE (Long-Term Evolution). Faster data rates, less latency, more capacity, and compatibility for a large variety of connected devices are the main features of 5G.

5.2.1 Key Features of 5G Networks

The essential features of 5G communication technology consist of three main categories, namely speed and capacity alongside low latency and frequency bands, while also incorporating massive MIMO (Multiple Input Multiple Output) and network slicing and security and standardization and deployment issues. 5G delivers dramatically faster data rates than previous networks because peak downloads operate at 10 Gbps, which equals 100 times more speed than 4G LTE [2]. The high-definition multimedia streaming maintained smooth performance through 5G technology, allowing users complete download speed enhancement and enhanced data transfer speed. The 5G network holds an impressive capability for conveying multiple devices at once without creating any network delays, as highlighted in the previous article. The data transmission stability and efficiency receive support from beamforming and massive MIMO technologies. Real-time applications, including industrial automation and virtual reality, utilize the 5G platform because its delay period falls below 1 millisecond [3].

The 5G network operates across different spectrum bands that utilize powerful high-band mmWave frequencies, although these frequencies require additional structural enhancements due to their restricted reach. The efficiency of networks gets enhanced by Massive MIMO through its ability to operate multiple antennas for handling simultaneous data streams. Network slicing emerges as a significant advancement of 5G technology because it lets operators optimize application-specific virtual networks through specialized industrial services.

5G technology incorporates security systems together with standardized practice methods, which provide enhanced cryptographic techniques for dealing with escalating cyber threats. The 3rd Generation Partnership Project (3GPP), alongside other organizations, initiates standardization strategies to establish worldwide network interoperability [4]. The implementation of 5G technology meets many technical barriers because it demands major infrastructure development along with new fiber-optic networks and small cells and antenna systems. Worldwide 5G network deployment requires government institutions to create rules and distribute frequency spectrum, which supports network implementation.

5.3 Drones in Transportation

Drones appear to be everywhere, and their applications are growing daily, taking on new and varied roles. Only a small portion of the market is occupied by the consumer-oriented drones that have gained so much popularity, and they only provide a portion of the controls and features that their business counterparts offer. Industrialized drones can be used for a wide range of commercial applications thanks to their improved features, which

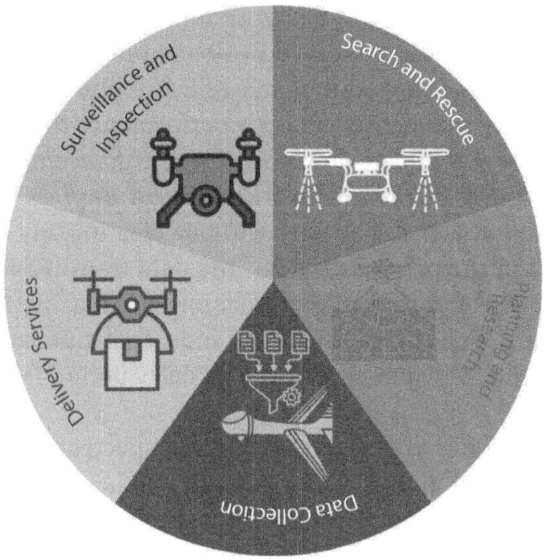

Figure 5.1 Applications of drones in transportation.

include bigger carrying capacities, GPS mapping and flight planning, accurate controls, and geofencing.

Transportation is one of the sectors with the most potential to use drone technology shortly. Drones have a lot of potential and might impact the industry in many ways, including traffic control, infrastructure development, and transportation. The various ways in which drones are used in transportation are shown in Figure 5.1.

5.3.1 Surveillance and Inspection

Unmanned aerial vehicles or drones have brought revolutionary changes to surveillance alongside inspection and transportation while enhancing operational effectiveness with increased security and lower expenses. Security forces, as well as law enforcement agencies, use drones to perform surveillance operations unparalleled by any alternative technology. The company delivers real-time aerial surveillance services that help secure borders and manage crowds together with protecting areas of large dimensions. During disaster response situations, drones enable quick assessments of affected areas which help teams locate survivors and estimate damage without endangering the rescue personnel [5]. The inspection of infrastructure demands drones as an indispensable tool. Such devices facilitate inspections of inaccessible zones, including bridges and dams, together with power lines and wind turbines. Drone-based devices with improved cameras and sensors detect both hardware flaws and signs of aging structures. The quick system allows immediate maintenance work that helps stop potential equipment failures. Spectral cameras mounted on drones enable farmers to detect crop conditions; thus, they can cure diseases while adjusting irrigation systems, thus enhancing both agricultural yield and resource distribution. Managers use drones for simultaneous animal welfare monitoring and cattle supervision, together with herding management responsibilities.

The drone industry produces a major shift in transportation operations. Delivery businesses actively use drone technology to develop quick product delivery services for densely populated areas through their e-commerce and retail operations. The implementation of drone systems has revolutionized the medical supply transportation of difficult-to-reach healthcare facilities and remote places, which has improved healthcare services. Warehouse inventory management benefits from drones through scanning and tracking technologies that assist with warehouse inventory management. Both factors reduce physical work requirements and lead to more exact process execution. The experimental phase of drones indicates their potential to change freight transportation through challenging areas and short distances [6].

Drones offer numerous advantages. A cost-effective answer exists because they reduce the need for expensive manned aircraft operations and ground checks. The drones extend their scope through fast and precise information delivery, allowing operators to achieve better performance. The dangerous work that drones accomplish ensures human safety at all times. Full benefits from drone systems require addressing legal issues while resolving concerns about privacy and overcoming technical problems with power systems, delivery systems, and environmental durability. The innovative drones provide secure solutions that enable the transformation of surveillance operations and inspection services while advancing transportation methods. Drones will achieve increased applications thanks to technological advancements and strengthened regulatory systems, which will produce additional industrial improvements through innovation.

5.3.2 Delivery Services

The delivery services area in the transportation business uses drones to develop innovative service delivery systems that provide quick and extensive reach together with efficiency. Leading companies in e-commerce and retail industries utilize UAVs, also known as drones, to transport merchandise directly to client residences. The technology will enable quick delivery reductions in city areas where traffic congestion creates significant delivery time delays. People who use drones for delivery benefit from quick service because these devices bypass road restrictions, which results in highly efficient delivery operators satisfying their customers.

Modern drones serve essential functions across the healthcare sector through logistical support of medical products together with vaccines and blood samples to distant locations that are challenging to reach [7]. Drones serve as a critical resource to transport vital medical goods into both disaster areas and disaster-affected zones and provide emergency support, which can result in saving lives. This delivery system function proves advantageous for remote locations apart from traditional methods [8]. The utility of drone delivery enables different advantages in delivering products when considered among other options.

The removal of delivery vehicles from operations allows the company to avoid building extensive truck and driver staff. These delivery vehicles contribute to environmentally friendly solutions because they leave behind fewer carbon emissions than other delivery trucks. The operational flexibility of drones allows them to perform delivery tasks in every kind of atmospheric situation along with diverse terrains with no hindrance from traffic conditions [9].

5.3.3 Current Challenges for the Usage of Drones in Transportation

The various significant obstacles related to drone implementation in transportation serve as barriers to reaching their full potential as a solution. The firm faces legal requirements as its main significant challenge. Such regulatory measures function as safety guidelines to monitor drone operations, particularly in commercial business engagements for maintaining population safety. Diverse foreign policies create obstacles for companies that try to follow these regulations. Organizations must acquire distinct approvals that come with restrictive operational protocols [10]. One main operational challenge exists because drone batteries have a restricted operational lifetime. Drone batteries cannot currently maintain flight time, which equals national parcel delivery duration or regional surveillance operations. The system needs frequent battery replacement or charging operations that remain challenging to perform. The system has a fundamental weakness regarding its maximum weight capability that represents an additional drawback. The existing drone technology features small devices that enable the freight of small packages or lightweight materials only. A drawback of weight restrictions limits drone expansion as a transportation system for larger loads, which thereby hinders industrial and commercial drone solutions. Maneuvering and collision also present other challenges, especially in traffic-dense environments such as urban settings. Drones presently integrate broad arrays of sensitive contact sensors combined with sophisticated mathematical operations to travel through various areas. The complex operational environments require robots to implement collision-prevention protocols that protect against buildings, trees, and all other obstacles encountered in urban airspace.

The navigation system development by fleet operators must achieve perfection for security in aviation operations and the health protection of human life. The growing number of drones in the airspace has made air traffic control issues more significant. Critical airspace procedural norms and authority-regulator communications are necessary for aerial drone integration in order to remove the risk of aircraft-drone collisions. The primary obstacles are security and privacy-related complications. Residents are exposed to safety risks and privacy problems due to a variety of drone modification tactics created by political entities.

Although there is a formal mechanism in place to protect private rights, the data collection carried out by surveillance drones may violate those rights. Drones have a lot to offer the transformation of transportation, but before they can be integrated, all particular issues must be resolved. Improved aircraft battery technology, increased capacity thresholds, sophisticated flight control systems, and enhanced security measures are all critical to the development of drone flight. Improved legal frameworks and increased innovation make it feasible to use drones for transportation on a large scale. Drones and air transportation networks work together to increase operational efficiency, reduce costs, and provide long-term advantages.

5.4 Drones and 5G Integration

Drones and 5G technologies together produce ground-breaking innovations for the contemporary technological realm, propelling both UAV system development and full industrial transformation. Better drone operating capabilities are produced by the fifth-generation wireless technology's large bandwidth and quick network capacity. Businesses can leverage improved drone functions through the network, increasing operational effectiveness across a variety of applications. Drone operations combined with 5G technology create a number of new opportunities that result in improved UAV capabilities and industrial revolution. Low-latency behavior emerges from 5G networks as their most advantageous aspect when used with this technology. Operators can now operate drones in real-time thanks to accurate navigation and quick environmental change response, which reduces latency times and allows them to execute difficult operations with accuracy. 5G network speed capability enables better drone operations through direct statement. Through drone integration, 5G technology enables the real-time transmission of enormous video data alongside various essential operational data. The precise delivery of crucial information through 5G enables efficient monitoring along with surveillance and inspection tasks that require prompt judgment decisions.

The introduction of 5G technology simplifies Beyond Visual Line of Sight (BVLOS) operations for drones, which were previously restricted to manual drone navigation. Drones allow remote long-distance operations through 5G technology because of their fast data transfer speeds and limited response delays. The outcome allows for new possibilities to develop applications in precision agriculture combined with infrastructure inspection and search and rescue operations using wide coverage. Drones obtain greater operational effectiveness from edge computing systems installed within 5G networks.

The real-time processing capabilities at the network edge of edge computing remove the requirement for moving all data back to central servers. New UAV systems based on these gains in speed will become more advanced and adaptable due to faster decision-making by drones. Networking support between 5G enables a seamless, coordinated operation among many drones. Real-time drone communication enables swarm intelligence applications to perform effective data collecting and surveillance operations along with collaborative tasks in disaster response situations [11].

The integration of drone technology with 5G networks offers multiple advantages to users, although it generates important problems related to protection rights and systems security. The correct use of 5G-connected drones requires robust cybersecurity measures and handling privacy concerns because of their technological advancement. The development of unmanned aerial vehicles enters a significant new phase because of the essential 5G and drone systems combination. Through this merging technology, companies can access fresh possibilities while drones attain better functionality.

5.5 Network Architecture for Drone and 5G in Transportation

UAVs present special challenges to network architecture, which requires drones and 5G transportation integration to work effectively in a strong system design. A system architecture built for drone communication needs a low-latency connection with high bandwidth, reliable functioning, and scalable capabilities to handle diverse transportation operations.

5.5.1 Infrastructure of 5G Network

Core Network (CN)
As the central intelligence center, the CN controls multiple operations, such as enforcing network policy, authenticating users, and managing network slices. Network slicing allows dedicated drone communication networks by guaranteeing peak efficiency and isolation [12].

Radio Access Network (RAN)
Droned-based traffic goes through Radio Access Network (RAN) to connect with 5G networks. Beamforming technology connects to small cells to provide substantial capacity along with minimal latency, especially in areas where numerous drones are present.

5.5.2 Edge Computing and Low Latency

Edge computing nodes get installed at specific strategic network boundary locations to lower latency values. The positioned edge nodes enable drones to execute data processing immediately while allowing them to make local decisions quickly. Real-time navigation systems, along with collision avoidance solutions, exist only when combined with edge computing technology coupled with minimal system delays [13].

5.5.3 Drone Traffic Management System

The designated airspace receives its drone traffic management duties from a dedicated server. Drone route coordination and safe separation gaps receive active support through this system, which also conducts real-time flight path modifications using live data. This server exchanges data with edge computing nodes as well as 5G core network elements [14].

A uniform communication protocol enables coordinated operation between drones as well as between drones and ground control stations and the traffic management system. A defined communication pipeline helps deliver current drone location information along with atmospheric conditions as well as airspace traffic reports between secure networks.

5.5.4 Authentication and Security

The security of drone and 5G network communication becomes possible through gateway authentication systems. These security components authenticate drones to confirm their identities while establishing protected communication route systems [15]. End-to-end encryption protects data movement between drones and ground control stations as well as between the core network infrastructure using encryption techniques.

5.5.5 Analytics and Network Monitoring

The functionality of real-time monitoring stations covers both 5G network performance alongside edge computing nodes as well as drone traffic. The analysis of advanced data patterns helps teams check for abnormal events while projecting network needs and enhancing drone command operations.

5.5.6 Ground Control Stations (GCS)

Operators gain remote drone control capabilities through dependable interface operation thanks to the 5G network connection to ground control stations [16]. User interfaces that use 5G capabilities for remote piloting allow operators to pilot drones with exact control through low-latency systems.

5.5.7 Regulatory Compliance

A specific module operates within the network architecture to maintain compliance with aviation rules and standards. The system possesses automation tools to execute height enforcement protocols alongside no-fly zone limitations besides meeting regulatory needs by simplifying contact with aviation control.

5.5.8 Integration with Other Transportation Systems

The network framework needs to facilitate straightforward partnership capabilities with linked cars and public transportation systems, as well as intelligent traffic signals for developing a cohesive transportation system.

5.5.9 Redundancy and Scalability

The architecture needs scalable capabilities that allow it to manage an expanding number of users together with drones [17]. The network scalability improves through the implementation of cloud-based solutions together with virtualization technology. The system must include redundancy mechanisms because they protect network dependability. The system maintains critical operations by using failover methods together with extra nodes and communication channels that provide backup support during a network outage.

The foundational structure of this network architecture serves as a foundation for functional integration between drone and 5G implementations in transportation systems. The framework provides specific drone operation solutions that ensure secure and low-latency connectivity with regulatory compliance for the development of an effective interconnected transportation system.

5.6 Security and Privacy Considerations

The technology which refers to drones has emerged rapidly as a transformative technology across diverse industries, including transport, because it operates under the name of unmanned aerial vehicles or UAVs. Transportation systems that integrate drones require the immediate resolution of vital privacy and security risks that continue to grow more prevalent. The paper investigates security and privacy issues that jeopardize safe drone integration into transportation systems and presents potential solution approaches.

5.6.1 Security Considerations

Data Security

Transmission and collection of position data alongside photos and videos represent frequent activities of drone-based transportation operations. Products need resistance-based encryption systems to defend against hacking incidents and data breaches. The security of sensitive data depends on implementing protected storage devices that use encryption while requiring secure communication pathways [18].

Authorization and Authentication

To control drones and their communications securely one must implement strong authorizing systems with proper access procedures. Security measures that use both biometric verification and multi-factor authentication work to control unauthorized access and stop abuse cases from happening.

Anti-Hacking Measures

The control and navigational systems of drones are at risk of cyberattacks that could lead to detrimental failures. Reduction of cyberattack danger becomes possible through the combination of safe coding practices with frequent software updates and intrusion detection systems [19].

Physical Security

The protection of drones must be established against any attempts of theft, vandalism, or physical contact events. Drones achieve better physical protection when kept in secure facilities alongside physical defenses as well as flight restrictions that enforce geographical boundaries [20].

5.6.2 Privacy Considerations

Data Minimization

The protection of privacy relies on two factors: first reduce the gathering of personally identifiable information (PII) and second acquire only essential data. The implementation of data minimization principles decreases both sensitive data misuse possibilities and unauthorized access incidents.

Privacy by Design

User privacy requirements must become an essential part of drone system development from the beginning. New design features and data anonymization methods that build privacy capabilities form the core of improving drone system privacy through technological advancement [21].

Transparent Policies

A company must build distinct and clear privacy instructions to handle data collection and data storage alongside information sharing. A drone system that prioritizes privacy must communicate to the public how data will be used and implement voluntary procedures and consent requirements for data processing.

Geofencing and No-Fly Zones

Geofenced technology establishes restriction areas around sensitive regions to stop drones from entering residential locales or vital infrastructure which helps maintain privacy protection for those regions.

Security together with privacy must receive priority attention when drones enter widespread transportation network use. A complete approach consisting of technological progress, legal standards, and public safety awareness must be developed to ensure both the safety and ethical integration of drones in transportation networks. Drone technology gives its greatest value when security data practices are combined with strong authentication mechanisms and unambiguous privacy protections, which regulatory entities must provide. Drones used in industrial transportation will acquire widespread adoption if corporations strike a balance between innovation and respect for personal rights.

5.6.3 Enhanced Security Features in 5G

5G is the fifth-generation wireless technology that enables high-speed communication networks and rapid processing of a variety of devices. Drone technology is evolving as a result of modern technological advancements, which mostly benefit the transportation sector. The secure operations and efficient execution of drones in transportation systems rely substantially on the improved security aspects provided by 5G networks. Network slicing technology is a core 5G capability utilized by operators to create virtual business-focused networks. To improve security, the system builds specialized drone-operation network slices that send encrypted communications without interference.

Real-time drone-control interactions rely on 5G transmission to safely deliver dynamic and emergency operations with ultra-fast, ultra-low latency replies. 5G demonstrates a high capacity for managing a large number of linked devices, making it easier to handle expanding drone activities within transportation networks. Network congestion avoidance capabilities help to reduce security vulnerabilities that arise when devices reach their operational capacity. The 5G network protects data during transfer with its powerful security protocol, which secures confidential information such as location and video feeds. Drones used for transportation require powerful encryption solutions to protect their networks from illegal digital attacks and ensure data confidentiality.

5.7 Regulatory and Ethical Considerations

Drone technology has been integrated into transportation networks as a vital component in the creation of cost-effective delivery solutions. Drone technology advancement necessitates a rigorous examination of the legal and ethical factors that impact its transportation use. The introduction of drone transportation systems necessitates both appropriate technology improvement and ethical compliance *via* suitable regulatory procedures.

5.7.1 Regulatory Considerations

Drones require official regulatory enforcement to ensure their operational safety and respect, including measures addressing aircraft permissions, airspace limits, weight regulations, and privacy protection norms. All drone operators must obtain suitable licensing, which demonstrates their technological ability, safety expertise, and skill competence. During the certification process, both ethical requirements and legal compliance standards must be given more importance. The current airspace rules serve as protective measures to prevent drone accidents with aircraft while also managing drone-aircraft interaction in shared airspace.

Operators set up no-fly zones around critical locations such as airports and infrastructure to safeguard airspace security and fight any threats [22]. Displaying explicit cargo payload constraints determines drone weight and allowed delivery contents during safety checks, particularly in delivery operations. Proper drone operations require rigorous privacy laws that govern both data gathering and storage processing. Drone data collecting activities require parental agreement as a precondition for protecting personal privacy while providing important services *via* drone capabilities.

5.7.2 Ethical Considerations

Ethical drone inspections confirm proper use by protecting privacy, assuring environmental safety, assisting disaster relief operations, and making machine operations plainly apparent. Individual privacy protection necessitates particular standards that establish data usage policies, surveillance limitation limits, and total monitoring visibility in order to maintain confidence and prevent unlawful surveillance. Assessing environmental drone effects should evaluate both their electricity utilization and their substances and sustainable characteristics. All manufacturing and operational aspects of ethereal drones need sustainable practices to minimize environmental destruction. The management of emergencies mandates that human beings should occupy the foremost safety priority position above all other elements.

Existing ethical codes need to define drone operational use to preserve rescue operation effectiveness while avoiding possible deterioration of the mission outcome. Every decision process of autonomous artificial intelligence drones needs absolute ethical transparency according to ethical guidelines. People need to comprehend the decision-making frameworks of these systems to make knowledgeable decisions about fairness and accountability issues in this technology [23]. Responsible network integration of drones depends on regulatory oversight in addition to ethical requirements that form an active part in this achievement. Drone technology draws its ethical framework from protecting privacy rights, environmental protection, and human safety principles while strict regulations guarantee safety standards and privacy protections. The expansion of drone industries in the future depends heavily on creating ethical as well as regulatory standards to gain public trust and unlock drone-based transportation possibilities.

5.8 Future Trends and Innovations

The integration of drones and 5G technology will cause remarkable evolutionary changes to be seen by all members of the transportation and mobility sectors. Self-piloted drone fleets derived from 5G and AI (Serverless) technology integration will boost delivery services and conduct area inspections and area monitoring duties. The unmanned vehicles using 5G technology will operate by themselves to move cargo and carry out surveillance tasks across locations. The implementation of this new development will boost network efficiency throughout final destination deliveries and transportation. A traffic management system featuring 5G drones works to deliver live transportation information to create another envisioned framework. This new technology will enhance traffic flow while optimizing routes and both these factors will decrease congestion while improving transportation performance. UAM urban air mobility lies directly ahead of the city because of current drone and 5G technology developments [24].

5G-connected cities will allow passengers to access drone and flying taxi transportation as an on-demand air-based service [25]. Accurate drone swarm operations for large group activities become feasible through improved swarm intelligence controls and communication methods. Well-coordinated 5G-enabled drone units will perform complex environmental inspections along with emergency assistance and large-scale agricultural surveillance because of this network's reliable performance and limited latency delay. The investment will support delivery services over long distances and remote infrastructure checks as well as agricultural field observation through this short-term spending.

The precise traffic predictions for this technology will become achievable through AI algorithms processing drone data obtained from the 5G network. 5G technology partnered with transportation authority enables organizations to improve their traffic patterns which results in better safety conditions during travel as well as minimized commuting times. Current drone development studies focus on creating solutions for environmental issues through better energy efficiency, longer operational periods, and reduced environmental effects. The combination of 5G technology with drones will boost their flight duration abilities through enhanced navigation and communication systems. 5G flying unmanned aerial vehicles facilitate proper transport management through drone-ground integration leading to airborne drone miles that result in the development of the real-cycle air-ground transport system. Drone operations will increasingly depend on AR systems to improve pilot situational understanding outside their aircraft. The new technology provides pilots with both collision warnings and path guidance because of 5G network connection capabilities.

Drones became common after people gained too much trust in dealing with large data amounts. Blockchain technology that integrates with 5G networks has proven to enhance airplane data security and drone transaction security. The overall development of blockchain applications will deliver equal advantages to every member of the transportation sector. During emergency response operations the salvation of human life remains an absolute priority. A formal rule should be implemented before drone deployment to establish ethical guidelines that will prevent drones from interfering with existing rescue activities since possible injuries would occur.

The next emerging technology known as drones collaborates with autonomy and algorithms to maintain ethical mapping operations. System operators need to establish why their protocols solely focus on impartiality, liability, and prejudice issues. Having a transparent algorithmic process through The Way ensures drones are operated faithfully and responsibly. Drones used in transportation systems merge ethical considerations along with legislative standards. Each domain that influences drone technology operates in the social world and natural environment and environmental ethics but legal regulations protect human rights and shield people from possible risks. The law and ethics perspective has established an ongoing drone processing strategy to benefit urban transport while technology acts as the key driver in this expanding industry.

5.9 Conclusion

The transportation business saw a revolutionary transition as a result of 5G communication technology and drone utilization, which provided amazing new capabilities for speed and electronic connectivity at previously unattainable levels. 5G communication technologies combined with drone capabilities produce extraordinary communication links, resulting in effective solutions for travel system administration. This development had a significant impact on the global transportation network. Network system adoption begins with considerable gains in reliability and efficiency performance following its introduction. Drones' operational range includes efficient and timely performance, as well as the ability to complete challenging complex jobs that exceed the accuracy limits of human hands and eyes.

Real-time traffic management, surveillance, and inspection rely on a wide range of services, including surveying and auditing. Advanced transportation systems better suited to complex modern times are the product of optimization systems powered by technology that efficiently governs resource allocation. The current uncertainty around new technology motivates people to develop better approaches for creating smart infrastructure and transportation services. The combination of technical advancement and network consolidation will result in various technological solutions that revolutionize commercial zones and produce new breakthroughs for surveillance systems, traffic management, and next-generation transportation development.

References

1. Gupta, A., Afrin, T., Scully, E., Yodo, N., Advances of UAVs toward Future Transportation: The State-of-the-Art, Challenges, and Opportunities. *Future Transp.*, 1, 326–350, 2021.
2. Attar, H., Issa, H., Ababneh, J., Abbasi, M., Solyman, A., Khosravi, M.R., 5G System Overview for Ongoing Smart Applications: Structure Requirements and Specifications. *Comput. Intell. Neurosci.*, 2022, 1–11, 2022.
3. Morgado, A., Huq, K.M.S., Mumtaz, S., Rodriguez, J., A Survey of 5G Technologies: Regulatory, Standardization and Industrial Perspectives. *Digit. Commun. Netw.*, 4, 87–97, 2018.
4. Khan, R., Kumar, P., Jayakody, D.N.K., Liyanage, M., A Survey on Security and Privacy of 5G Technologies: Potential Solutions, Recent Advancements, and Future Directions. *IEEE Commun. Surv. Tutor.*, 22, 196–248, 2020.

5. Barmpounakis, E.N., Vlahogianni, E.I., Golias, J.C., Unmanned Aerial Aircraft Systems for Transportation Engineering: Current Practice and Future Challenges. *Int. J. Transp. Sci. Technol.*, 5, 111–122, 2016.
6. Yadav, A., Goel, S., Lohani, B., Singh, S., A UAV Traffic Management System for India: Requirement and Preliminary Analysis. *J. Indian Soc. Remote Sens.*, 49, 515–525, 2020.
7. Prasanth, A., Lakshmi, D., Dhanaraj, R.K., Balusamy, B., Technological Advancement in Internet of Medical Things and Blockchain for Personalized Healthcare: Applications and Use Cases, in: *CRC Press*, pp. 1–290, 2024.
8. Li, X., Tupayachi, J., Sharmin, A., Martinez Ferguson, M., Drone-Aided Delivery Methods, Challenge, and the Future: A Methodological Review. *Drones*, 7, 191, 2023.
9. Aggarwal, S., Gupta, P., Mahajan, N., Balaji, S., Singh, K.J., Bhargava, B., Panda, S., Implementation of Drone-Based Delivery of Medical Supplies in North-East India: Experiences, Challenges and Adopted Strategies. *Front. Public Health*, 11, 1–15, 2023.
10. Zieliński, T., Challenges for Employing Drones in the Urban Transport Systems. *Saf. Def.*, 2, 1–8, 2022.
11. Chen, J., Wang, W., Zhou, Y., Ahmed, S.H., Wei, W., Exploiting 5G and Blockchain for Medical Applications of Drones. *IEEE Netw.*, 35, 30–36, 2021.
12. Nidhi, D., Mihovska, A., Prasad, R., Overview of 5G New Radio and Carrier Aggregation: 5G and Beyond Networks, in: *23rd International Symposium on Wireless Personal Multimedia Communications (WPMC)*, pp. 1–6, 2020.
13. Ge, X., Ultra-Reliable Low-Latency Communications in Autonomous Vehicular Networks. *IEEE Trans. Veh. Technol.*, 68, 5005–5016, 2019.
14. Prabhakar, P., Yokesh, V., Aruchamy, P., Nanthakumar, S., Artificial Intelligence-Enabled Fully Echoed Q-Routing and Adaptive Directional Medium Access Control Protocol for Flying *Ad-Hoc* Networks. *Int. J. Commun. Syst.*, 38, 1–28, 2025.
15. Aydin, Y., Kurt, G.K., Ozdemir, E., Yanikomeroglu, H., Group Authentication for Drone Swarms, in: *IEEE International Conference on Wireless for Space and Extreme Environments (WiSEE)*, pp. 72–77, 2021.
16. Haque, S.R., Kormokar, R., Zaman, A.U., Drone Ground Control Station with Enhanced Safety Features, in: *2nd International Conference for Convergence in Technology (I2CT)*, pp. 1207–1210, 2017.
17. Winkler, S., Zeadally, S., Evans, K., Privacy and Civilian Drone Use: The Need for Further Regulation. *IEEE Secur. Privacy*, 16, 72–80, 2018.
18. Yang, D., Zhao, Y., Yi, Z., Yang, D., He, S., A Design Scheme of Data Security for Unmanned Aerial Vehicles, in: Advanced Hybrid Information Processing, *Lecture Notes of the Institute for Computer Sciences, Social Informatics and Telecommunications Engineering*, W. Fu and L. Yun (Eds.), p. 468, Springer, Cham, 2023.
19. Zhi, Y., Fu, Z., Sun, X., *et al.*, Security and Privacy Issues of UAV, in: *Computer Networks*, p. 224, 2020.

20. Tedeschi, P., Al Nuaimi, F.A., Awad, A.I., Natalizio, E., Privacy-Aware Remote Identification for Unmanned Aerial Vehicles: Current Solutions, Potential Threats, and Future Directions. *IEEE Trans. Ind. Inf.*, 20, 1069–1080, 2024.
21. Altawy, R. and Youssef, A.M., Security, Privacy and Safety Aspects of Civilian Drones: A Survey. *ACM Trans. Cyber-Phys. Syst.*, 1, 1–25, 2016.
22. Tran, T.H. and Nguyen, D., Management and Regulation of Drone Operation in Urban Environment: A Case Study. *Social Sci.*, 11, 1–13, 2022.
23. Wang, N., Christen, M., Hunt, M., Ethical Considerations Associated with Humanitarian Drones: A Scoping Literature Review. *Sci. Eng. Ethics*, 27, 51–57, 2021.
24. Ahmed, F., Mohanta, J.C., Keshari, A., Recent Advances in Unmanned Aerial Vehicles: A Review. *Arabian J. Sci. Eng.*, 47, 7963–7984, 2022.
25. Cao, H., Wu, Z., Yu, W., Drone for Intelligent Traffic Monitoring: Current Status and Future Trends. *Comput. Exp. Simul. Eng. Mech. Mach. Sci.*, 146, 1133–1150, 2023.

6

Impact and Assessment of Artificial Intelligence-Enabled UAV for Real-Time Data Streaming Application

Gobinath C.[1]*, Thanjaivadivel M.[2], Sarathkumar Rangarajan[3] and K. Kalaivanan[4]

[1]School of Computing and Information Technology, Department of AIML, REVA University, Bangalore, India
[2]Department of Computer Science and Engineering, Nandha Engineering College, Erode, India
[3]Victorian Institute of Technology, School of IT and Systems, Melbourne, Australia
[4]Department of Computer Science and Engineering, SRM Institute of Science and Technology, Trichy, Tamil Nadu, India

Abstract
This chapter describes the utilization of AI-powered UAVs for streaming live data with the potential to boost perception, decision-making, and communication using advanced algorithms. The UAVs process large volumes of data from onboard sensors and external sensors using computer vision, machine learning, and deep learning for the analysis of the world in real time. The UAVs automatically stream data from different sources such as aerial images, video feeds, environmental sensors, and geospatial maps. The AI onboard processing computes data in real time and supplies useful information and actionable intelligence to users. Applications of AI-powered UAVs have extensive ramifications in industries and are revolutionizing industries like agriculture, disaster relief, surveillance, and environmental monitoring. The industries obtain increased efficiency, data-based decision-making, and optimization of resources and therefore AI-powered UAVs are an important asset for today's needed applications with efficient and rapid processing of data.

***Keywords*:** Unmanned aerial vehicles, artificial intelligence, precision agriculture, disaster management

**Corresponding author*: gobinath0119@gmail.com

A. Prasanth, Rajesh Kumar Dhanaraj, Munish Sabharwal, Vandana Sharma and Seifedine Kadry (eds.) Drones for Transportation Logistics and Disaster Management, (115–144) © 2026 Scrivener Publishing LLC

6.1 Introduction

The technology for real-time streaming of data has been progressing continually with the convergence of Artificial Intelligence (AI) and Unmanned Aerial Vehicles (UAVs). The technology framework is centered on the critical aspect of AI-equipped UAVs to collect, analyze, and disseminate data in industries. These systems, by combining AI algorithms and UAV operations, can retrieve precise data from distant and remote locations with a level of unprecedented efficiency. The application aids decision-making in real-time and cost-effective utilization of resources [1].

Artificial intelligence continues to revolutionize business and the way human beings are engaging with technology, leading to innovation across several industries. To continue enhancing real-time data stream solutions, UAVs have been most in demand specifically. The combination of AI and UAV technology holds promise in surveillance, disaster relief, agriculture, and conservation [2]. Integration has been able to combine UAVs with big data systems, leading to a new revolution in data gathering and analysis.

UAVs are not new in themselves but offshoots from their military beginnings. Technology in UAVs has, however, developed significantly over the past two years, progressing from its classical use in surveillance and reconnaissance to a multi-tool for civilian and commercial usage. Meanwhile, UAVs are able to fly smarter and autonomously with the introduction of AI. Due to this complementarity, UAVs are now able to offer advanced data-streaming systems, which were unattainable a few years ago.

The origin of this change is real-time streaming of data, which facilitates real-time access of decision-makers to raw data and supports rapid and well-informed decision-making. Methods of collecting data conventional in nature with respect to fixed infrastructure and manpower are, in most cases, not sufficient to offer real-time intelligence. AI-powered UAVs, on the other hand, offer an elastic, mobile, and dynamic solution that can stream data in real-time from remote and inaccessible areas [3]. Besides easing operations, the technology makes predictive and prescriptive analytics more convenient, facilitating proactive response to unfolding situations.

The goal of this deep research is to study the Impact of AI-based UAVs for Real-Time Data Streaming Applications. This deep research will cover all the aspects ranging from the technology used in AI-based UAVs, UAV uses in different industries, the influence on data analysis, challenges, and limitations to be overcome, and the ethics and regulatory aspects in their applications [4].

The remainder of this chapter will provide a full overview of these problems, and a full description of the potential of AI-based UAVs and the general reasons for their use in real-time stream data technology.

6.2 Overview of UAVs Powered by AI

6.2.1 Understanding AI-Enabled UAVs: Definition and Features

Unmanned aerial systems (UAS) or unmanned aerial vehicles (UAVs), also known as drones, have been transformed in the last few decades with the incorporation of artificial intelligence (AI) in them. The integration of AI and robotics makes the UAS intelligent and autonomous, and hence the aerial vehicles are responsive and efficient. This chapter puts the features of AI-based UAVs and technology drivers into perspective.

Autonomous drones as AI-controlled UAVs have the capability to fly, execute operations, and make decisions independently without any direct human control. A sensor suite, onboard processing, and AI-controlled software enable such UAVs to perceive, respond to changes in the environment, and execute operations for the achievement of established objectives [5]. From consumer-class drones of small sizes to sophisticated military-class UAVs, all utilize AI technology for precision improvement, responsiveness, and operation efficiency.

6.2.2 Capabilities of AI-Enabled UAVs

Autonomous AI-based UAVs are capable of mapping and recognizing flight paths on their own. They can even recognize obstacles and adapt to the environment through changing behavior, as depicted in Figure 6.1. This is a very important characteristic of precision agriculture and search and rescue [6].

Data Collection and Analysis: They are equipped with cameras, LiDAR sensors, and other vision sensors. They collect data in real time. AI processes and analyzes data onboard, and one can receive useful information and insights.

Machine Learning and Decision Making: AI drones utilize machine learning to learn and make decisions. They are able to, for instance, recognize objects, recognize patterns, or recognize crop health in agriculture.

Data Streaming and Communication: It is possible to stream the data in real-time to ground stations or embedded devices. The real-time streaming is specifically useful for applications like enabling prompt decision-making in emergency situations and surveillance of locations.

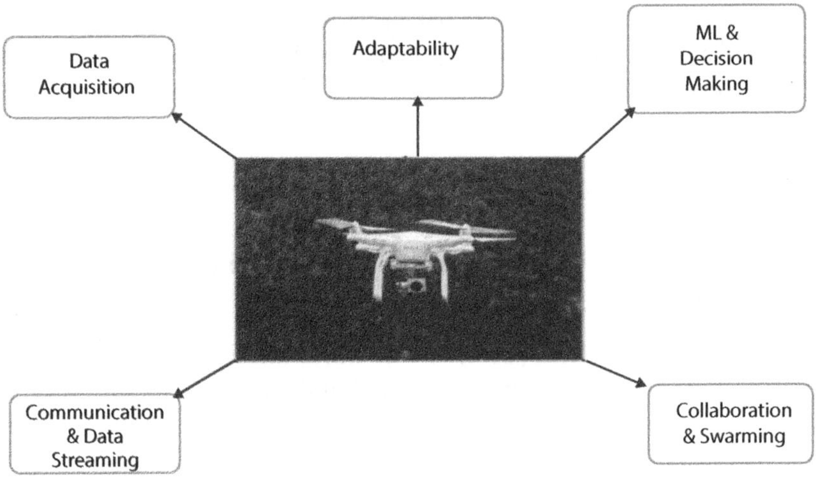

Figure 6.1 Basic elements of AI-enabled UAVs.

Flexibility: Dynamic condition-based flexibility can be achieved for AI-driven UAVs, and hence they are more efficient in situations where quick decision-making is worth the effort. For instance, while studying their surroundings, they can modify flight patterns according to varying atmospheric conditions [7].

Collaboration and Swarming: Some artificial intelligence-powered drones can fly in groups, called a swarm, in order to collaborate and complete an activity. It is applied to aerial monitoring and mapping, in which various angles are highly valuable.

6.2.3 Advantages of Using UAVs for Data Streaming

UAVs are becoming part of an ever-growing list of industries due to the many benefits of utilizing UAVs combined with AI for real-time data streaming. The significant benefits of utilizing AI-based UAVs for real-time data streaming operations are outlined in this section.

6.2.3.1 Quick and Effective Information Gathering

One of the most powerful features offered by artificial intelligence-based unmanned aerial vehicles, or UAVs, is the enormous potential for rapidly collecting valuable information from distant or even totally remote locations. In contrast to conventional means of data collection, normally plagued by geography-based limitations or processes requiring much physical exertion and toil, AI-based UAVs' sophisticated design and technology allow them to experience the luxury of flying over such difficult terrain without any kind of restriction. This makes them capable of collecting valuable information at a very high rate of efficiency and transmitting it in real-time for additional analysis and assessment [8]. In other uses like disaster relief missions, environmental monitoring operations, and infrastructure inspection missions, where timely and accurate information is of the first order, the unmatched accessibility and high speed of such UAVs are of the most value and utility.

6.2.3.2 *Improved Safety Features and Higher Cost-Effectiveness*

For such purposes such as structural inspection, subjecting human labor to dangerous or hard-to-reach areas may be expensive and dangerous. AI-powered UAVs can perform the task independently in an attempt to reduce the risk of human injury and reduce the cost of operation [9]. AI-wearable UAVs can also perform the inspection more frequently and thus identify defects even before they lead to serious problems, maximizing economy and safety.

6.2.3.3 *Accuracy and Precision*

The fusion of sensor technology and sophisticated algorithms of artificial intelligence in drones facilitates precise data collection and sophisticated analysis. From building structural integrity checks to crop health checks in agriculture, AI-driven UAVs facilitate highly precise and detailed data [10]. The precision helps in effective decision-making in industries, thus improving productivity, safety, and overall operational efficiency.

6.2.3.4 *Real-Time Decision Support*

Where there is a pressing requirement for decision-making, such as in the case of disaster relief or law enforcement, the real-time broadcast of information from Unmanned Aerial Vehicles (UAVs) is of utmost importance. Such UAVs provide immediate feedback to decision-makers, allowing for rapid and informed decisions, which in turn facilitates resource allocation, risk mitigation, and saving lives.

6.2.3.5 *Access to Remote Regions*

Artificial intelligence-driven unmanned aerial vehicles (UAVs) can reach remote or far-off places, for example, disaster sites, forests, and coastlines.

This capability is needed in other uses, for example, wildlife census, environmental research, and search and rescue, where it is not possible to offer ground access [11].

6.2.3.6 *Enhanced Monitoring and Surveillance*

Real-time monitoring and surveillance by the use of AI-powered UAVs is invaluable in operations that need to be monitored in real time. UAVs are able to monitor moving targets, monitor deviations, and monitor real-time continuously defined spaces. These are beneficial in uses such as wildlife conservation and border surveillance, where real-time monitoring offers security, prevents crime, and allows proactively informed decision-making.

6.2.3.7 *Scalability and Flexibility*

Artificial intelligence-driven unmanned aerial vehicles (UAVs) are highly programmable and scalable and are thus highly suitable to be used in any profession. They can be easily equipped with numerous different sensors and artificial intelligence programs which can be individually customized to their function to suit different professions. They are thus a type of universal tool for different industries.

6.2.4 Most Important Technologies and Components Utilized

AI-based UAVs possess basic assembly units and advanced technology in a bid to fly and process information by themselves and in real-time. The most fundamental units needed by a UAV for it to fly are addressed in this chapter, including sensors, communications units, and onboard computer hardware.

Sensors: Sensors are the sensing organs of AI-driven UAVs, which allow them to perceive and interpret environmental information. Sensors are utilized at scale:

Cameras: UAVs utilize various configurations of cameras, from RGB to multispectral to thermal in attempting to acquire high resolution visible data. Cameras are a key product in service from aerial surveillance to farm surveillance and high resolution images for inspection.

LiDAR (Light Detection and Ranging): LiDAR sensors shoot laser pulses to create accurate 3D models of the terrain. LiDARs are the centerpiece of the usage of topography mapping, forestry surveying, and infrastructure inspections. Though expensive, their accuracy makes them invaluable [12].

GPS (Global Positioning System): GPS receivers are used to give accurate positioning and navigation so that UAVs can be programmed to follow pre-determined paths and return home with ease.

Inertial Measurement Units (IMUs): IMUs monitor the movement and direction of a UAV and are crucial for stability and accurate navigation.

Environmental sensors are instruments utilized to measure key environmental parameters, such as temperature, humidity, and air quality [13]. Their use is of utmost significance in the areas of environmental monitoring and scientific research.

Onboard Processing Units: UAVs have internal computers installed inside that carry the potential for conducting sophisticated computations at lightning-fast speed. The onboard processing units ranging from embedded computers to special-purpose designed Graphics Processing Units (GPUs) support the processing of data, identification of objects, decision-making operations, and control of autonomous flight routines. Equipped with onboard AI algorithms, UAVs are able to process sensor data effectively, thereby enabling them to fly with greater autonomy and flexibility.

Communication Systems: AI-driven UAVs carry communication modules onboard to facilitate real-time streaming of data. Hardware employed could be radio transceivers, satellite link, or cellular network, depending upon range requirements and coverage area. Data sent out by the UAV is received and decoded by ground stations or remote stations.

AI Algorithms: The heart of AI-driven UAVs is the AI algorithms that perform the data processing and enable the decision-making. These can range from a huge collection of computer vision and machine learning approaches such as:

Object Detection and Recognition: The machine learning algorithms can detect and recognize objects within the range of the UAV, beneficial for use in such fields as surveillance and wildlife tracking [14].

Path Planning and Navigation: UAVs utilize intelligent algorithms to dynamically plan and adjust flight paths, providing flexibility to accommodate mission needs as well as flight conditions [15].

$$f(n) = g(n) + h(n) \tag{6.1}$$

where:

- $f(n)$ is the total estimated cost to reach the target,
- $g(n)$ is the cost from the start node to the current node,
- $h(n)$ is the heuristic estimate from the current node to the goal.

This Equation (6.1) ensures optimal route planning by UAVs in the face of obstacles and mission constraints.

Data Analysis and Insights: Machine processing provides valuable insights, which can be utilized in processes like tracking the health of crops in agriculture and infrastructure risk assessment.

Machine Learning for Autonomy: Advanced models can better understand sensor information so that UAVs can respond with data-driven solutions in real time for maximum effectiveness and autonomy.

6.3 Applications of AI-Enabled UAVs in Real-Time Data Streaming

6.3.1 Environmental Monitoring and Conservation

AI-based drones have revolutionized environmental monitoring and surveillance with live streaming capabilities that otherwise are not possible. Some of the uses consisted of a series of environmental tasks ranging from counting animals to observing ecosystem and climate change.

6.3.1.1 Wildlife Conservation

Drones powered by artificial intelligence are employed in wildlife monitoring and conservation in their natural environment. With high-definition cameras, alongside cutting-edge image recognition technology, the drones can monitor and follow endangered species, count the number of the animals, and identify possible danger in the form of poachers. Live streaming of information enables conservationists to respond in real time to the illegal activities, maintaining biodiversity.

6.3.1.2 Forestry and Land Governance

UAVs equipped with LiDAR and multispectral cameras are capable of collecting enormous data related to land surfaces and forests. UAVs enable forest management through precise tree counting, disease outbreak identification, and vegetation health monitoring. Conservationists can make decisions based on real-time data in forest conservation, deforestation, and forest fires [16].

6.3.1.3 Marine and Coastal Surveillance

Artificial intelligence-guided unmanned aerial vehicles, or just UAVs, equip highly sophisticated waterproof sensors specifically tailored to survey all the components of marine environments and coastlines. These high-tech gadgets also possess the ability to effectively scan and analyze the rise and fall of sea levels, the erosion process, as well as the level of pollution in such environmentally fragile locations. Moreover, in real-time, these highly effective UAVs possess the ability to transmit valuable information to scientists, thereby assisting them in making very sound decisions in their bid to safeguard sensitive coastal ecosystems as well as marine habitats [17].

6.3.1.4 Environmental Research

Environmental scientists use AI-based UAVs to collect data regarding air quality, soil composition, etc. UAVs can perform remote sensing activities to acquire knowledge in the form of environmental change and ascertain the impact of climate change in real time.

6.3.2 Disaster Management and Response

Real-time streaming of data is of the highest priority in disaster response and disaster management. AI-based UAVs are of the highest priority to offer situational awareness and relief in case of natural and man-made disasters.

6.3.2.1 Search and Rescue Operations

Disaster-hit zones get surveyed and salvaged by AI-equipped UAVs. They survey large areas in a few seconds, identify survivors, and measure intensity of damage. Real-time streaming of data provides rescue teams minute-by-minute updates, enabling coordination and saving lives [17].

6.3.2.2 Damage Assessment

Following natural disasters like earthquakes, hurricanes, or floods, remotely deployed aerial vehicles with high-resolution cameras and LiDAR scanning technology can inspect damaged infrastructure, identify areas needing assistance, and allocate resources accordingly. The real-time streaming capability provides decision-makers with the most up-to-date, accurate information, thus facilitating effective resource allocation [18].

6.3.2.3 Hazard Monitoring

Unmanned Aerial Vehicles, or UAVs, with very sophisticated and advanced sensors, i.e., gas sensors or radiation sensors, but not restricted to them, can be employed to carry out surveillance in risky zones possibly formed after such enormous accidents like industrial accidents or chemical leaks. Streaming in real time offers a window of opportunity to carry out an analysis of the situation at hand in real time, to carry out a rapid assessment, and to implement the necessary safety measures in real time to provide safety to people and the environment [18].

6.3.2.4 Evacuation Planning

In regions of frequent disasters, the use of artificial intelligence-boosted unmanned aerial vehicles, or UAVs, is a viable means of assessing and analyzing possible threats from numerous emergencies. These boosted UAVs can be used for collecting and transmitting real-time information relevant to the constantly changing situation on the ground, thus helping decision-makers and authorities make informed decisions on the best time and precise locations to evacuate people in need.

6.3.3 Precision Agriculture and Crop Monitoring

The agricultural sector benefits immensely from AI-enabled UAVs with real-time data streaming capabilities, enhancing crop management and improving overall yields.

6.3.3.1 Crop Health Assessment

Multispectral and thermal sensor drones can monitor crop health by identifying stress factors such as disease, pests, or nutrient deficiencies. Through real-time data streaming, farmers can respond rapidly to issues, optimizing crop yields [19].

$$NDVI = \frac{(NIR - RED)}{(NIR + RED)} \tag{6.2}$$

where:

- NIR is the near-infrared light reflected by vegetation,
- RED is the red light absorbed by plants.

This helps in detecting plant stress, disease, and nutrient deficiencies in real-time.

6.3.3.2 *Irrigation Management*

UAVs can quantify the content of soil moisture and show where irrigation is needed. With real-time information, UAVs help for accurate irrigation management, water conservation, and increased efficiency of agricultural processes [20].

$$SMI = \frac{(\theta - \theta_{min})}{(\theta_{max} - \theta_{min})} \tag{6.3}$$

where:

- θ is the observed soil moisture,
- θ_{*min*} and θ_{*max*} are the minimum and maximum soil moisture values, respectively.

This helps in precise irrigation planning and water conservation.

6.3.3.3 *Yield Forecasting*

Machine learning capabilities of drones can read crop data and forecast yields. Farmers will get real-time data about projected harvests, which will help them plan distribution and marketing more efficiently.

6.3.3.4 *Soil Analysis*

Soil sampling UAVs collect data on soil quality and composition. This data is streamed in real-time to inform farmers about optimal planting strategies and fertilizer applications, resulting in sustainable and efficient farming practices [21].

6.3.4 Infrastructure Inspection and Maintenance

AI-driven UAVs are transforming infrastructure inspection and maintenance in many industries. Real-time data streaming from these UAVs improves safety, lowers costs, and reduces downtime.

6.3.4.1 *Bridge and Building Inspections*

High-definition cameras and LiDAR-enabled UAVs can perform complex bridge, building, and structure inspections. Live streaming in real time enables inspectors and engineers to detect and locate structural defects and plan maintenance or repair operations [22].

6.3.4.2 Power Line and Utility Inspections

Utility firms use AI-powered UAVs to inspect power lines, windmills, and other assets. The UAVs report real-time equipment conditions, and the utility companies use this to pre-emptively repair the equipment and avoid outages [23].

6.3.4.3 Monitoring Oil and Gas Facilities

In the energy industry, gas sensors and thermal imaging cameras mounted on UAVs are employed to monitor oil and gas facilities. Streaming of real-time data helps in leak and other safety threat detection, reducing environmental risks and labor safety [23].

6.3.4.4 Railway and Pipeline Inspections

AI-powered UAVs are used to track railroads, pipelines, and other related infrastructure. Real-time streaming of information helps in detecting potential issues, like tracks being damaged or pipeline leakage, so that it can be repaired or serviced instantly [24].

6.3.5 Surveillance and Public Safety

AI-powered UAVs are transforming surveillance and improving public security through the provision of real-time data streaming capabilities. The police and law enforcement agencies employ UAVs in surveillance and monitoring activities. UAVs are applied in tracking suspects, monitoring public crowds, and responding to emergency situations in real-time. Real-time data streaming provides officers with information that is vital to decision-making and situational awareness [25]. UAVs are utilized in border surveillance to identify and prevent illegal crossing. Real-time data streaming provides border patrol officials with the ability to respond immediately to illegal movement and maintain border security [26]. In big public gatherings or protests, AI-powered UAVs can provide aerial surveillance to monitor crowd movement and identify potential security threats. Real-time data streaming provides assistance to the authorities in effective crowd management and maintaining public security. In firefighting operations, thermal-imaging UAVs can provide real-time data regarding the location and movement of the fire. The data is utilized by the firemen in the planning of containment operations and evacuation of priority locations [27].

6.4 Case Studies: Real-World Implementations

6.4.1 AI-Enabled UAVs for Monitoring Wildlife Populations

AI-powered UAVs have revolutionized wildlife conservation and monitoring. They offer real-time data to optimize management and protection. Case Study 1 (Endangered Species Monitoring): AI-powered cameras on UAVs monitor endangered species, such as rhinos in Africa, to recognize individuals and monitor movement to react immediately to the danger of poaching. Case Study 2 (Marine Life Monitoring): In the marine environment, UAVs monitor whales, study sharks, and monitor the health of coral reefs. Their real-time data streaming offers marine biologists an immediate look at marine ecosystems.

6.4.2 UAV-Based Aerial Imagery for Disaster Assessment

UAVs have revolutionized disaster assessment and management. UAV AI systems play a critical role in relief planning, survivor detection, and dam age assessment. Case Study 1 (Earthquake Response): Following a devastating earthquake, UAVs employed LiDAR and high-resolution cameras to estimate damage, and real-time data informed search and rescue [28]. Case Study 2 (Hurricane Damage Assessment): UAVs assessed damage and facilitated recovery in hurricane-stricken areas. UAVs surveyed the affected areas and took pictures of destruction, allowing authorities to quickly assess and distribute resources.

6.4.3 Precision Agriculture Using AI Algorithms and UAVs

UAVs and AI algorithms transformed precision agriculture to improve crop yield and management. Case Study 1: Smart farming using UAVs driven by AI An AI-equipped UAV with multispectral sensors was employed by a farm in an AI-driven case study to provide real-time crop health feedback. Farmers received real-time fertilizer, pest, and irrigation feedback, improving yield and resource optimization. Case Study 2: Vineyard farming Grapevine cultivation in grape farming is regulated through precision agriculture. AI-equipped UAVs captured high-resolution images of vineyards, and their algorithms scanned grapevines' health in real-time continuously. It improved grape quality and wine yield by allowing vineyard managers to make timely decisions on pruning, harvesting, and disease control [29].

6.4.4 Automated Infrastructure Inspection with UAVs

AI-powered UAVs are transforming infrastructure inspection with enhanced safety, cost savings, and improved efficiency. Case Study 1 (Bridge Inspection): Camera- and LiDAR-equipped UAVs inspected the structural condition of a large bridge. They inspected the bridge, taking high-definition images and identifying faults. Engineers scheduled maintenance with real-time information, cutting costs and downtime. In Case Study 2 (Power Line Inspection), utility operators utilized AI UAVs to inspect. UAVs inspected power line and tower conditions, indicating equipment status and fault detection to achieve timely maintenance and lower outage risk.

6.4.5 UAVs for Surveillance and Emergency Response

AI-piloted UAVs are central to emergency response and surveillance, providing real-time information to law enforcement and first responders. Case Study 1 (Law Enforcement Operations): In one case, UAVs tracked a mass gathering, enabling agencies to monitor crowd behavior and locate security threats. The real-time intelligence enabled officers to maintain public safety. Case Study 2 (Search and Rescue Operations): AI-driven UAVs are essential to search and rescue operations. UAVs were used to locate a missing hiker in a remote location, quickly scanning the region and providing real-time intelligence to the rescue team, leading to a successful rescue operation.

6.4.6 Machine Learning and Deep Learning Techniques

Machine learning and deep learning algorithms are the fundamental pillars on which AI-driven data analysis has been developed [30]. During the past few years, these incredible technologies have accelerated and exponentially developed, leading to remarkable improvements that deliver greater accuracy, greater efficacy, and the ability to scale to handle any form of data analysis process more effectively.

6.4.6.1 Advancements in Machine Learning

Machine learning, as a general and continuously expanding field of inquiry, has experienced a series of major advances and breakthroughs over the course of its history:

Ensemble Learning, or the deliberate mixing of numerous models with various ensemble methods, such as Random Forests, Gradient Boosting, and AdaBoost, has played a major role in enhanced predictive accuracy and general robustness. Due to these advances, ensemble models have become highly capable tools in tackling complex and complicated real-world problems that demand highly advanced solutions.

6.4.6.2 *Reinforcement Learning (RL)*

Reinforcement learning, the exciting computer science discipline that enables machines to learn a series of sequential choices with the passage of time, has taken monumental strides in the last two years. This innovative paradigm has been employed in varied industries, from autonomous systems to robots and even gaming, with exceptional capabilities to maximize and enhance decision-making.

Reinforcement learning is applied in UAV training for autonomous decision-making. The Q-learning equation is:

$$Q(s,a) = Q(s,a) + \alpha[r + \gamma maxQ(s',a') - Q(s,a)] \tag{6.4}$$

where:

- $Q(s,a)$ is reward by taking action a in state s,
- α is the learning rate,
- r is the reward received,
- γ is the discount factor for future rewards,
- $\max Q(s',a')$ is the maximum expected reward for the next state s'.

This helps UAVs learn optimal flight and decision strategies through trial and error.

6.4.6.3 *Transfer Learning*

The idea of transfer learning has now been widely accepted in the area of machine learning, and it is a stark departure from the way we train models. With this new method, models that have been specially trained to perform a specific task can now be transferred with ease to perform a different task. This aspect not only makes machine learning models more flexible but also greatly minimizes the amount of large training data required and the computational resources needed to train such models.

6.4.6.4 *Interpretable Models*

There has been increasing interest in interpretability of machine learning models or Explainable AI (XAI). The reason is in explaining AI decisions and making them more understandable because AI decisions have caused concern about black-box deep learning models, particularly high-risk domains like healthcare and finance.

6.4.6.5 *Advancements in Deep Learning*

Deep learning, being part of the much larger field of machine learning and more narrowly concerned with the implementation and creation of neural networks, is currently a leading driver of incredible advancements and discoveries in an enormous variety of disparate fields [31]:

6.4.6.6 *Convolutional Neural Networks*

Convolutional Neural Networks, otherwise known as CNNs, have actually revolutionized the domain of image and video processing in extraordinary manners. Such revolution has brought massive impacts on several applications, such as object detection, image classification, and even face recognition, which are now executed with greater accuracy and efficiency than before. The deep structure typical of such networks, and more importantly the convolutional layers, is particularly extremely proficient in processing complicated and advanced visual information. Hence, CNNs allow a vastly enhanced and quicker image and video interpretation, thus emerging as a very valuable resource in computer vision [32].

6.4.6.7 *Recurrent Neural Networks (RNNs) and Long Short-Term Memory (LSTM)*

RNNs and LSTMs are an absolute requirement nowadays to handle sequential data, used in speech recognition, natural language processing, and time series forecasting. Since they are capable of remembering temporal relations, they are able to learn and process based on time patterns, and therefore are used in a huge number of applications where context-sensitiveness and sequential knowledge are required.

6.4.6.8 *Generative Adversarial Networks (GANs)*

GAN has largely unlocked new and promising frontiers in numerous applications, especially image generation, style transfer, and the now common construction of deep fakes. These new networks have made extensive use of the incredible power of adversarial training methods to create high-quality, incredibly real images and data that are frequently indistinguishable from their real counterparts [33].

Transformers have certainly become the de facto standard natural language processing architecture. Among other models, BERT, GPT-3, and T5 are great examples that have attained state-of-the-art performance on a vast range of tasks such as, but not limited to, language comprehension, translation, and generation of text. Self-attention mechanisms, which form a part of transformers, have revolutionized the field of sequence-to-sequence tasks to a great extent, and thus there has been phenomenal progress in the way these tasks are addressed and performed.

The rapid progress of machine learning and deep learning techniques reflects the immense scope of AI in data analytics in healthcare and finance to autonomous systems and creative content creation.

6.4.7 Object Detection and Recognition

Object recognition and object detection, both of which are extremely important areas in the domains of computer vision and artificial intelligence, have witnessed unprecedented growth in recent years. These unprecedented growths have significantly improved numerous applications, ranging from but not limited to self-driving vehicles, sophisticated surveillance systems, and rich augmented reality experiences.

6.4.7.1 Single-Stage Object Detectors

The past couple of years have witnessed a revolutionary change in object detection, thanks primarily to the creation of algorithms like YOLO, or You Only Look Once, and SSD, or Single Shot MultiBox Detector. These new methods have come a long way in making real-time object detection a reality. They are best suited to situations where the primary concerns are speed and accuracy, both of which they can balance in perfect harmony to the requirements of different applications.

6.4.7.2 Two-Stage Object Detectors

The two-stage detector class, of which high-profile examples such as Faster R-CNN and Mask R-CNN are extremely prominent, has developed over the years quite well in their ability to deliver higher levels of accuracy and flexibility in object detection tasks. These sophisticated systems not only perform excellently in performing standard object detection, but are also able to handle more complex challenges, such as instance segmentation, and thereby enable a much deeper and richer visual data analysis [34].

6.4.7.3 *Optimal Object Detection*

The pursuit of optimality has been the overarching goal, resulting in the development of highly optimized object detection models that are defined by record levels of performance. Not only are these models outstanding in what they can accomplish, but they also demonstrate extensive applicability in resource-constrained devices. Additionally, they are highly efficient in real-time processes, where speed and accuracy are of utmost importance.

6.4.7.4 *Fine-Grained Object Recognition*

Today's recognition algorithms have made tremendous progress, so much so that they can recognize fine-grained details with incredible accuracy, whether it is about different bird species or different breeds of dogs. All these developments have far-reaching real-world practical uses, ranging from wildlife watching, the very successful e-commerce industry, and many more applications that involve being able to distinguish between classes at very fine and subtle levels.

Recent research has further highlighted *few-shot and zero-shot learning*, which enables models to recognize objects with fewer or even no examples. This phenomenal flexibility provides AI systems with the capability to deal with new, unseen objects with more flexibility.

$$IoU = \frac{|A \cap B|}{|A \cup B|} \tag{6.5}$$

where:

- A is the predicted bounding box,
- B is the ground truth bounding box.

IoU measures how well a detected object matches the actual object in AI-driven UAV imagery.

Breakthroughs in object detection and identification have greatly pushed the boundaries of AI to comprehend and interpret visual information, with vast implications for applications such as autonomous driving, security, and retail.

6.4.8 Image and Video Processing

The field of image and video processing has determined innovative changes determined by the development of AI algorithms. It enables the actual processing of visual information and improves healthcare, entertainment and surveillance applications.

Super Resolving Algorithms have unique features to improve the quality of your photos and videos, making them clearer and more detailed. Many applications, such as video streaming, surveillance, and medical imaging, where image quality is very important, have overlooked many applications, such as video streaming, surveillance, and medical imaging.

6.4.8.1 Real-Time Image Segmentation

Image Segmentation fields can now identify object restrictions and forms in real-time in images and videos. This technology is very involved in augmented reality, self-driving cars, and medical imaging, where precise object boundaries are required.

6.4.8.2 Image and Video Signature

AI algorithms can now generate descriptive captions for images and videos. This innovation provides the multimedia content to the visually impaired and also improves search engine optimization.

6.4.8.3 Video Analysis

Video analysis algorithms can process and analyze video content in real time. They have applications in many different applications, from security systems where facial recognition, traffic observation, and decision-making processes are of utmost importance in real time.

6.4.8.4 Model Generation for Creativity and Art

Model generation such as deep neural networks can generate art, music, and other creative content. Such models have given rise to new forms of artistic creation and content generation and have opened new applications in the creative industry.

For segmentation of images, UAV employs Softmax activation function in deep learning:

$$\sigma(z_i) = \frac{e^{z_i}}{\sum_{j=1}^{n} e^{z_i}} \tag{6.6}$$

where:

- z_i is the input to the neural network,
- n is the number of possible classes.

This equation facilitates accurate object classification in UAV imagery.

The rapid development of image and video processing has revolutionized the processing and manipulation of visual data in medicine and entertainment, education, and advertising.

6.4.9 Sensor Fusion and Data Integration

Sensor fusion and data integration are important activities of data analysis in the contemporary world. Sensor fusion means the capability to merge one or more sources of sensors into a richer, broader view of the environment. Sensor fusion and data integration are applied in numerous industries, including robotics, healthcare, environmental monitoring, and smart cities.

6.4.9.1 Multimodal Sensor Fusion

Multimodal sensor fusion integrates data from more than one sensor—cameras, LiDAR, radar, GPS, etc. Such a total combination allows us to comprehend the surroundings in detail and hence is of utmost importance for applications such as autonomous cars and drones, wherein heterogeneous data fusion is crucial to make smart choices.

6.4.9.2 Healthcare Data Integration

Integration of healthcare data is required to develop an even more informative patient profile. The electronic health records, wearable, imaging, and genomic data are completely analyzed and integrated. The integrated data are used by artificial intelligence algorithms to improve patient diagnosis, treatment planning, and patient care.

6.4.9.3 *Environmental Monitoring*

Environmental monitoring relies on integrated data from various sources, i.e., satellites, surface weather stations, IoT sensors, and remote sensors. The integrated data streams provide real-time information regarding environmental conditions, which is utilized in applications like weather forecasting, climate monitoring, and prediction of disasters.

6.4.9.4 *Industrial Automation*

Sensor information from manufacturing and industrial equipment, robots, supply chain factories, and quality control factories is collected and aggregated to automate. Artificial Intelligence software interprets the aggregated information to improve manufacturing productivity, maintenance scheduling, and product quality.

6.4.9.5 *Smart Cities*

Smart city projects are highly reliant on information convergence from different sources, including traffic cameras, environmental sensors, citizen feedback, and others. The combined information presents an end-to-end snapshot of the city, enabling improved urban planning, transport management, and public services.

The achievements realized in data integration and sensor fusion are the building blocks of data-driven decision-making in applications that need end-to-end understanding of sophisticated environments and systems.

6.5 Challenges and Considerations in AI-Enabled UAVs for Real-Time Data Streaming Applications

The use in real-time data streaming application areas is of highly valuable potential, yet simultaneously raises a full set of serious issues and problems. It is particularly their requirement that must be examined in depth as to enable adequate guarantees for their proper, secure, and ethical application in a variety of marketplaces. In what follows, we will elaborate in more detail on the most important issues and problems raised by application areas of AI-capable UAVs in real-time data streaming.

6.5.1 Data Privacy and Security

Privacy and data security are major concerns in the use of AI-powered unmanned aerial vehicles (UAVs) for live data streaming. These are issues of protecting sensitive data collected in the skies through the employment of the UAV and making such data secure and confidential.

Data Encryption: There must be strong encryption processes implemented for real-time data streamed. Data encryption helps guard data from being intercepted or manipulated by other unauthorized users and preserves data confidentiality and integrity.
Data Storage: Data collected by UAVs should be stored securely to preserve the integrity of the data. Proper backup data and recovery measures should be implemented so as to recover data in case of any technical failure or cyberattack. Data-at-rest encryption safeguards data on UAVs or on ground stations.
Compliance with Data Protection Law: Operators of UAVs must comply strictly with data protection law like that of the European continent's General Data Protection Regulation (GDPR). Law requires compliance with data protection law and privacy law while handling sensitive and personal information. Lawful compliance is required to avoid legal sanction.
Cybersecurity Controls: The UAV systems should have strong cybersecurity controls to prevent unauthorized access and cyberattacks. The controls are in the form of the installation of intrusion detection systems, frequent security audits, and vulnerability assessment to help detect and neutralize threats.

6.5.2 Regulatory Frameworks and Airspace Management

Effective integration of AI-enabled UAVs into the airspace requires a proper regulatory environment and efficient airspace management policies for their legal and safe operation.

Licensing and Certification
UAV operators and pilots must pass air regulators' licensing and certification requirements. They are compulsory as a measure aimed at ensuring safe and prudent operation of UAVs. They are meant to verify that the operators are qualified and capable of ensuring safe UAV operation.
Airspace Integration
There must be effective airspace management in the balance of manned air traffic and UAV flying. Air traffic management systems must be modified to accommodate the requirements and operational behaviors of UAVs

that can be unique compared to manned air vehicles. Integration processes must be safe and should not result in any air traffic delay or crash.

Geofencing

Geofencing technology is utilized to establish no-fly zones and exclusion zones that restrict UAVs from entering the airspace, such as airports, military installations, and sensitive government sites. Geofenced areas are used to restrict access.

Remote ID and Tracking

Remote identification and tracking of drones are crucial in the provision of accountability and safety. Remote identification technology can empower governments to identify the activities of drones and understand who the operators are, further enhancing aggregate management and safety in shared airspace.

6.5.3 Technical Constraints and Limitations

While they have a greater capability of functionality, AI drones are plagued by numerous technical issues that can interfere with their operations, especially when they are used to stream data in real-time.

Low Flight Duration

UAVs have limited battery life with low mission duration. Streamed mission data needs to provide for constant recharging or battery replacement, particularly for high-endurance missions. Effective power management techniques have to be utilized in order to maximize time.

Payload Capacity

Payload capacity of UAVs is naturally limited, and use of high-end sensors and AI processing further reduces their load-carrying capacity. Payload weight, flight stability, and mission performance have to be balanced in a very fine manner, which is the main technical challenge. Payload optimization needs to be implemented to enable the UAVs to perform their task assigned to them without compromising on performance.

Communication Range

Performance of real-time data transmission by UAVs is dependent on ground station to a UAV communication range. If the range is inadequate, operational coverage can limit operations, and therefore it is challenging to achieve a link under stable conditions. Satellite data links and network relays offer advanced communication media with longer distances and stable, high-quality signal. Environmental Factors Adverse weather conditions like strong winds, rain, and poor visibility can adversely affect UAV operations.

Environmental Conditions
Inclement weather like windy conditions, rain, and low visibility can harm UAV flight. AI software needs to be designed in such a way to consider environmental conditions and ensure stable and secure data streaming. Real-time weather data and flight control flexibility can save from destruction under adverse weather conditions.

6.5.4 Ethical Considerations and Societal Impact

The application of artificial intelligence-based unmanned aerial vehicles is morally wrongful and highly influential in society and hence needs thorough scrutiny and regulation.

Privacy Concerns
UAVs equipped with high-definition cameras and sensors do pose the negative effect of invading people's private space and violating their privacy. Norms and morality must be established to regulate UAVs traversing open as well as closed spaces. Privacy by design principles must be incorporated in the UAV system.
Data Fairness and Bias
The AI algorithms used in UAVs must be trained from a representative and diverse data set so that they are not subject to bias and discrimination during decision-making. Steps for reducing bias need to be incorporated in a manner that equity is provided in data analysis, especially in such mission-critical operations as surveillance and police work.
Job Displacement
Part of the employment automation with UAVs could displace jobs, especially in areas such as surveillance and agriculture. In the effort to create space for social and economic effects, labor retraining and transition is required.
Community Acceptance
UAV flights over populated areas can be resisted by society on the basis of noise pollution, visual intrusiveness, or perception of surveillance. Information campaign and community relations are essential to gain community acceptance and transparency of UAV operations.

6.6 Future Directions and Opportunities

Future Directions and Opportunities in AI-Enabled UAVs for Real-Time Data Streaming Applications.

6.6.1 Integration of AI and UAV Technologies

The partnership of artificial intelligence and unmanned aerial vehicle technology is spearheading the innovation in sectors. The partnership has many opportunities that are expected to revolutionize the use of UAVs and their impact on applications involving real-time data transmission.

AI for Increased Autonomy
Artificial intelligence algorithms are capable of enabling UAVs to autonomously decide within intricate surroundings. They can learn and adapt to changing conditions, calculate the shortest route to avoid obstructions, and maximize flight efficiency. AI autonomy is needed in applications like search and rescue, precision agriculture, and bridge inspection.
AI-Augmented Sensing
New computer vision and sensor technologies, both motivated by AI, will make the UAVs more sensitive perceptually. They will identify and classify objects, sense environmental features, and offer more precise and stable real-time data stream.
Swarm Intelligence
Swarm intelligence and AI will facilitate collective action in UAV fleets. They will work together to protect remote regions, communicate effectively, and perform cooperative missions, such as environmental monitoring, disaster relief, and surveillance.
Edge Computing
Edge computing and AI will make it possible to compute data onboard by UAV, thereby decreasing data to be uplinked to ground station. Real-time processing of this nature is crucial in the event of applications that require low-latency decision-making.
AI-Optimized Hardware
Hardware devices used for AI processing in UAVs will be brought into focus. AI chips and processors will be specially designed to make UAVs operate with advanced AI models, allowing them to be more effective and versatile.

6.6.2 Collaborative Research and Development

Collaborative research and development is key to achieving the potential of AI-capable UAVs. Interdisciplinary collaboration, collaboration between academia and industry, and international collaboration will spur innovation.

Interdisciplinary Research
Interdisciplinary studies among AI, robotics, aerospace engineering, and data science experts will result in advancements in UAV technology. Research will be directed towards incorporating artificial intelligence into unmanned aerial vehicle operations and creating new streaming data applications.
Academic and Industry Partnerships
Collaboration of schools with industrial stakeholders will enable the translation of leading-edge research into deployable technologies. The collaboration will speed up both the development and incorporation of artificial intelligence-based unmanned aerial vehicle technology.
Testbeds and Prototyping
The development of testbeds and prototyping centers for AI-based UAVs will be essential in testing and validation of emerging technology. The centers will be essential for iterative development and thorough testing of UAV systems.
Cross-Sector Collaboration
Interdisciplinary collaboration among sectors including healthcare, agriculture, disaster management, and environmental monitoring is anticipated to facilitate the development of multifunctional UAV solutions. Such adaptable UAVs possess the capability to be utilized across a range of applications involving real-time data transmission.
International Collaboration
Global cooperation is in the research and development of UAVs and will promote innovation and knowledge sharing. Global cooperation can tackle global problems such as disaster relief and monitoring of the environment on a grand scale.

6.6.3 Policy Implications and Standardization

The integration of AI-powered UAVs into society needs competent policies, regulation, and standards. These are the factors that will determine the future of the AI-UAV technology.

Regulatory frameworks
The current regulations and laws must be updated with the AI-enabled UAVs by the governments and aviation authorities. Safety concerns, privacy concerns, and airspace management concerns need to be addressed through regulations. Safety and innovation need to be encouraged on a priority basis.

Privacy and Data Management
There should be stricter privacy regulations put in place to safeguard the population against pervasiveness of UAV surveillance. The regulations should include data property rights, right to consent, and data storage rights.
Airspace Integration
Rules for safe coexistence of UAV and manned aircraft in the same airspace have to be framed. Air traffic management infrastructure must be developed to facilitate safe coexistence.
Liability and Insurance
There should be liability provisions in case of accidents or data loss in connection with UAVs. There should be special insurance policies for UAV operation to cover the risk completely.
Ethical Principles
Ethical frameworks for AI-driven UAV flight need to be built in a way that UAV flight can be harmonious with values and human rights. They need to counterbalance against considerations such as ethics of surveillance and ethics of information collection.

6.6.4 Advancements in Hardware and Software Solutions

New hardware and software technologies are pushing AI-enabled UAVs to capability and autonomy. Longer flight times and downtime will be provided by increased battery life, with improved UAV performance on long-endurance missions. Fewer and lighter components will result in faster UAVs with more payload. LiDAR and hyperspectral sensors will be more sensitive and sophisticated at detecting environmental data, with uses in precision agriculture and disaster relief. Enhanced AI processors will have UAVs processing data in real-time, minimizing their reliance on ground-based processing. Enhanced software tools will simplify the development, testing, and deployment complexity, while edge computing will allow UAVs to react instantaneously with no need for a continuous ground connection. In addition, AI-driven user interfaces will make the use of UAVs easier, and advances in communications, such as 5G, will provide reliable high-speed data streaming between UAVs and ground stations to facilitate seamless UAV-ground station integration.

6.7 Conclusion and Future Enhancements

The advent of AI-powered UAVs has made possible, for the purposes of real-time data streaming, a new and expanded universe of technological possibility and revolutionary potential. In the remainder of this concluding section, we shall discuss at length the drastic implications of AI-enabled UAVs for real-time data streaming, envision the futures of expansion and advancement in the field, and underscore the vital necessity of using this technology for ethical and moral purposes. AI-powered UAVs have transformed industries and applications, leaving an irreversible impact on society, with their real-time data streaming applications having a profound impact on environmental monitoring and conservation, disaster response and management, precision agriculture and crop monitoring, infrastructure inspection and maintenance, and surveillance and public safety operations; their transformative power permeates various industries, such as healthcare, search and rescue, construction, and entertainment, transforming potential and opening new frontiers of innovation and development through innovation in data streaming and analysis.

The prospects for development and growth in the field of AI-driven UAVs for streaming data in real-time are favorable. Some of the ongoing development includes improving cutting-edge AI algorithms for greater autonomy of UAVs, multimodal sensor fusion to provide richer understandings of the environment, work on swarm technologies for efficient group use, optimizing energy efficiency and sustainability, simplifying human-machine interfaces for broader UAV accessibility, ongoing regulatory reform to address needs for safety and innovation, technological harmonization *via* international cooperation, and projected economic growth and jobs in the developing AI-driven UAV industry.

References

1. Abrahams, M., Sibanda, M., Dube, T., Chimonyo, V.G., Mabhaudhi, T., A systematic review of UAV applications for mapping neglected and underutilised crop species' spatial distribution and health. *Remote Sens.*, 15, 19, 4672, 2023.
2. Adil, M., Song, H., Mastorakis, S., Abulkasim, H., Farouk, A., Jin, Z., UAV-assisted IoT applications, cybersecurity threats, AI-enabled solutions, open challenges with future research directions. *IEEE Trans. Intell. Veh.*, 9, 4, 4583–4605, 2023.
3. Aggarwal, S., Kumar, N., Alhussein, M., Muhammad, G., Blockchain-based UAV path planning for healthcare 4.0: Current challenges and the way ahead. *IEEE Netw.*, 35, 1, 20–29, 2021.
4. Babel, L., Flight path planning for unmanned aerial vehicles with landmark-based visual navigation. *Rob. Auton. Syst.*, 62, 2, 142–150, 2014.

5. Brighenti, F., Carnemolla, F., Messina, D., De Guidi, G., UAV survey method to monitor and analyze geological hazards: the case study of the mud volcano of Villaggio Santa Barbara, Caltanissetta (Sicily). *Nat. Hazards Earth Syst. Sci.*, 21, 9, 2881 2898, 2021.
6. Cao, W., Qing, H., Xu, X., Liu, C., Chen, S., Zhong, Y., Liu, Q., Pre-archaeological investigation by integrating unmanned aerial vehicle aero-magnetic surveys and soil analyses. *Drones*, 6, 9, 243, 2022.
7. Chao, H., Gu, Y., Gross, J., Guo, G., Fravolini, M.L., Napolitano, M.R., A comparative study of optical flow and traditional sensors in UAV navigation, in: *2013 American Control Conference*, pp. 3858–3863, IEEE, 2013.
8. Creswell, A., White, T., Dumoulin, V., Arulkumaran, K., Sengupta, B., Bharath, A., A. Generative adversarial networks: An overview. *IEEE Signal Process Mag.*, 35, 1, 53–65, 2018.
9. Curcio, A.C., Peralta, G., Aranda, M., Barbero, L., Evaluating the performance of high spatial resolution UAV-photogrammetry and UAV-LiDAR for salt marshes: the Cádiz Bay study case. *Remote Sens.*, 14, 15, 3582, 2022.
10. Davies, L., Vagapov, Y., Grout, V., Cunningham, S., Anuchin, A., Review of air traffic management systems for UAV integration into urban airspace, in: *2021 28th International Workshop on Electric Drives: Improving Reliability of Electric Drives (IWED)*, IEEE, pp. 1–6, 2021.
11. Dawkins, J., Unmanned combat aerial vehicles: examining the political, moral, and social implications. Monograph, School of Advanced Air and Space Studies, 2005.
12. Del Cerro, J., Cruz Ulloa, C., Barrientos, A., de León Rivas, J., Unmanned aerial vehicles in agriculture: A survey. *Agronomy*, 11, 2, 203, 2021.
13. Erdos, D., Erdos, A., Watkins, S.E., An experimental UAV system for search and rescue challenge. *IEEE Aerosp. Electron. Syst. Mag.*, 28, 5, 32–37, 2013.
14. Garg, P.K., Unmanned aerial vehicles: An introduction, Mercury Learning and Information, 2021.
15. Gonzalez, L.F., Montes, G.A., Puig, E., Johnson, S., Mengersen, K., Gaston, K.J., Unmanned aerial vehicles (UAVs) and artificial intelligence revolutionizing wildlife monitoring and conservation. *Sensors*, 16, 1, 97, 2016.
16. Hermand, E., Nguyen, T.W., Hosseinzadeh, M., Garone, E., Constrained control of UAVs in geofencing applications, in: *2018 26th Mediterranean Conference on Control and Automation (MED)*, IEEE, pp. 217–222, 2018.
17. Idachaba, F.E., Monitoring of oil and gas pipelines by use of VTOL-type unmanned aerial vehicles. *Oil Gas Facil*, 5, 01, 47–52, 2016.
18. Janiesch, C., Zschech, P., Heinrich, K., Machine learning and deep learning. *Electron. Markets*, 31, 3, 685–695, 2021.
19. Jat, D.S. and Singh, C., Artificial intelligence-enabled robotic drones for COVID-19 outbreak, in: *Intelligent systems and methods to combat covid-19*, pp. 37–46, 2020.
20. Jenssen, R. and Roverso, D., Intelligent monitoring and inspection of power line components powered by UAVs and deep learning. *IEEE Power Energy Technol. Syst. J.*, 6, 1, 11–21, 2019.

21. Jian, L., Li, Z., Yang, X., Wu, W., Ahmad, A., Jeon, G., Combining unmanned aerial vehicles with artificial-intelligence technology for traffic-congestion recognition: electronic eyes in the skies to spot clogged roads. *IEEE Consum. Electron. Mag.*, 8, 3, 81–86, 2019.
22. Kim, K. and Kang, Y., Drone security module for UAV data encryption, in: *2020 international conference on information and communication technology convergence (ICTC)*, IEEE, pp. 1672–1674, 2020.
23. Koslowski, R. and Schulzke, M., Drones along borders: Border security UAVs in the United States and the European Union. *Int. Stud. Perspect.*, 19, 4, 305–324, 2018.
24. Mohammed, F., Idries, A., Mohamed, N., Al-Jaroodi, J., Jawhar, I., UAVs for smart cities: Opportunities and challenges, in: *2014 international conference on unmanned aircraft systems (ICUAS)*, IEEE, pp. 267–273, 2014.
25. Mohd Noor, N., Abdullah, A., Hashim, M., Remote sensing UAV/drones and its applications for urban areas: A review, in: *IOP conference series: Earth and environmental science*, vol. 169, IOP Publishing, p. 012003, 2018.
26. Nawaz, H., Ali, H.M., Laghari, A.A., UAV communication networks issues: A review. *Arch. Comput. Methods Eng.*, 28, 3, 1349–1369, 2021.
27. Nedjati, A., Vizvari, B., Izbirak, G., Post-earthquake response by small UAV helicopters. *Nat. Hazard.*, 80, 1669–1688, 2016.
28. Rahman, S., Khan, S., Porikli, F., A unified approach for conventional zero-shot, generalized zero-shot, and few-shot learning. *IEEE Trans. Image Process.*, 27, 11, 5652–5667, 2018.
29. Restas, A., Forest fire management supporting by UAV based air reconnaissance results of Szendro fire department, Hungary, in: *2006 First International Symposium on Environment Identities and Mediterranean Area*, IEEE, pp. 73–77, 2006.
30. Kaur, N., Kaur, G., Aruchamy, P., Chaudhary, N., An Integrated Approach Based on Fuzzy Logic and Machine Learning Techniques for Reliable Wine Quality Prediction. *Procedia Comput. Sci.*, 252, 613–622, 2025.
31. Prasanth, A., Ragavi, B., Jayachitra, S., Kuntavai, T., Automated Detection of Oropharyngeal Carcinoma Using Super-Resolution Generative Adversarial Network and Convolutional Neural Network, in: *Proceedings of IEEE International Conference on Smart Electronics and Communication Systems*, pp. 1–6, 2025.
32. Aruchamy, P., Sundarrajan, M., Choudhry, M.D., Jothi, A., Enhancing security of medical images using code-based intermittent encryption and convolutional neural network. *Comput. J.*, 130, 1–22, 2024.
33. Balasubramaniam, S., Kadry, S., Prasanth, A., Dhanaraj, R.K. (Eds.), *Generative AI and LLMs: Natural Language Processing and Generative Adversarial Networks*, Walter de Gruyter GmbH & Co KG, 2024.
34. Tang, L. and Shao, G., Drone remote sensing for forestry research and practices. *J. For. Res.*, 26, 791–797, 2015.

7

Blockchain-Based Security and Privacy Solutions for Drones Systems

Sugandhi Malhotra[1]*, Gaganjot Kaur[2], K. Saranya[3] and Anu Sayal[4]

[1]HOD, CSE-APEX Chandigarh Group of Colleges, Jhanjeri, Mohali, Punjab, India
[2]Department of Computer Science Engineering, Raj Kumar Goel Institute of Technology, Ghaziabad, India
[3]Department of Electronics and Communication Engineering, Dr. Mahalingam College of Engineering and Technology, Tamil Nadu, India
[4]School of Accounting and Finance, Taylor's University, Subang Jaya, Selangor, Malaysia

Abstract
In today's fast-paced world the exponential rise in the use of drones requires a lot of improvements and stability in drone communication as they are playing a major role in many societal and military applications. The development of Unmanned Aerial Vehicles (UAVs) is accelerating and is promoting several civic uses. They continue to be vulnerable to several security vulnerabilities, which endangers everyone's safety. When they are connected to the Internet, security challenges multiply since their data stream is vulnerable to intrusions. Drones have demonstrated their capacity to offer cost-effective, real-time solutions for a variety of applications, including healthcare, monitoring of smart grids, surveillance in smart cities, and border security. Despite its numerous security and privacy concerns, researchers from all over the world have offered a wide range of ways to safeguard drone communication against cyber-attacks.

The majority of these solutions, which rely heavily on computation, were based on cryptographic methods. There aren't many blockchain-based solutions and those that do have expensive transaction storage costs as well as bandwidth and communication dependability problems. Motivated by these facts, the authors introduce the article that outlines the structural framework governing the secure communication of drones, coupled with the integration of blockchain technology. Subsequently, an evaluation of prevailing blockchain-driven resolutions is

**Corresponding author*: mailmetech@gmail.com

A. Prasanth, Rajesh Kumar Dhanaraj, Munish Sabharwal, Vandana Sharma and Seifedine Kadry (eds.) Drones for Transportation Logistics and Disaster Management, (145–180) © 2026 Scrivener Publishing LLC

conducted, alongside an examination of extant regulations about data privacy that demand consideration during the formulation of blockchain-centric remedies. Ultimately, this leads to the recognition of plausible hurdles and unresolved matters, which are intended to steer forthcoming research endeavors.

***Keywords*:** Blockchain, drones, Ethereum, security, privacy, unmanned aerial vehicles (UAV), cryptographic methods

7.1 Introduction

The internet is succeeded by an ever-changing way of sending data, information, files, images, and videos. But the hot and major issue while transmission has always been "Secure Asset Transfer." Due to this ongoing issue, several researches have been made in the past. The most promising technology that evolved to support this secure transfer is Blockchain. Moving the content or most appropriately asset transfer *via* Blockchain guarantees security, transparency, ease, and low cost. With an introduction to Blockchain, now any person anywhere around the world with a simple device phone can transfer securely peer to peer.

Blockchain security is confirmed by its proven facts. In its 8 years of creation, the Bitcoin Blockchain commonly known as public Blockchain has never been successfully hacked. Each transaction is recorded and maintained on thousands of computers making it impossible for anyone to fake the transfer which in turn also makes tampering transparent. If anyone is trying to tamper with the data, it merely becomes impossible for the intruder to change the data at all the sites simultaneously. It helps in detecting tampering easily on the Blockchain network. Blockchain is aptly termed as an edge technology to defeat cyber-crime. Blockchain drones not only hold their application in financial and currency transfers but also in other asset transfers like land titles, vote casting, medical data transfer, etc. Everyone is blessed with this technology to transfer their assets securely with great transparency. It is rightly said that the only permanent change is Blockchain.

Blockchain is a chain of Blocks which in turn is a piece of data structure responsible for handling transactions. Each block in the blockchain is connected with 2 other blocks i.e., the previous and next block except the first and last block. Each block data structure contains transactions and related data. Every time a new transaction arrives or happens, a new block is created. This new block becomes part of the ledger. Every block is uniquely identified by one of the very important properties i.e., Block Number.

Another important property of each block is its hash. Each Block contains a unique set of transactions. After applying the hashing algorithms to each block's data, a unique hash is generated. Each Block contains the Previous Block Hash (PBH) except the first block in the chain. The First Block in the Blockchain is termed as Genesis Block and the Block Number defines the position of the Block from the Genesis Block as depicted in Figure 7.1.

One of the most popular Block Explorer etherscan.io maintains a complete set of data available on Blockchain. Each block is linked with another through PBH which creates a linkage between all the different blocks. Ledger is the term given to the collection of blocks. An example of etherscan.io is shown in Figure 7.2. Block Height and Block Number are the same in this case.

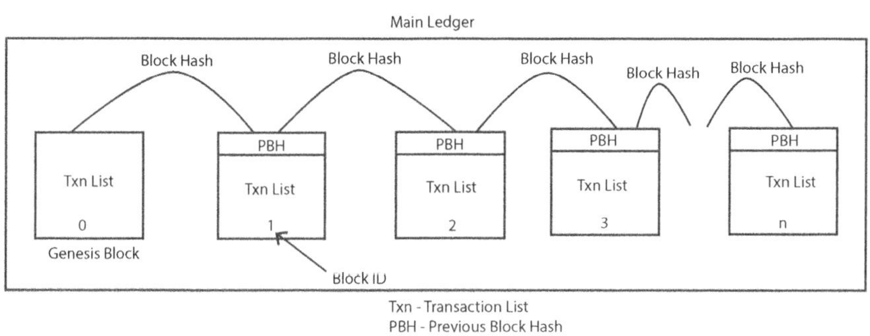

Figure 7.1 Blockchain – chain of blocks.

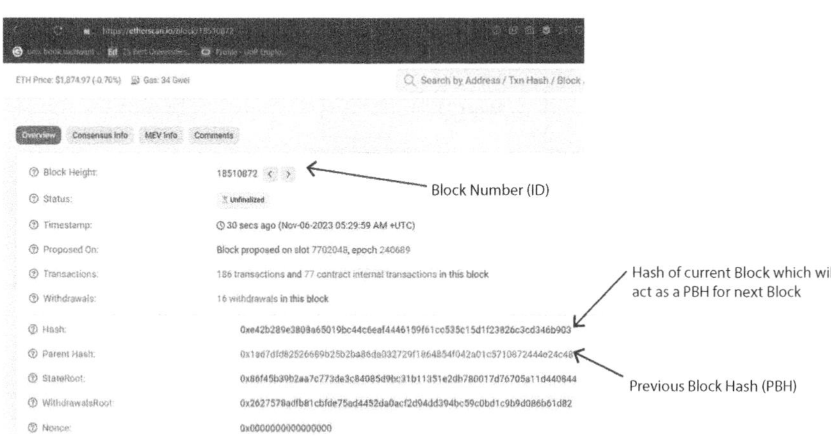

Figure 7.2 Ethereum block parameters.

7.1.1 Principle Functionality of the Blockchain

Transaction is the basic unit of any Blockchain network. How the transaction is initiated and how it lands in the main Blockchain ledger is the fundamental part of the transaction life cycle in Blockchain. All the transactions generated by different people reside in the main pool. The main pool is the local storage area of all the miners for all the unvalidated and unconfigured transactions.

Every time the transaction initiates, it resides in the local copy of the miner. A miner is a type of node. The core responsibility of a miner is to validate a transaction and create its block out of those transactions. Once it's validated, miners in turn receive some reward. The candidate block is a miner block which is the candidate to be the part of the main ledger but not yet chosen to be part of the main ledger. Candidate blocks compete with blocks of other miner blocks; the one who wins the race gets the opportunity to be part of the main block ledger.

This competition is all about. Every miner on the network is assigned a puzzle to solve. This process is automated and the responsibility of launching a puzzle is assigned to a piece of software known as Consensus Algorithm. A puzzle is launched by the Consensus algorithm after a certain interval of time known as ΔT. Each independent miner is supposed to solve the puzzle in a specific period known as ΔT1. The miner who is not able to solve the puzzle in the specified time automatically goes out of the race. The one who solves the puzzle in the least time gets a chance to publish the block in the main ledger. Transaction Flow in the Blockchain network is depicted in Figure 7.3.

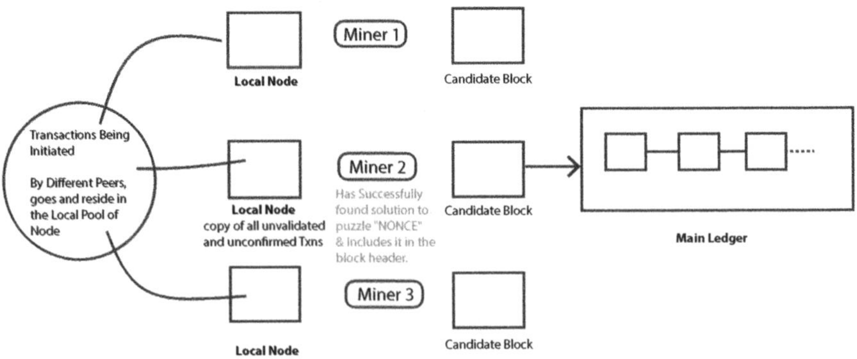

Figure 7.3 Transaction flow – blockchain network.

In the absence of central authority, Consensus algorithms must manage everything on the network. It is a Consensus Algorithm that ensures smooth functioning of the blockchain network and keeps everything intact.

A question always comes to mind: how do miners select transactions and what do they get in return? Transaction owners pay fees for transaction handling. This fee is not standard. Each transaction owner may pay a different amount for the same transaction. For example, the 1st Transaction owner may pay 0.1 coins and the 2nd Transaction owner may pay 0.2 coins for the same transaction. The fee is one of the criteria for why fewer transactions get faster validation. The same transaction may be picked by many miners. Let's assume the case where miner 1 and miner 2 pick the same set of transactions, In this case, if miner 2 transactions get validated early then the consensus algorithm automatically removes the same transactions allocated to miner 1.

A consensus algorithm can be one of the 4 types:

i. Proof of Work (POW)
ii. Proof of Stake (POS)
iii. Proof of Authority (POA)
iv. Proof of Elapsed Time (PoET)

POW is simple. Miner creates a block, validates the transaction, and solves the puzzle. It is about you doing some work, proving that you have done the work and you get the reward. It is always up and consumes huge amounts of electricity without any need or usage. It is all about stealing the job.

In the case of POS; every network delegate has a minimum value of stake known as a token in the network. Based on what value you have staked in the network; you are given network power. The more ethers you have staked, the more chances to validate a transaction and create a block. For example, a miner with 100 ethers stake will get less chance as compared to one with 1000 ethers. The best thing is you won't steal the task but you will be given the job and you get the reward.

In POA, a certain number of authoritative people are selected. They are responsible for validating the transaction and creating a block. POS is under the umbrella of POA. The authoritative people are chosen on the basis of various factors like expertise, type of project, technical knowledge, type of project, etc.

PoET is most preferred in private blockchains. Hyper Ledger (Private Ledger) is inside an organization. No stake no reward concept is used as no 3rd party is involved in this case. Blocks are created automatically after a certain interval of time and are not dependent on any authority. It is automated.

The major reason for using hashing in the Blockchain is it's an irreversible process i.e., it converts plain text to cipher. But at any point, the cipher cannot be decoded back to plain text. Blockchain uses Asymmetric cryptography. The public key of the Receiver is used for Encryption and the private key of the Receiver is used for Decryption. Blockchain recognizes you on the network with a public key. Blockchain does not take any of your personal information, so it can not generate private keys for you if you lose it. This is majorly for ensuring security. It is not very user-friendly but provides the utmost security. One needs to make sure that they do not lose their private keys. It is just like a greater number of doors and windows; more chances of thieves breaking in. So, no way of ensuring the duplicity of private keys on the network. Figure 7.4 shows the graphical representation of address generation in the case of Ethereum. This figure elaborates on the process of generating the private and public keys in the case of Ethereum.

The process flow of generating Private Key and Public Key in Ethereum is defined in the following steps:

1. The Seed Phrase acts as a Passphrase. It is like a raw material that is processed further to generate the product that you are trying to manufacture. As shown in Figure 7.4,

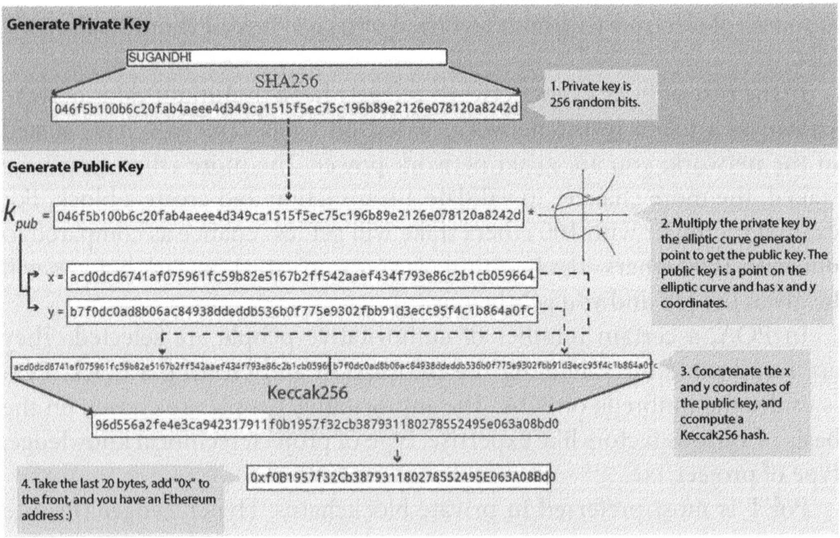

Figure 7.4 Process flow private and public key generation – Ethereum.

"SUGANDHI" is acting as a seed phrase. The difference between the actual process and this example is here, the user is allowed to enter the seed phrase but in real time the seed phrase is generated automatically. Users are not allowed to type passphrases as it is easily recognizable. For the same phrase, every time the same private key will be generated. So, the Pass Phrase is randomly generated for each transaction. This is due to maintaining security. If two users enter the same seed phrase then the same public and private key will be the outcome. And anyone can access another account. So, the generation of seed phrases is automated to secure the transaction.

2. The SHA 256 algorithm is applied to the seed phrase and a real-time private key is generated.
3. Elliptic Curve Generator works on the real-time private key; which further divides it into public a and y coordinates.
4. Public x and y coordinates are later concatenated.
5. On concatenated coordinates; later Keccak26# is applied to generate the public key.

Public keys are generated from private keys only. Once one gets the public key, one does not apply any hash on it to generate an address. The last 20 bytes of it (public key) are accepted and prefixed with 0x.

Another important concept to understand in the Blockchain is Wallets. Wallets act as a gateway as depicted in Figure 7.5, providing access to the ledger. It establishes a connection with the Blockchain network.

An important question that comes into mind is why can't we ignore wallets to connect to the Blockchain network and directly connect with it? The answer to this is Wallets have become a mandatory point because of the

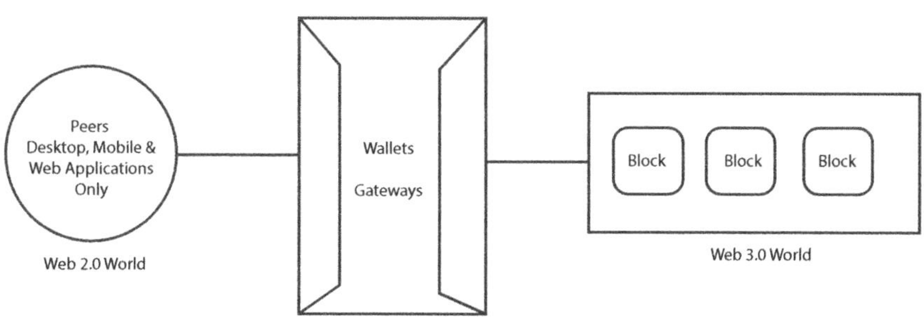

Figure 7.5 Blockchain wallets.

transition from Web 2.0 to Web 3.0. In Web 2.0, to and from communication between UI/UX (client) and Server worked well for web, mobile, and desktop applications as depicted in Figure 7.6.

AS world entered the Web 3.0 space with the innovation of Blockchain Technology. Additional libraries were required to connect with the Blockchain technology as depicted in Figure 7.7. These additional libraries act as middleware. Web 2.0 was responsible for creating and targeting only those applications that were not distributed and decentralized. But with the advent of Blockchain technology, Era is mainly dealing with decentralized applications (DAPPs). Additional Libraries are required to communicate with the Blockchain network and to get information from there. Wallets are middleware agents to connect Web 2.0 with Web 3.0 space. Blockchain is sitting in the Web 3.0 world and users as Clients are sitting in the Web 2.0 world.

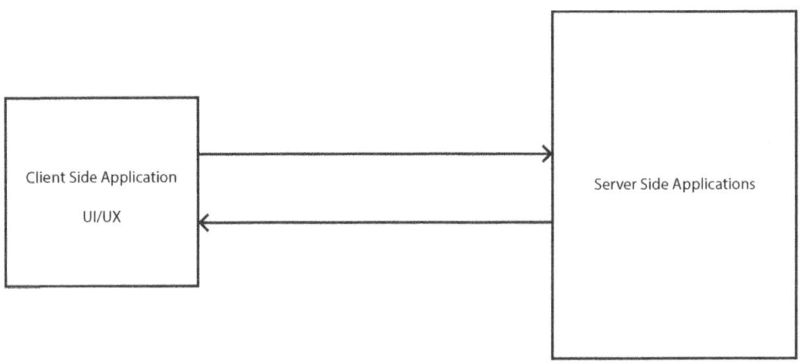

Figure 7.6 Web 2.0.

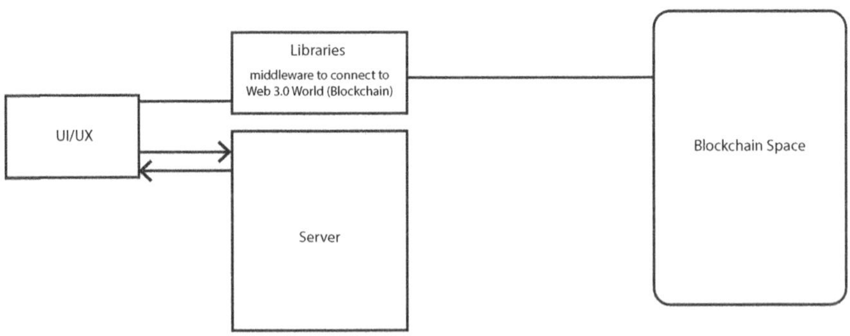

Figure 7.7 Web 3.0.

Wallets connect with one of these blockchain nodes so that they can move data to and from the blockchain. Nodes act as service providers and make money. Wallets can create their own node and connect to the Node through a Remote Procedure Call (RPC). There are 6 types of wallets as mentioned below but every wallet performs the same task i.e., to connect to the Blockchain network and secondly store private and public keys.

i. Web Wallet
ii. Mobile Wallet
iii. Desktop Wallet
iv. Hardware Wallet
v. Paper Wallet
vi. Multi-Signature Wallet

The first three types of wallets are free and easy to use. They are hosted on a centralized server and can be accessed through web, mobile as well as desktop applications. Hardware wallets look like pen drives. Hardware wallets are more secure as web/Mobile/Desktop wallets are hosted on a centralized server where private and public keys are stored as depicted in Figure 7.8. Hackers can hack or steal it easily. Hardware wallets are more secure but one is compromising on user-friendliness. Hardware wallets are capable of holding cryptocurrency. Hardware wallets are also sometimes referred to as Trezor Wallets.

Paper Wallets are cheaper versions of hardware wallets. You create your wallet using web, mobile, and desktop but you do not allow them to save your private or public keys. You put it on a piece of paper or notepad. Types 1 to 3 are known as hot wallets as they are always connected to the internet and Type 4 to 6 are known as cold wallets.

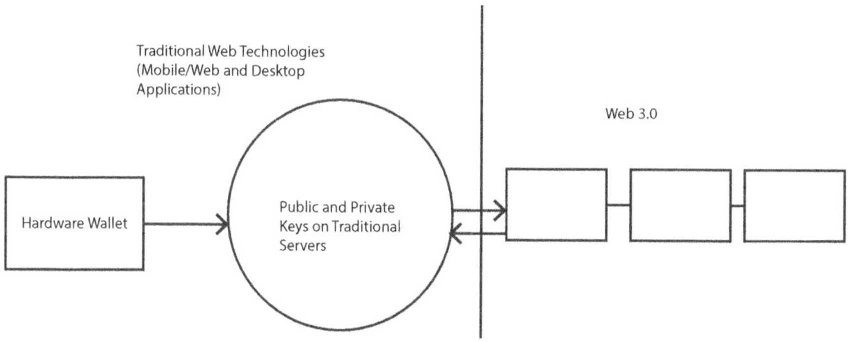

Figure 7.8 Hardware wallets.

Many questions come to the mind of the user before selecting the wallet as it stores private and public keys like

a) Any hacker can hack or steal private or public keys
b) What if my hardware wallet damaged
c) What if I lose my paper?

Organizations want to become automated and decentralized. Funds are managed in the form of cryptos. If the organization maintains one official account with 1 private key and 1 public key with 3 owners as depicted in Figure 7.9. Then it will become difficult to identify who did the transaction because all are using the same private and public key. This is where multi-signature wallets come into the picture. It will generate 3 different keys for 3 different owners as depicted in Figure 7.10.

The multi-signature wallet can be configured in any of the 3 ways like 1/3, 2/3, or 3/3 account; It means that at least 1 owner should sign the transaction, 2 owners should sign the transaction or 3 owners should sign the transaction respectively. In case of need, for the same owner also 3 different private keys can be generated for better security. This is the concept of multiple private keys. Here 3 is a random number, users can generate a different number of private keys as required.

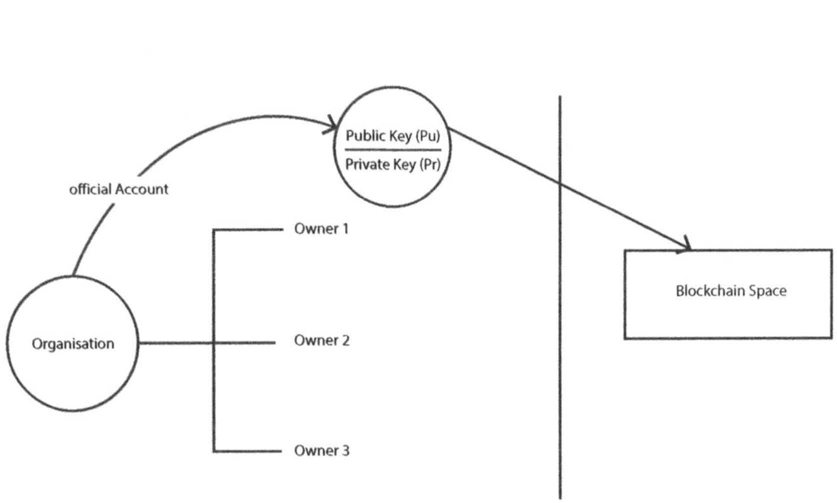

Figure 7.9 Decentralized organization with one private and public key.

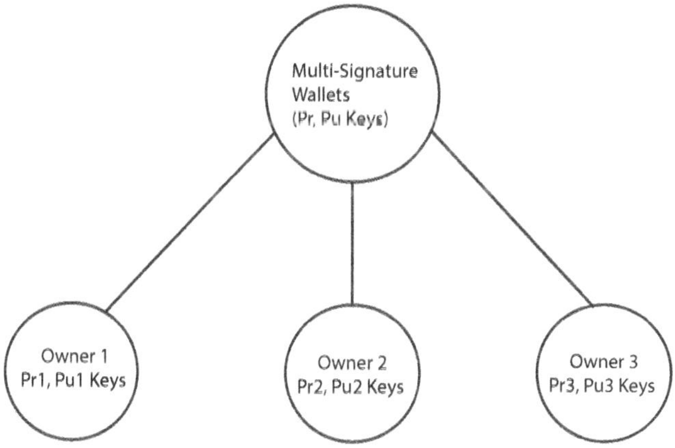

Figure 7.10 Multi-signature wallets – 3 separate private and public keys.

7.1.2 Foundation of Drones

A UAV, an acronym denoting Unmanned Aerial Vehicle., is an aircraft capable of autonomous navigation without a human flyer operating the system. Originally conceived as a cost-effective substitute for manned military aircraft, UAVs are poised to persist into the future, driven by technological enhancements. Analogous to the evolution of the internet and Global Positioning System i.e. Global Positioning System (GPS) tracking, UAVs are transcending their initial defense-oriented projects and evolving into valuable tools for various civilian enterprises, owing to recent strides in their functionality, network infrastructure, communication systems, and manufacturing techniques [12–17]. This shift has led to their integration into governmental and commercial services [18], consequently generating a substantial market opportunity within the industry.

Various communication technologies are employed within devices based on Virtual Circuit (VC) principles, together with Wireless Personal Area Networks based on Bluetooth [19], WPAN based on Zigbee, Wireless LAN Area Networks, Low Power Wide Area Networks, cellular networks based on technologies like Long Term Evolution, advanced Long Term Evolution, 5G, and even awaited 6G [20]. These technologies serve as essential underpinnings for multimedia streaming surveillance in the context of dynamic network segments about architectures involving drones, UAVs, and similar Internet of Things (IoT) devices [21–23]. Nevertheless, just like every coin can be flipped for the 2nd side, illustrations of misuse in such applications exist, posing significant concerns and serving as motivational factors for the current study.

In the contemporary landscape, the significance of VC-based applications has experienced a substantial upswing due to their inherent systems supporting high speed and altitude, markedly contributing to their adoption rates. When these technologies are judiciously deployed and managed, they hold the potential to emerge as dependable and cost-effective solutions for wireless communication, addressing a wide array of real-world challenges [24]. For instance, many governments predominantly employ UAVs.

The potential and ability of UAVs to fly for a long duration with a controlled speed and height have made them suitable for the aviation industry. Originally UAVs simply referred to as drones were built for military usage. During the First World War, Britain and America tested their pilotless aircraft Aerial Targets and Aerial Torpedo respectively. Both showed promising results but neither was used in war. During the interwar period, the British launched several radio-controlled aircraft. The term 'drone' was used at this time.

Beyond their military origin, drones have captured the consumer and commercial market. They are widely used in various sectors like filming, photography, agriculture, healthcare, climate monitoring, launching missiles and satellites, etc. The most well-known use of drones was the 9/11 terrorist attack at the United States World Trade Center. After realizing the key potential of UAVs, the government of Telangana state in India has established Telangana Drone City to establish it as India's leading UAV test and business center.

7.1.3 Blockchain and Drones Security

An early application of drones in the military made it a secure application due to several clearance checks involved at the government level. Back then, fewer drones were operating which was the main reason for the secure implementation of drones. But as the applications of drones grew vastly with Industry 4.0, it has given rise to a major issue of security. Drones carry sensitive data. There exists a challenge to ensure the credibility and integrity of messages being passed *via* drones to UAVs. Nowadays, drones are widely used to carry atomic bombs, weapons, and so on. An attacker can easily get access to drones and can coordinate them to target explosive infrastructure such as gas stations or public events, gatherings, and societies. Not only this, drones are prone to various cyber-attacks as listed in Table 7.1.

Blockchain features gain traction to fight against cyber-attacks. Added information that is collected and maintained in the Blockchain helps to

Table 7.1 Cyber security attacks on drones and their impact.

Cyber security attack on drone	Affected security parameter	Impact
Jamming [37]	Availability: Capturing complete control of drones and disrupting their accessibility by overloading the network with traffic.	This may lead to crashes or unavailability of drone services.
Tampering [28, 38]	Integrity: System coordinates can be changed. Data can be altered.	May lead to mission failure. Disruption and failure of the assigned tasks. Altering the destination address may lead to data leakage.
Collision [39]	Accessibility and Availability: Launching 2 or more drones operating at the same frequency will create passing information and information collisions may happen.	This may lead to the failure of assigned jobs.
Selective Forwarding Attacks [27, 40]	Reliability: Introducing a malicious drone in the network acting as a normal drone node.	This may lead to a higher packet drop rate and make the network unreliable.
Flooding Attacks [26, 40]	Availability and Accessibility: Introducing malicious drones in the network causes overutilization of the network.	This will lead to high traffic and affect the throughput of the network.
Man in the Middle (MITM) Attack [36, 37]	Integrity and Confidentiality: Placing a malicious drone between two authorized nodes.	This will lead to packet sniffing and eavesdropping resulting in privacy breaches.
Denial of Service (D0S) Attack [39, 40]	Availability: Bombarding drones with several requests from illegitimate users.	This will lead to the unavailability of services to legitimate users.

detect different types of attacks. Blockchain not only helps to detect different types of attacks but also helps you to take timely actions to prevent them from occurring. Blockchain has the potential to achieve greater network security visibility and accelerate the pace of implementing new security services. After proven results of Blockchain to ensure security; users can combine Blockchain with drones to make the communication of drones more secure, transparent, reliable, and safe.

Distributed Ledger feature in Blockchain is utilized to ensure that data captured *via* drones kept on Blockchain is tamper-proof. Drone operators can be easily identified through the smart contracts feature of Blockchain. Communication can be secured with UAVs and drone devices using the key concept of Blockchain. Using keys shared by UAVs; drones can encrypt the data at their end before transmitting it to UAVs. The UAVs can also use hash functions to ensure the secure forwarding of data to servers.

7.1.4 Related Work

Much work and research have been done in the areas of Blockchain and drones due to the application of these technologies in diverse fields. The motivation of combining Blockchain with drones has come from the contribution of Blockchain in the Secure Communication World and the vulnerability of drones in maintaining privacy and data security. Table 7.2 lists the various attacks which are common with drones and their solution mechanisms proposed or implemented by various contributors in their research work along with the scope of using Blockchain over proposed techniques.

7.2 Drones – The New Network Architecture

7.2.1 Components and Parameters

Drones are versatile aerial vehicles that can be remotely controlled or operated autonomously. They consist of numerous mechanisms and machineries that work collectively for enabling flight and functionality. This report delves into the major components and parameters of a typical drone. Figure 7.11 depicts the types of communication in drones.

(a) Frame: The frame is the structural backbone of the drone, providing support for all other components. Frames are

Table 7.2 Motivation and contribution – leveraging blockchain for secure communication.

Drone attacks	Attack solution methods	Conclusion
Right to Privacy is vague [1]: Civilians use drones with cameras at weddings to capture images, Google uses drones to facilitate Google Street services, and so on. The major question here is "Will the use of drones/ UAVs preserve privacy or jeopardize it?"	Techniques to protect privacy while collecting, processing, and storing data. Privacy by Design (PbD) measures to be taken that camera installed on drones not targeting or targeted towards personal spaces like windows. Another solution given is legislative.	PbD acts only as a preventive measure but the UAV's operation is different and can be targeted manually by the operator. Legislative solutions are after-operation solutions to control the effect of privacy breaches. However, these both solutions do not guarantee privacy and confidentiality.
Data Privacy and Data Protection [2]: It majorly focuses on the behavioral privacy aspects i.e. collecting someone's personal data without consent or adding misinterpretation comments on that.	Legal reforms exist to protect privacy. It provides regulatory impact on the usage of drones.	Laws sometimes act weak due to their ill-fitness to the situation and need at hand.
Privacy, Data Protection, and ethical risks concerns [3]	It offers two techniques: Regulatory enforcement and Privacy by Design. Usage restrictions are also another criterion that is enforced by legislative solutions.	Privacy, data protection & ethical concerns are specific to UAVs and their operators. It is limited to following fairness principles by the operator and taking the required actions by the government if any breach happens to privacy.

(*Continued*)

Table 7.2 Motivation and contribution – leveraging blockchain for secure communication. (*Continued*)

Drone attacks	Attack solution methods	Conclusion
Privacy and Security [4] attacks on the level of safety and confidentiality of information to society interest.	Due to the huge number of operations performed on different types of drones to cater the security needs, they have divided the operations into 3 categories – open, specific, and certified. Open category operations do not require any authorization; specific operations require authorization limited to specific functioning, and certified operations require complete authorization for operations.	Operation categorization is based on the risk analysis and mitigation required to solve it. It can be biased based on risk assessment by the authorization party.
Privacy Concerns [5]: Drones deployment on 5G base stations can be compromised or eavesdropped by the attacker drones.	A Blockchain-based identification management of drones should be maintained to differentiate between authorized drones and unauthorized ones.	The expiration of drone ID is difficult to handle. Even if this is handled perfectly, it may affect trajectory privacy by violating access authorizations.

(*Continued*)

Table 7.2 Motivation and contribution – leveraging blockchain for secure communication. (*Continued*)

Drone attacks	Attack solution methods	Conclusion
Privacy of Drone Big Data [6]: Drone big data is tightly associated with society's privacy issues.	Adopting blockchain to ensure that big data storage is tamper-proof, traceable, and secure.	Numerous applications of Blockchain and its security have given rise to the usage of Blockchain in the highly insecure drone operations world.
Conventional Cyber Attacks [7]: Attackers can launch several traditional active or passive attacks to hamper the data privacy.	Deep Neural Networks (DNN) have been combined with Blockchain to secure data communication.	Blockchain is one of the proven solutions to protect the data over any network.
Privacy Attack [8]: Authorization and Authentication is at stake while using drones in IoT devices.	Leveraging the solutions of Blockchain to implement a secure authentication model.	Blockchain is the key solution to all the aspects of security and privacy in today's technological glitches and attacks.
Privacy and Data Security [15]: Drones are being used by hackers and attackers for performing malicious activities. STRIDE (Spoofing, Tampering, Repudiation, Information Disclosure, Denial of Service, Elevation of Privileges) is a common category of attacks on drone networks.	Intrusion detection like Rule based, Signature based and Anomaly based can be implemented to detect the unauthorized entities.	Several Attack detection mechanisms exist but there exists a need to prevent the attacks.

(*Continued*)

Table 7.2 Motivation and contribution – leveraging blockchain for secure communication. (*Continued*)

Drone attacks	Attack solution methods	Conclusion
Privacy and Authentication is invaded [38] via Man in the Middle Attack (MITM) and Denial of Service (DoS) attack	Blockchain ensures immutability, transparency, and decentralization in the network which makes it utmost secure to handle the attacks on the data in the drone network.	Blockchain is the present and future of the information and data security world.

usually constructed from carbon fiber, plastic, or metal materials. They come in various configurations, such as quadcopters having four rotors mechanism, hexacopters with six rotors mechanism and octocopters i.e. eight rotors mechanism.

(b) Motors and Propellers: Motors generate thrust to lift the drone off the ground. Each motor is paired with a propeller that converts rotational motion into thrust. The number of motors and propellers depends on the drone's configuration. Brushless motors are commonly used for their efficiency and durability.

(c) Electronic Speed Controllers (ESCs): ESCs govern motor velocity by managing the electrical power fed to them. Instructions transmitted from the flight controller are received by the ESCs, prompting them to modulate the motor's revolutions per minute in accordance with the received commands.

(d) Flight Controller: The controller is the brain of the drone and is responsible for processing all their received sensor data as well as translating the pilot inputs or autonomous commands into motor control signals. It typically integrates an Inertial Measurement Unit for measuring the orientation as well as all the gyroscope data, while ensuring stable flight.

(e) Batteries: The drones are pumped and powered by rechargeable lithium-polymer (LiPo) batteries, whereas

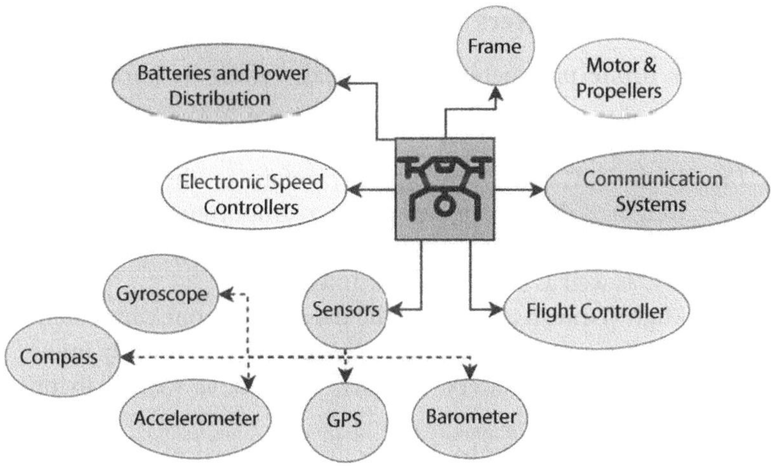

Figure 7.11 Parts of a drone.

these batteries furnish essential power to power up the motors, drive the electronics, as well as energize various constituents. The power distribution board typically manages the distribution of power from the battery to various components of the drone.

(f) Sensors: The drones use various sensors to collect and gather the data related to their environment. The common set of sensors include:
   - GPS, to provide the positioning data for navigation as well as geolocation.
   - Barometer, to measure the atmospheric pressure and estimate altitude.
   - Compass, to basically determine the drone's heading.
   - Accelerometer, typically used to measure the acceleration in various directions.
   - Gyroscope, to measure the angular velocity as well as to maintain stability.

(g) Communication Systems: The drones typically require communication systems to transmit as well as receive multiple sets of data. The remote controllers allow manual control by the operator, while the wireless communication protocols like Wi-Fi enable the smooth message flow between the drone as well as the ground level base station.

7.2.2 Drone Parameters

The various drone parameters that typically affect its performance are listed below:

(a) Payload Capacity: It typically refers to the maximum weight a drone can carry. It's an important and critical parameter for applications like aerial photography, delivery, as well as scientific instruments [34].
(b) Flight Time: It is basically the duration a drone can remain airborne on a single battery charge. This is typically influenced by the battery capacity, payload, as well as flying conditions.
(c) Maximum Altitude: It's the highest vertical distance a drone is estimated to reach above its take-off point. As per the regulatory restrictions, it is often a limited parameter to ensure aviation safety.
(d) Range: It basically refers to the maximum distance between the drone as well as the remote controller or ground station. It usually depends on the communication system as well as potential signal interference.
(e) Maximum Speed: The maximum speed indicates the highest velocity a drone can achieve in flight, and it varies as per the drone's design, motor power, as well as aerodynamics.
(f) GPS Accuracy: This is a critical parameter for drones that use GPS for navigation. It basically determines how precisely the drone can determine its position, affecting flight stability as well as accuracy.

In basic terms, understanding the components as well as all the related parameters of drones is essential for the professionals involved in drone operation as well as development. Notably, drones have revolutionized industries by providing new perspectives as well as capabilities, and besides, a solid understanding of their technical aspects is fundamental for safe as well as effective operation.

7.2.3 Drones Network Topology Architecture

As per the literature, the drones have revolutionized and developed in terms of technicalities and across various industries with aerial capabilities, necessitating advanced communication systems to confirm the optimal operational efficiency. The basic and typical configuration of a drone network's topology architecture substantially impacts the effectiveness of communication, scalability, as well as resilience [35]. The selected structure must be in line with the specific needs of the mission as well as the operational limitations.

Out of all, the one common arrangement is the star architecture, where the drones establish a direct communication channel with a central ground station. This centralized type strategy streamlines data management as well as operational control, and is specifically suitable for instantaneous decision-making scenarios. Besides, it also introduces a potential vulnerability by concentrating the network's functionality at the ground station, thereby risking the system's overall integrity [36].

On the contrary, the *ad-hoc* mesh type of architecture embodies a decentralized concept [37] where the said drones function as individual nodes, forming flexible links with neighboring drones. This self-organizing approach enhances system overall robustness, as drones can relay information throughout the network, allowing smooth and easy data transmission even if certain nodes experience disruptions. This type of architecture is particularly advantageous when communication links are inconsistent or are typically subject to frequent changes, as seen in search as well as rescue missions conducted in remote or crisis-stricken regions.

As the third type, the hybrid topology combines the different structural approaches to balance their respective advantages. This advantageous arrangement has the ability to incorporate elements of both stars as well as mesh topologies, providing a flexible and robust solution that can adapt to a wide range of mission requirements. For instance, in a scenario involving a swarm of drones, the mesh type of network could be employed for communication within the swarm, while a star network might be employed for interaction with a central control station to relay critical directives. Thus, it is notable that every topology has implications regarding communication delay, scalability, as well as energy usage.

The basic network topology architecture plays a pivotal and critical role in establishing the organization as well as connections among these UAVs within a communication framework. This architecture typically governs the flow of data as well as information between drones as well as ground stations, highly impacting the efficiency as well as dependability of the entire system. One prevalent topology is the star configuration, wherein as explained before, the drones establish direct communication channels with a central ground station, resembling a hub-and-spoke pattern. This typical setup definitely allows for centralized control as well as data aggregation, simplifying management but on the other side, potentially introducing a sole failure point. Alternatively, the *ad-hoc* mesh based topology involves drones as nodes that can communicate directly with each other, forming a self-organizing network and basically decentralized in nature. Although the said approach offers enhanced redundancy as well as adaptability, which proves crucial in situations where communication links might be disrupted, it is a little complicated in development.

Figure 7.12 reveals the basic understanding and planning for collection of data as well as managing the drone components for future use. The hybrid topology amalgamates overall diverse architectural models to satisfy the various operational demands within a singular network. The choice of architecture impacts multiple factors such as communication delay, scalability, as well as energy efficiency, all of which critically and typically influence the performance as well as dependability of drone missions, spanning from large level surveillance as well as reconnaissance to collaborative swarm operations. Comprehensively, the star architectures minimize delay due to direct connections, although they might lack scalability. In contrast to this, the mesh architectures perform better with enhanced performance in withstanding faults as well as scaling, yet they may introduce higher latency as data travels between multiple sets of nodes.

Selecting an appropriate drone network topology architecture holds immense significance, directly affecting the success of the missions.

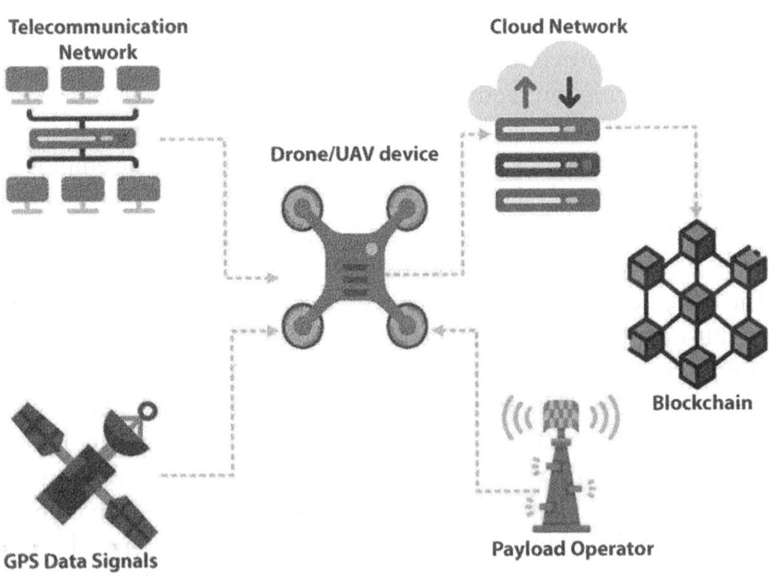

Figure 7.12 General architecture.

Besides, the decision hinges on the operational environment, mission goals, known communication prerequisites, as well as the desired level of fault tolerance. By comprehending and understanding the overall merits as well as limitations of various architectures, the typical drone operators can understand and alter networks that optimize communication efficiency as well as heighten the overall effectiveness of their missions.

7.3 Drones Privacy Solutions

The number of detractors of drones have articulated multiple types and quantity of concerns that the aggregation of lawful imagery as well as videos could possibly facilitate extensive surveillance, granting drone operators (whether the official governmental bodies and/or the private entities) the capability to ascertain individuals' activities across multiple set of timeframes, without adhering to the customary data disclosure protocols expected if the data sets were acquired through alternative methods. With their remarkable capabilities, the drones have introduced transformative possibilities across various sectors as referred in [1], but this technical and fruitful advancement has also sparked a number of concerns about privacy as well as data protection. Integrating camera systems as well as sensors in drones raises significant privacy issues, as these devices can typically capture sensitive information from individuals as well as their surroundings. To address such levels and types of concerns, several technical solutions have been developed to uphold privacy while maximizing the benefits of drone technology, as described in [2].

Out of many, the one approach is the anonymization of the overall collected data. This typically involves the steps of removing and/or obfuscating personally identifiable information from captured images as well as videos, and thus, ensuring that individuals cannot be identified from the recorded data. Besides, the advanced techniques, such as blurring and/or pixelation, are basically and often employed by the concerned to achieve this.

Geo-fencing: It is another privacy-based solution that typically creates the virtual boundaries in geographical areas. The drones are programmed to avoid entering these restricted zones, typically including private properties and/or sensitive areas. Geo-fencing basically helps and thus, prevents unintentional intrusion into private spaces as well as reduces the overall risk of unauthorized surveillance. In this case the encryption plays a pivotal role in securing data during transmission as well as storage. Besides, implementing the strong encryption protocols ensures that intercepted data remains incomprehensible to unauthorized entities. This typical measure safeguards sensitive information, while maintaining data privacy even if the drone's communication channels are compromised.

Data minimization is another effective strategy that basically involves the steps of collecting only the necessary information to fulfill the drone's purpose. The potential and possible privacy risks are inherently minimized by reducing the amount of data collected, processed, as well as stored, as cited in [3]. Additionally, the ideology of privacy by design emphasizes integrating privacy considerations from the inception of drone systems. This basically involves the steps of implementing privacy-enhancing technologies as well as practices during the overall development process, while ensuring that privacy is a fundamental aspect of the drone's functionality [4]. An UAV is a type of aircraft designed to function autonomously, devoid of an on-board human pilot. Such drones form a critical component of an unmanned aircraft system, encompassing both the UAV and also a ground-based controller, as well as an interconnected communication setup between all the said components.

The operation of UAVs can encompass varying levels of self-governance, involving either remote manipulation by a human operator or self-directed navigation executed by onboard computational systems. The different types of drone communication systems concerning possible aspects are presented in Figure 7.13.

- Inter-Drone Communication: In the current scenario, standardization for communication between drones has not yet

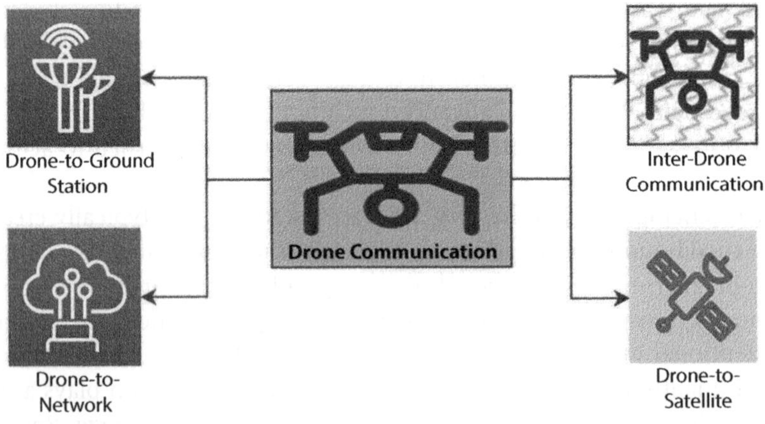

Figure 7.13 Drone communication.

been established. Interestingly, the application of Machine Learning can be exploited to create and enhance an intelligent wireless communication system utilizing UAVs [25]. In many instances, interactions between drones can be likened to Peer-to-Peer (P2P) communication, which opens up vulnerabilities to a range of P2P attacks, encompassing tactics like jamming such as Distributed Denial of Service (D-DoS) and Sybil attacks [26–28].

- Communication from Drone to Ground Station: This mode of transmission hinges on well-established and acknowledged industrial protocols firmly rooted in wireless communication principles, encompassing technologies like Bluetooth and Wi-Fi 802.11. These protocols encompass frequency bands spanning both 2.4 GHz and 5 GHz. Nevertheless, most drone-to-ground communications are public and lacking in robust security, opting for single-factor authentication that can be effortlessly breached. Therefore, as a result, these connections are typically susceptible to both passive threats like eavesdropping as well as the active attacks like man-in-the-middle breaches.
- Drone-to-Network Communication: This type of communication segment basically provides the flexibility to opt for a network based on the desired security level. It may also encompass the typical utilization of cellular communication methods, relying on frequency bands such as 3 GHz, 4 GHz, 4G+ (LTE), as well as 5 GHz, while ensuring the robust security measures for these wireless communication networks as a critical aspect when they are deployed.
- Drone-to-Satellite Communication: This type of communication mode becomes very critical when transmitting the real-time coordinates using the GPS. This typical functionality facilitates the ability to recall a drone to its initial position in instances where it surpasses the designated operational area and/or ventures beyond the line of sight. The satellite-based communications are generally regarded as secure as well as reliable; however, in some instances, they entail significant costs as well as maintenance requisites. AS a result of that and consequently, they find extensive usage predominantly within the domain of military applications around the globe.

- Drone-to-Network communication provides a versatile approach where the selection of a network is typically and critically influenced by the desired security level. This can basically encompass utilizing existing wireless communication protocols like Bluetooth as well as Wi-Fi 802.11. The cellular communication, operating in frequency bands such as 3 GHz to 5 GHz, can also be integrated at this level. The security aspect of these communication networks is paramount, as data integrity as well as confidentiality must be definitely maintained to prevent unauthorized access and/or data breaches. Additionally, implementing the robust encryption as well as authentication mechanisms is essential to protect the sensitive information transmitted between the drone as well as the network.
- The overall Drone-to-Satellite communication offers a reliable and convenient solution for real-time coordinate transmission through GPS. This capability becomes invaluable for scenarios where the drones need to be properly tracked as well as controlled from a remote location. While satellite communications are known for their reliability, their adoption is expensive and requires continuous commitments. They are particularly predominant in very specific military applications due to the stringent security requirements as well as the quality and ability to operate in remote and/or challenging and/or harsh environments.

Figure 7.14 typically demonstrates the overall categorization of drones based on communication systems. IN typical understanding, the Drones are categorized into distinct classes according to their intended applications.

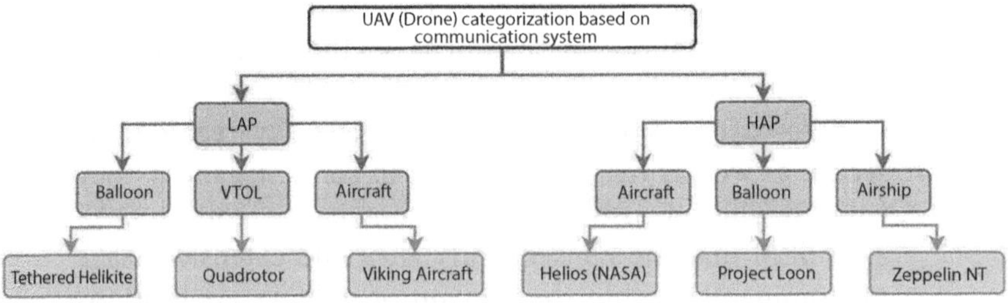

Figure 7.14 Overall categorization of drones based on the different communication systems.

The range and number of applications is wide-ranging and, from a design standpoint, it technically and primarily falls within three categories: the military, the industrial (enterprise), as well as lastly, the commercial.

Figure 7.15 presents the details related to the four different types of drone communication. Regulatory frameworks, such as the European Union's General Data Protection Regulation (GDPR), also play a critical and typical role in shaping drone privacy solutions. These regulations develop data collection, processing, as well as storage guidelines, ensuring that individuals' rights to privacy are respected. In addition to this, the ideology of a "Blockchain of Internet of Drones" represents a transformative paradigm where the convergence of blockchain technology as well as the IoT converge to revolutionize and develop the drone industry. In this groundbreaking framework, drones are equipped with IoT sensors as well as connected to a decentralized blockchain network, yielding security, accountability, as well as operational efficiency benefits. This type of interconnected network enables real-time tracking as well as recording of crucial drone data, including flight paths, locations, as well as operational status.

By immutably storing this data on the blockchain, unparalleled transparency as well as trust are established, fostering more enhanced and better collaboration among stakeholders such as regulatory bodies, drone operators, as well as the general public. Furthermore, this innovation enhances safety by facilitating collision avoidance as well as airspace traffic

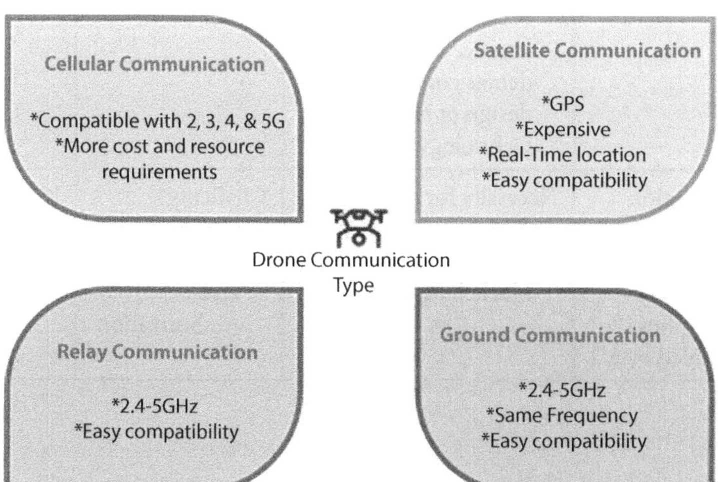

Figure 7.15 Detailed four drone communication types.

management through decentralized coordination, reducing and/or mitigating the overall risk of accidents. Smart contracts embedded within the blockchain enable automated actions as per the predefined conditions, while adding a layer of autonomy to drone operations. With its capacity to validate compliance with regulations as well as securely manage drone identities, the Blockchain of the Internet of Drones reshapes the drone landscape, ensuring secure, efficient, as well as responsible drone operations while moving towards a new era of aerial innovation. Focusing on this, Table 7.3 presents a tabulated survey of the present state-of-the-art literature related to blockchain of the internet related to drones.

Table 7.3 Blockchain of internet of drone.

Application	Weakness	Lack of security requirement	References
Communication for drones empowered by Privacy-Preserving 5G Technology	Insufficiently conducted experiments within the design	Privacy concerns related to drones and user information.	[5]
UAV big data privacy protection	Centering on the utilization of homomorphic encryption with NTRU and the absence of pertinent details concerning the design of blockchain technology	Authorization and automaticity	[6]
Identification and Flight Mode Detection of Drones Enabled by 5G	Necessity for potent drones and dearth of experiments in blockchain design	Challenges in Privacy, Automation, and Authorization	[7]

(*Continued*)

Table 7.3 Blockchain of internet of drone. (*Continued*)

Application	Weakness	Lack of security requirement	References
Secure authentication model	Employing a public blockchain diminishes both privacy and performance	Automaticity, authorization, and privacy	[8]
Dynamical cache data for offloading	Absence of Experiments in Communication and Encryption Complexity	Privacy and tolerance issues	[9]
Drones to manage a public blockchain	Necessity for High-Powered Drones in Blockchain Maintenance	Automaticity, privacy, authorization, and authentication	[10]
Ensuring Secure Data Management in AI-Powered Pandemic Scenarios	A complex system featuring multiple light chains alongside a main chain necessitates the use of robust drones	Privacy issues and authorization	[11]

7.4 Integration of Blockchain Functioning with Drones

Blockchain, Artificial Intelligence (AI), and UAVs find applications across diverse fields such as healthcare, defense, smart cities, and the intelligent power grid. Although the amalgamation of drones with blockchain, AI, and fifth-generation wireless technology (5G) offers numerous benefits, several challenges still necessitate resolution. A few of the issues as listed in Figure 7.16 are discussed below with suggested strategies:

- Privacy Concerns: In order to safeguard the privacy of participants in the proposed scheme, a publicly accessible blockchain is employed. This approach enhances security and communication performance for UAVs but simultaneously exposes sensitive user information, giving rise

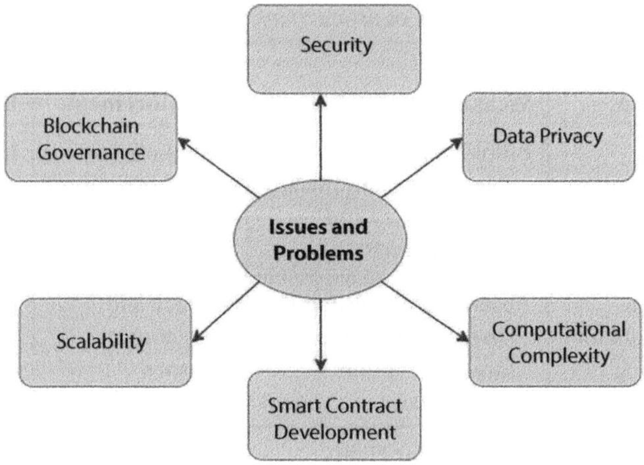

Figure 7.16 Major issues.

to apprehensions about data privacy on the blockchain. Consequently, corporations are imposing limitations on blockchain usage, spurring a high demand for privacy-preserving technologies within the blockchain ecosystem.

- Computation Challenges: The incorporation of AI as well as 5G in the overall proposed system enhances UAV communication security as well as network performance. However, the uncertain nature of these technologies results in significant data generation, where 5G's rapid data speeds as well as ultra-low latency lead to a data deluge. Besides, the reduced and limited computational capacity of edge-AI algorithms poses a hurdle in effectively managing the large amounts of generated data.
- Smart Contract Development: The immutability of smart contracts necessitates meticulous programming to prevent the overall flaws and/or vulnerabilities that could compromise the integrity of the blockchain network. Multiple and more often mistakes in smart contracts can severely harm the overall blockchain network, emphasizing the need for security as well as vulnerability assessments to ensure its overall safety and protection.
- Scalability Challenges: Scaling up the blockchain-enabled UAV network is a typical, critical and paramount concern. While a blockchain processes an average of 12 transactions

per second, this rate is sometimes insufficient for the demands of a 5G network. In contrast to this and as per the social media platforms that handle millions of transactions per second, the blockchain's performance falls short at some instances. Addressing this issue requires enhancing the scalability of the blockchain network.

- Decentralized Governance: The intricacies of deploying as well as managing a distributed ledger, even in a public blockchain network, often pose multiple types of challenges. The decisions surrounding oversight, management, troubleshooting, creation, as well as deployment of smart contracts, dispute resolution, as well as governance norms increase at a rapid rate. Thus, a robust governance framework is imperative to effectively manage these complexities for smooth system operation.
- Quantum Computing Threat: The protection and safety of the within a blockchain network through public-key cryptography faces a critical threat from quantum computing. Predictions indicate that by the year 2027, overall quantum computing could compromise blockchain security by decrypting public key encryption. Besides, the emergence of a quantum-secure blockchain network becomes imperative to mitigate this kind of possible and potential risk.
- Data Processing Delays: The multiple type sensors on UAVs generate substantial amount of data, potentially leading to suboptimal solutions due to the UAVs' battery limitations. This consideration typically requires efficient data processing strategies to overcome the overall set of delays.
- Blockchain Standardization: The absence of consensus on blockchain technology's standardized usage effects and harms the establishment of a real-time UAV network. This overall lack of clarity impedes blockchain's seamless and smooth integration as well as basic deployment within the UAV ecosystem.

7.5 Security Challenges and the Road Ahead

The development of the new technologies has typically brought a number of fourth challenges as well as potential risks concerning the most critical aspect, i.e., protecting the personal data as well as privacy. These critical concerns have prompted the need and establishment for proactive measures, not only within the regulatory framework at the EU level, where updates to personal data protection laws have been made but also, within the realms of self-regulation as well as co-regulation. To add on and complement existing legislation as well as enhance adherence as well as awareness, international as well as European efforts have led to the overall development of technical standards, guidelines, codes of conduct, as well as certification protocols.

Supporting such initiatives and endeavors, the GDPR focuses and highlights the importance of standardization as well as certification initiatives within the data protection domain. Standards are often viewed as a means to ensure security, interoperability, transparency, as well as the establishment of trust among multiple individuals whose data is being managed. These typical efforts collectively aim to address the multifaceted challenges posed by the adoption of novel technologies, particularly within the context of privacy as well as data protection [29].

Recent strides in drone technology are ushering in fresh prospects and applications across diverse domains, particularly with the emergence of compact drones [30, 31]. However, these strides are concurrently presenting novel hurdles relating to security, adaptability, and uniformity. This study delves into the realm of drone technology, its realms of application, its multifaceted utility for citizens, security aspects including safeguarding and confidentiality concerns, prevailing threats and vulnerabilities faced by drones, existing methodologies for fortifying drone cybersecurity, as well as the sources of information tapped for the existing literature review and work done in this field.

Compact UAVs are charting a fresh trajectory for both civilian and armed sectors, introducing unexplored avenues. Nonetheless, these small drones grapple with inherent architectural complexities and the delineation of safety and protection parameters [32, 33]. The way of evolution or in other words the exponential evolution of the IoT has augmented the scopes as well as measurements of drone technology, while concurrently introducing new perils. These diminutive airborne intelligent devices often pose formidable challenges and issues to the security as well as privacy of collected data. The typical design of these compact drones is still evolving and in the development stage to meet the requirements of their operational domain. Besides, the overall fundamental design intricacies necessitate the incorporation of security measures, privacy safeguards, as well as data transformation protocols [38–40].

7.6 Conclusion

To conclude, the steps of integrating security as well as privacy solutions based on Blockchain technology presents a promising and critical source and place for enhancing the foundational aspects of drone systems. The convergence of Blockchain as well as drone operations very effectively tackles the complex security, privacy, as well as data integrity challenges, ushering in a novel era characterized by heightened trust as well as enhanced accountability across the system. Blockchain's inherent attributes, encompassing immutability, decentralization, as well as transparency, typically target the vulnerabilities inherent in drone systems. Besides, given the step of establishing an unmodifiable ledger encompassing transactions as well as interactions, Blockchain effectively protects the integrity of crucial data, thereby thwarting endeavors aimed at manipulating data, spoofing, and/or unauthorized access.

In trend and based on the newfound level of transparency, it fosters trust among various stakeholders, facilitating hiked collaboration amongst drone operators, regulatory bodies, as well as end-users. Besides the Blockchain's capabilities in executing smart contracts typically offer the potential for automating drone operations, which basically streamlines adherence to regulations, airspace restrictions, as well as privacy policies. This framework reduces the overall risk of human error as well as enforces adherence to predetermined protocols, thereby bolstering safety as well as security measures. However, implementing Blockchain-based solutions requires grappling with challenges like scalability, interoperability, and integration into existing systems. Moreover, the resource-intensive nature of Blockchain networks necessitates optimization to effectively cater to the real-time demands of drone operations. To surmount these challenges, it becomes imperative for technology developers, regulatory bodies, and industry stakeholders to collaborate. As drone technology continues to gain prominence across logistics, surveillance, and disaster response sectors, the incorporation of Blockchain-based security and privacy solutions evolves.

References

1. Volovelsky, U., Civilian uses of unmanned aerial vehicles and the threat to the right to privacy an Israeli case study. *Comput. Law Secur. Rev.*, 30, 306–320, 2014.

2. Clarke, R., The regulation of civilian drones' impacts on behavioral privacy. *Comput. Law Secur. Rev.*, 30, 286–305, 2014.
3. Finn, R., Privacy, data protection and ethical risks, in: *Civil RPAS operations*, 2014.
4. EASA, Concept of Operations: A risk-based approach to regulation of unmanned aircraft, 2015, https://www.easa.europa.eu/sites/default/files/dfu/204696_EASA_concept_drone_brochure_web.pdf.
5. Wu, Y., Dai, H.N., Wang, H., Blockchain-based privacy preservation for 5G-enabled drone communications. *IEEE Netw.*, 35, 1, 50–56, 2021.
6. Lv, Z., Qiao, L., Shamim, M., Jun, B., Analysis of using blockchain to protect the privacy of drone big data. *IEEE Netw.*, 35, 1, 44–49, 2021.
7. Gumaei, A., Al-Rakhami, M., Mehedi, M., Pace, P., Deep learning and blockchain with edge computing for 5G-enabled drone identification and flight mode detection. *IEEE Netw.*, 35, 1, 94–100, 2021.
8. Yazdinejad, A., Parizi, R.M., Dehghantanha, A., Karimipour, H., Srivastava, G., Enabling drones in the Internet of things with decentralized blockchain-based security. *IEEE Internet Things*, 8, 8, 6406–6415, 2021.
9. Luo, S., Li, H., Wen, Z., Qian, B., Morgan, G., Blockchain-based task offloading in drone-aided mobile edge computing. *IEEE Netw.*, 35, 1, 124–129, 2021.
10. Islam, A., Rahim, T., Masuduzzaman, M., A blockchain-based artificial intelligence-empowered contagious pandemic situation supervision scheme using the internet of Drone Things. *IEEE Wirel. Commun.*, 28, 4, 166–173, 2021.
11. Chang, Z., Guo, W., Guo, X., Chen, T., Min, G., Blockchain-empowered drone networks: Architecture, features, and future. *IEEE Netw.*, 35, 1, 86–93, 2021.
12. Sachs research, Drones reporting for work, 2020. https://www.goldmansachs.com/insights/technology-driving-innovation/drones/.
13. Sharma, A., Vanjani, P., Paliwal, N., Basnayaka, C.M., Jayakody, D.N.K., Communication and networking technologies for UAVs: A survey. *J. Netw. Comput. Appl.*, 168, 1–12, 2020.
14. Noor, N.M., Abdullah, A., Hashim, M., Remote sensing UAV/drones and its applications for urban areas: A review. *In IOP Conf. Series Earth Environ. Sci.*, 169, 1–9, 2018.
15. Yaacoub, J.P., Noura, H., Salman, O., Security analysis of drones systems: Attacks, limitations, and recommendations. *Internet Things*, 1–10, 2018.
16. Mogili, U.R. and Deepak, B.B.V.L., Review on application of drone systems in precision agriculture. *Procedia Comput. Sci.*, 133, 502–509, 2018.
17. Outay, F.H., Mengash, A., Adnan, M., Applications of unmanned aerial vehicle (UAV) in road safety, traffic and highway infrastructure management-Recent advances and challenges. *Transp. Res. Part A Policy Pract.*, 141, 116–129, 2020.

18. Zhang, L. and Bai, J., An Ethereum-based wind power energy network contract management solution. *Int. J. Commun. Netw. Distrib. Syst.*, 28, 1, 43–60, 2022.
19. Mozaffari, M., Saad, W., Bennis, M., Nam, Y.H., Debbah, M., A tutorial on UAVs for wireless networks: applications, challenges, and open problems. *IEEE Commun. Surv. Tutor.*, 21, 3, 34–60, 2019.
20. Fan, C., Bao, S., Tao, Y., Li, B., Zhao, C., Fuzzy reinforcement learning for robust spectrum access in dynamic shared networks. *IEEE Access*, 1258, 27–39, 2019.
21. Aruchamy, P., Balraj, L., Sowndarya, K., An energy-aware link fault detection and recovery scheme for QoS enhancement in Internet of Things-enabled wireless sensor network. *Comput. Electr. Eng.*, 1–20, 2025.
22. Jeyaselvi, M., Dhanaraj, R., Sathya, M., Memon, F.H., A highly secured intrusion detection system for IoT using EXPSO-STFA feature selection for LAANN to detect attacks. *Cluster Comput.*, 559–574, 2022.
23. Prasanth, A., Lakshmi, D., Dhanaraj, R.K., Technological Advancement in Internet of Medical Things and Blockchain for Personalized Healthcare: Applications and Use Cases, in: *CRC Press*, pp. 1–216, 2020.
24. Lu, Y., Security in 6G: The Prospects and the Relevant Technologies. *J. Ind. Integr. Manage.*, 271–89, 2020.
25. Chen, M., Challita, U., Saad, W., Yin, C., Debbah, M., Machine learning for wireless networks with artificial intelligence: a tutorial on neural networks. *arXiv Preprint arXiv*:1710.02913, 2017.
26. Midha, S., Kaur, G., Tripathi, K., Cloud deep down—SWOT analysis, in: *2nd International Conference on Telecommunication and Networks (TEL-NET)*, Noida, India, pp. 1–5, 2017.
27. Midha, S., Sharma, M., Tripathi, K., Practical Implications of Using Dockers on Virtualized SDN. *Webology*, 312–330, 2021.
28. Midha, S., Verma, S., Verma, K., Mittal, M., Jhanjhi, N., Masud, M., AlZain, M.A., A Secure Multi-factor Authentication Protocol for Healthcare Services Using Cloud-based SDN. *Comput. Mater. Contin.*, 3711–3726, 2022.
29. Mushtaq, M.F., Jamel, S., Deris, M.M., Triangular Coordinate Extraction (TCE) for Hybrid Cubes. *J. Eng. Appl. Sci.*, 2164–2169, 2017.
30. Rowaihy, H., Enck, W., McDaniel, P., La Porta, T., Limiting sybil attacks in structured p2p networks, in: *Proceedings of the IEEE INFOCOM 2007—26th IEEE International Conference on Computer Communications*, Anchorage, AK, USA, pp. 2596–2600, 2007.
31. Naoumov, N. and Ross, K., Exploiting p2p systems for ddos attacks, in: *Proceedings of the 1st International Conference on Scalable Information Systems*, Hong Kong, China, pp. 47–53, 2006.
32. Podkorytov, A. and Ryapukhin, A., The reduction of computational cost in GNSS data network processing. *AIP Conf. Proc.*, 2467, 030028, 2022.

33. Burova, A. and Ryapukhin, A., Reduction of the number of multiplication operations in digital signal processing algorithms by classical methods of discrete Fourier transform. *AIP Conf. Proc.*, 2402, 040002, 2021.
34. Wazid, M., Bera, B., Mitra, A., Das, A.K., Ali, R., Private blockchain-envisioned security framework for AI-enabled IoT-based drone-aided healthcare services, in: *2nd ACM MobiCom Workshop on Drone Assisted Wireless Communications for 5G and Beyond, DroneCom '20*, London, UK, pp. 37–42, 2020.
35. Allauddin, M.S., Kiran, G.S., Kiran, S.R., Srinivas, G., Mouli, U.R., Prasad, P.V., Development of a surveillance system for forest fire detection and monitoring using drones, in: *IEEE International Geoscience and Remote Sensing Symposium, IGARSS '19*, Yokohama, Japan, pp. 9361–9363, 2019.
36. Mishra, Garg, B.D., Narang, P., Mishra, V., Drone-surveillance for search and rescue in natural disaster. *Comput. Commun.*, 156, 1–10, 2020.
37. Midha, S. and Tripathi, K., Extended Security in Heterogeneous Distributed SDN Architecture, in: *Advances in Communication and Computational Technology*, vol. 116, Lecture Notes in Electrical Engineering; Springer Nature, Singapore, 2021.
38. Midha, S. and Tripathi, K., Remotely Triggered Blackhole Routing in SDN for Handling DoS, in: *Proceedings of the International Conference on IoT Inclusive Life*, vol. 116, Lecture Notes in Networks and Systems; Springer Nature, Singapore, 2020.
39. Midha, S. and Tripathi, K., Data hiding based PKI Authentication Protocol. *SDN Test Mag. J.*, 82, 15848–15853, 2020.
40. Midha, S. and Tripathi, K., Extended TLS security and Defensive Algorithm in OpenFlow SDN, in: *Proceedings of the 2019 9th International Conference on Cloud Computing, Data Science & Engineering (Confluence)*, Noida, India, pp. 141–146, 2019.

8

Design and Development of Modular and Multifunctional UAV for Amphibious Landing and Surround Sense Module

Surendar G.[1], Balasubramanian E.[2*], Choi Jae-Sung[3] and C.H. Nirmal Prabhath[1]

[1]Centre for Autonomous System Research, Vel Tech University, Chennai, India
[2]National Institute of Technical Teachers Training and Research, Chennai, India
[3]Sun Moon University, Asan-Si, Chungcheongnam-Do, South Korea

Abstract

Unmanned aerial vehicles (UAVs) are versatile tools used in various situations and industrial applications. This article focuses on the invention of a modular UAV with an amphibious landing capability and a surround sense module for monitoring water quality in remote water bodies. The computed aided modeling and detailed structural and airflow features analysis are carried out. It is configured to achieve strong and lightweight features with amphibious features, fair stability, and good buoyancy for a fair landing and navigation through water bodies. The transition between aerial operation and water-based operation is seamless. The integrated Surround Sense Module (SSM) gives situational awareness in real-time through Sonar and Light Detection and Ranging (LiDAR) systems. Field tests are done rigorously to check the performance, reliability, and safety of the UAV being developed. With adaptability, modularity, and advanced sensing features, this UAV has the potential to transform roles from surveillance to disaster response in aquatic missions.

***Keywords*:** Amphibious drone, surround sense module, water quality, sensors integration, data acquisition, disaster response

**Corresponding author*: esak.bala@gmail.com

A. Prasanth, Rajesh Kumar Dhanaraj, Munish Sabharwal, Vandana Sharma and Seifedine Kadry (eds.) Drones for Transportation Logistics and Disaster Management, (181–198) © 2026 Scrivener Publishing LLC

8.1 Introduction

UAVs have been transforming the world since their inception in many engineering applications, and their usefulness is more promising in solving industrial problems. They are considered versatile tools [1] in adapting various payloads and environmental conditions for deployment in real-world industrial sectors. UAVs have been well utilized in diverse applications, including environmental monitoring, agriculture, mining, infrastructure assessment, and bridge inspection, and they play a vital role in addressing many challenges [2]. For the past and present decades, researchers have concentrated on lightweight UAV structures to optimize the body's mass and effectively utilize high strength-to-weight ratio materials to increase the system's endurance [3, 4]. Many pioneering works have focused on microelectronic mechanical systems (MEMS) based on the miniaturization of sensors [5] and enhancing the communication aspects of UAVs [6].

The continuous demands of UAVs are significantly exploited by integrating high-definition cameras, cutting-edge sensors and other noteworthy payloads such as thermal imaging, hyperspectral cameras, Radars and LiDARs. It broadened their scope, and they are a viable tool in solving day-to-day life problems [7]. The evolution of UAVs in many industrial and commercial sectors paved the way for the development of various UAVs. They are designed in such a way that the traditional operational difficulties can be overcome, and they can be deployed in diverse environments. However, one of the key aspects of UAV design is the concept of modularity, where users can save time in assembly and disassembly according to the mission requirements, and transportation of these kinds of UAVs would be straightforward and effective.

Modular design can help integrate various sizes and shapes of payloads, which is the need of the hour to satisfy the industrial requirement for minimizing the cost and human efforts [8]. The UAVs can be customized and reconfigured as needed, which is of paramount interest to the UAV sector for ease of assembly and deployment. It enables us to perform various tasks without difficulty and address the challenges imposed on us [9]. UAVs are much better as modular systems, which are responsive to operational needs and flexibly adapt to changes in the mission and the environment within which they have to operate, especially when they have to traverse different and complex terrains that include land-water interfaces [10]. They can perform a wide range of functions, from environmental monitoring and disaster response to surveillance and reconnaissance, thanks to their ability to customize their payloads to specific mission requirements [11].

Recently, the operational capabilities of UAVs have changed paradigms with the addition of amphibious landing capabilities. The ample capacity of a UAV to shift smoothly from aerial flight to water-based operations is termed "amphibious landing" [12, 13]. That capacity becomes more of an essence when sometimes the use of UAVs over water bodies in disaster relief maneuvers in flooded land or coastal surveillance is needed. Specific engineering challenges are considered to achieve a good amphibious landing. The UAV has to be designed robustly and lightly to ensure buoyancy and stability in the water [14]. In order to provide optimal maneuverability and control during take-off and landing sequences, the propulsion system has to allow for a seamless transition between aerial and water-based modes [15]. However, such amphibious landing capability is challenging to realize but essential for maximizing the utilization of UAVs in dynamic land-water interface conditions.

The surround-sense modules will greatly enhance the effectiveness of UAVs in various environments [16]. Such a system comprises sensors and imaging technologies, including IMU, ultrasonic sensors, radar, lidar, and cameras, which will provide a full perspective of the surroundings of the UAV. From this point of view, the core of modular and multifunctional UAVs shall be derived from the surround sense module for real-time situational awareness. This module gathers information about the environment, improving mission planning in unknown territories while providing valuable control inputs. This surround sense module shall serve as the sensory system for the UAV and, therefore, crucial for perception and navigation in more than one environment—both aerial and aquatic.

8.2 UAV Design Considerations

Making a UAV that can land on land and water requires great care, as it is not easy to switch from flying to moving in water. Many important points must be considered to bring this ability together, including the build, shape in the air, and how it works.

8.2.1 Design Parameters and Features

The structure of the UAV should be designed to achieve a compromise between strength and lightweight. A strongly built but lightweight frame is needed to resist all the forces that may be brought about by amphibious landings, while optimal aerodynamic performance is achieved during flight [17, 18]. Buoyancy is a very important factor for a UAV designed for amphibious operation. The design should, therefore, consider the UAV's weight and payload, with factors that invoke stability in water landings and take-offs [19, 20]. Floatation devices or enhanced landing gear can be added for better stability at water landing. Critical components, including electronic systems and propulsion parts, shall use water-resistant materials to prevent water ingress during operations conducted over water [21]. The landing gear has to be suitable for both land and water surfaces, provide stable landings, and can effect a seamless transition between aerial and aquatic environments.

8.2.2 Aerodynamic Characteristics

The wings should be designed to achieve adequate lift in the air, and stability is obtained over water landing [22, 23]. Hydrodynamic principles should be employed in UAV design to reduce drag and turbulence during water operations [24]. Streamlined shapes and optimized surfaces contribute towards aerodynamic efficiency and stability [25]. Control surfaces, like the rudder, must be designed to control water-based maneuvers, efficiently maintaining stability in water operations.

8.2.3 Power Requirements

The UAV should feature two power mechanisms optimized for flying and aquatic activities, comprising traditional propellers for flying and aquatic propulsion systems for amphibious maneuvers [26]. Considering that water operations pose extra difficulties, the propulsion system should be designed to achieve the best energy efficiency to support longer possible endurance periods in operations.

8.3 Development of Modular UAV

The design concept is that of a modular UAV with multirotor and amphibious system configurations. This novel configuration enables the UAV to hover and maneuver in the air with the multirotor system and navigate over the water surface using the float configuration when performing missions on water. The transitions between these modes according to the mission requirements and environmental conditions are well accommodated in-flight maneuvering. In this case, the multirotor system weighs 3 kg and acts as a quadcopter to provide VTOL capabilities. This system ensures stability and maneuverability in aerial data collection; hence, accurate positioning can be achieved in various applications. The UAV consists of an amphibious system configuration for waterborne operations composed of two parallel hulls with a deck connection, all weighing 1.5 kg. The design shown in Figure 8.1 ensures stability and buoyancy, allowing the drone to float on the water surface.

The propulsion system is selected based on the estimated weight, with a thrust-to-weight ratio 1.8. Brushless DC motors and propellers are used for achieving VTOL and aerial maneuvers. Over-deck propulsion systems integrated into the middle of the amphibious system, such as rear propellers, provide thrust for navigation and stability in waterborne mode. The motor selection and thrust requirements are detailed in Tables 8.1 and 8.2.

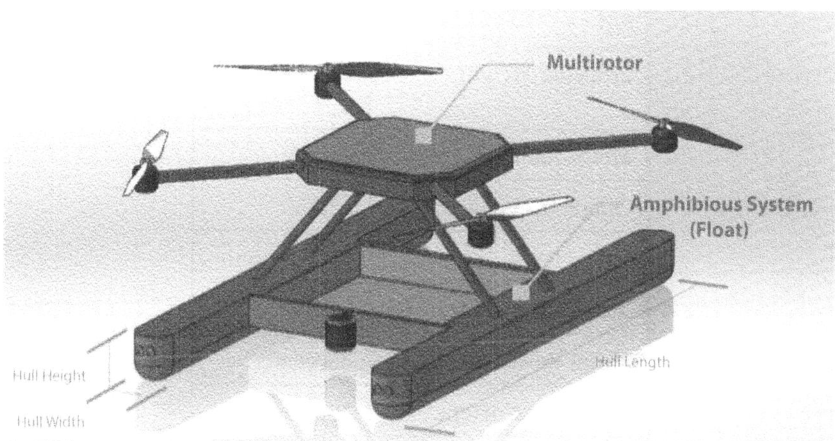

Figure 8.1 Conceptual model of modular UAV.

Table 8.1 Multirotor thrust calculation.

Parameters	**Values**	**Empirical relation**
Total mass	4.5 kg	Total mass = structural weight (3 kg) + subsystems (1.5 kg)
maximum take-off weight	44.1 N	Multirotor + Amphibious System
Thrust to Weight Ratio	1.8	T/W
Total Thrust Required	79.38 N	Thrust required = T/W
Thrust Available	98.0 N	Thrust Available = Thrust of the Multirotor' Motors

Geometrical terminologies used in the design calculations are shown in Figure 8.1, while design parameters for the multirotor and amphibious systems are presented in Tables 8.2 and 8.3. Other design considerations include power source, payload capacity, communication system, data processing and storage, navigation, and environmental resistance.

Table 8.2 Design parameters of modular UAV (Multirotor).

Parameters	**Values**
Propeller Diameter (d)	15 in
Propeller Pitch (p)	5 in
p/d	0.33
Propeller Weight	0.028 Kg
Motor Rating	400 kV
Operating Voltage	22.2 V
Motor RPM	6500 rad/min
Motor Thrust	24.5 N
Motor Power	850 watts
Operating Current	30 Amps
Motor Weight	0.156 Kg

Table 8.3 Design parameters of amphibious system (Floats).

Parameters	**Values**
Total mass	1.5 kg
Length of the hull (LH)	0.64 m
Length of the waterline (LWL)	0.5 m
Length to Beam Ratio (LBR)	8
Draft (d)	0.10 m
Beam of the Hull (Bwl)	0.12 m

High-capacity lithium polymer batteries, specifically the TATTU (6s) - (5200mAh) Li-po battery, are chosen to power drones for extended flight and waterborne operations. Table 8.4 details the estimated power requirements for the multirotor system and other components. The vehicle's center of gravity is calculated based on the arrangement of electronic components, control boards, batteries, and accessories to maintain stability.

Developing a reliable and efficient modular UAV with an amphibious landing mechanism is a complex engineering feat requiring careful consideration of structural, hydrodynamic, and operational aspects. The computational simulations are conducted using ANSYS CFD to evaluate the performance of the amphibious landing mechanism in airborne and aquatic environments. These simulations help to identify potential challenges and refine the design before building physical prototypes. Simulations are carried out using the modular UAV created in CATIA and loaded into ANSYS Fluent. A computational domain should be ten times the vehicle's original size. With an arm length of 0.5 m, the floating drone's actual domain measures about 0.64 x 0.64 m. The quality of the mesh is determined with skewness and orthogonality checks. The Angle of Attack (AoA) is varied between 0° and -10° while relative air speeds are kept between 3 m/s and 10 m/s for the CFD analysis. Figure 8.2 shows the pressure contour and

Table 8.4 Power requirement of modular UAV.

Parameters	Values
Required battery output voltage	0.322 V
Single LiPo cell output voltage	4.2 V
mAh	52000 mAh
Battery weight	0.7 kg

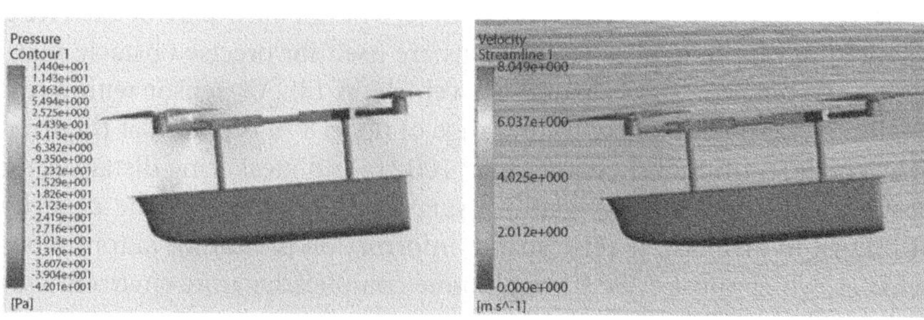

Figure 8.2 Pressure contour and velocity streamlines of modular UAV at -5° AoA.

velocity streamline simulations for an AoA of -5°, which yields the optimal drag coefficient of 0.58.

The structural components of the UAV are built from lightweight carbon fiber and aluminum materials. This makes the UAV lighter in weight and helps it remain buoyant during water landings and take-offs while the strength necessary for aerial flight is retained. The landing mechanism is a critical component for land and water based operations, which plays a crucial role in achieving buoyancy during water landings. A foamed material that is polymer-based absorbs the impact forces when landing on hard ground, becomes stable in water, and contributes to gains in efficiency. Reinforcement adaptability applies to the choice of these floats. All electronic components including the UAV flight controller, navigation instruments, and communication systems are waterproofed against water vulnerability during operation in aquatic environments. Water ingress protection is catered for in the design through sealing mechanisms and protective enclosures. A dual propulsion system is usually used, for example, separate propellers for flying and navigating in water for amphibious missions. Both systems are thereby coordinated to yield maximum thrust and control. Aquatic conditions make the control surfaces rudder adjustable. The configuration gives precise water navigation control using information from surrounding sense environmental conditions to evaluate heading conditions for creating an appropriate heading approach.

8.4 Surround Sense Module

The design and development of the Surrounding Sense Module (SSM) featuring LiDAR forward technology and Ping Sonar, as illustrated in Figure 8.3, has metamorphosed into a milky way of business transformation towards autonomous vehicle safety and autonomy. This module will be the first attempt to provide vehicles with an advanced sensory apparatus to understand all surroundings in real-time. The forward part of LiDAR is essential in this system, as laser beams are used for precise obstacle mapping of the environment around the vehicle. A LiDAR sensor with a resolution of 1 cm with a maximum range of up to 8 m is placed at the front of a vehicle, continuously scanning at 100 Hz and measuring distances for obstacle detection and navigation using data makes sense. The capacity of LiDAR to give fast, precise spatial information is vital for autonomous vehicles, which must drive through quite complex dynamic environments with much confidence.

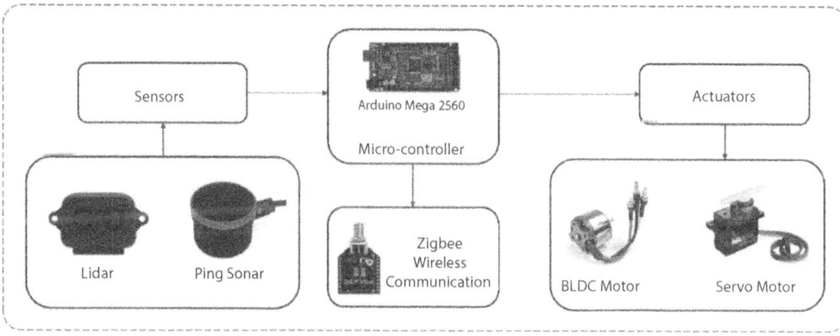

Figure 8.3 Block diagram of surround sense module.

LiDAR technology and sonar, such as the Ping Sonar module, add a layer of environmental awareness. Inspired by underwater sonar systems, Ping Sonar uses sound waves to measure distances and detect objects. Positioned strategically around the vehicle, 1 cm resolving power sensors from Ping Sonar, with a depth rating of 100 m, emit ultrasonic pulses at 115 kHz to create a detailed spatial map of the environment. This technology works best when normal vision cannot be used, for example, during adverse weather or darkness, adding significant value to the SSM. Considerations in design were cautious when integrating LiDAR and Ping Sonar technologies into the SSM. The placement and orientation of the sensors are designed to reduce blind spots and maximize coverage. Advanced algorithms are developed to handle the enormous volume of data these sensors generate towards real-time responsiveness and accurate obstacle detection. The SSM represents a significant step forward in safety and a key facilitator of driverless features. Vehicles will become much safer since the system will make them aware of their surroundings, thus preventing collisions and finding a secure path. As technology progresses, the design and development of such innovative sensing modules will play an increasingly important role in making that future one where vehicles can operate independently and more safely come to fruition.

In the perspective control systems technique, the Surround Sense System used during this experiment integrates LiDAR and Ping Sonar sensors to provide real-time environmental feedback as shown in Figure 8.4. The Arduino Mega microcontroller is the central processing unit, efficiently managing and coordinating inputs from the dual LiDAR sensors and the Ping Sonar. Its computational prowess enables real-time sensor data analysis, allowing the vehicle to make informed decisions based on its surroundings. Accordingly, a BLDC motor of the forward thrust system is

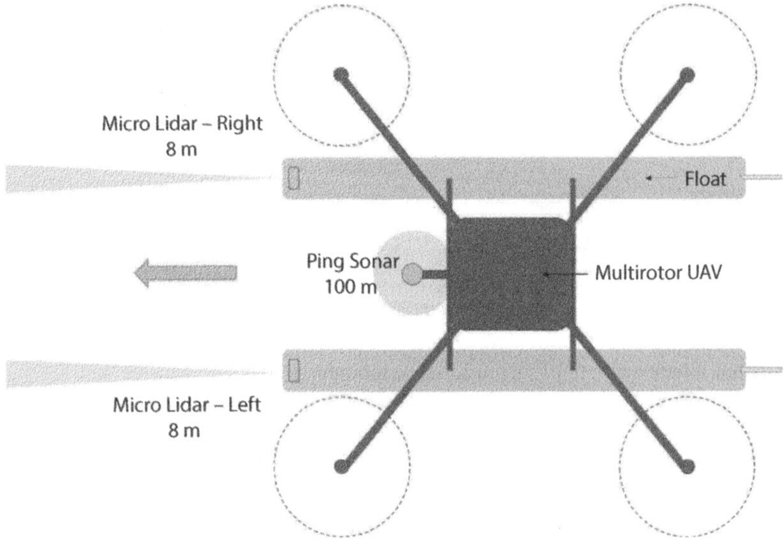

Figure 8.4 Surround sensors of UAV for environmental perception.

adjusted for propulsion, and two servo motors are used to steer the rudder for directional control. The designed orchestrated control mechanism ensures the vehicle's adaptability, allowing it to navigate complex terrains while responding efficiently to obstacles above and below the water surface.

8.5 Integration and Testing

Integrating an amphibious system and a SSM (Figure 8.5) into UAVs represents a ground breaking convergence of technologies, pushing the boundaries of UAV capabilities in both aerial and aquatic domains. This integration will widen the scope of UAV operations by allowing them to switch from flying to aquatic operations, thus opening more opportunities for environmental monitoring, search and rescue, and reconnaissance missions. An amphibious system will allow UAVs to land on water and take off from there, increasing the deployment scenarios for UAVs in maritime and aquatic environments. With floatation devices, these UAVs can conduct operations over remote water bodies, ensuring the collection of valuable data like water quality, bathymetry and surveillance capabilities in areas that cannot be accessed by conventional means using UAVs. At the same time, integrating an SSM will improve UAV situational awareness even further. In aqueous environments, LiDAR will perform scanning to detect all

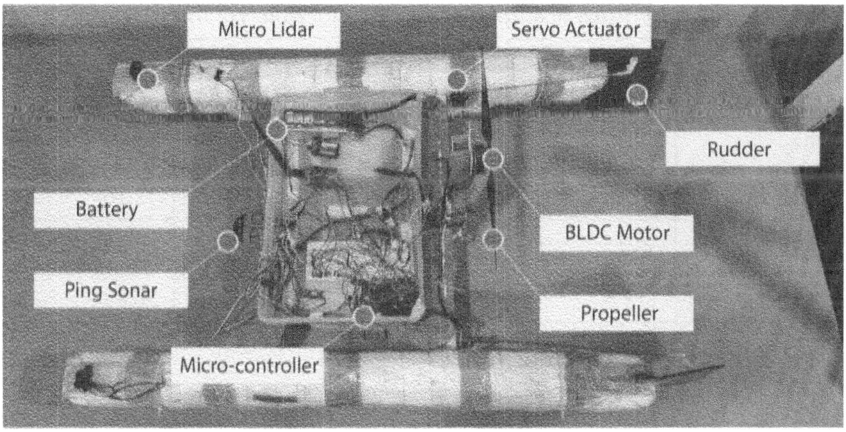

Figure 8.5 Integration of amphibious system and surround sense module.

surrounding objects for obstacle detection. When the UAV navigates in water, the Ping Sonar system uses sound waves to assess underwater environments, detect submerged obstacles, and navigate precisely. An onboard microcontroller Arduino Mega coordinates the amphibious system control and SSM. The developed intelligent controller manages the transition between flight and aquatic modes, interprets data from the SSM and executes responsive actions, ensuring a seamless and adaptive UAV operation.

The laboratory testing of the forward thrust system, considering rudder deflection and throttle impact, is crucial for optimizing power consumption and enhancing the amphibious system's performance. Understanding the entire system behavior helps engineers design more efficient propulsion systems for modular UAVs. The propeller at the rear generates the necessary thrust for forward motion, evaluated under different throttling conditions using a test rig. Results and corresponding maximum thrust values (Table 8.5) predict surplus thrust and enable us to understand the maximum operating state of thrust of the developed system. The current drawn during testing determines the battery size, regulating UAV for forward motion by controlling the thrust motor.

The prototypes undergo extensive field trials to prove their functionality and understand their performance, including simulated amphibious landings in controlled water environments. These trials evaluate the UAV's stability, buoyancy, and control, providing valuable data for further refinement and optimization. The developed systems are tested in open environment conditions on a tank of still water at two stages. In the initial stage test of flying mode, Figure 8.6 shows the demonstrated flight of a UAV in

Table 8.5 Forward thrust of amphibious system for different throttle positions.

Throttle position	Power (watts)	Voltage (V)	Current (A)	Thrust (N)
25%	12.3	12.71	0.95	7.28
50%	27.1	12.44	2.17	12.16
75%	50.4	12.38	4.08	24.73
100%	113.5	12.24	9.25	36.08

Figure 8.6 UAV in water landing and take-off.

water landing and take-off, which is performed to understand the stability of the developed UAV in water based operations. During this successful test flight, the SSM module was not incorporated to minimize the complexities at the primary test level.

To understand the buoyancy and controls of the amphibious system in aquatic mode as tested which is shown in the Figure 8.7 for the fixed distance of 7 m in a tank. The obstacle was placed at the end of the path, and the sensor of SSM was programmed for various detecting ranges (D).

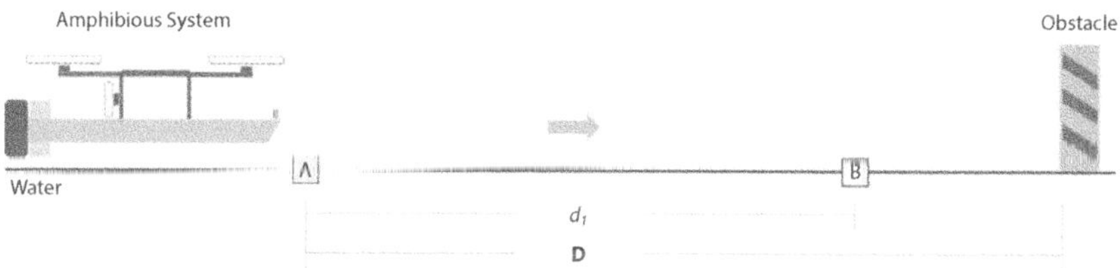

Figure 8.7 Performance testing of the surround sense module.

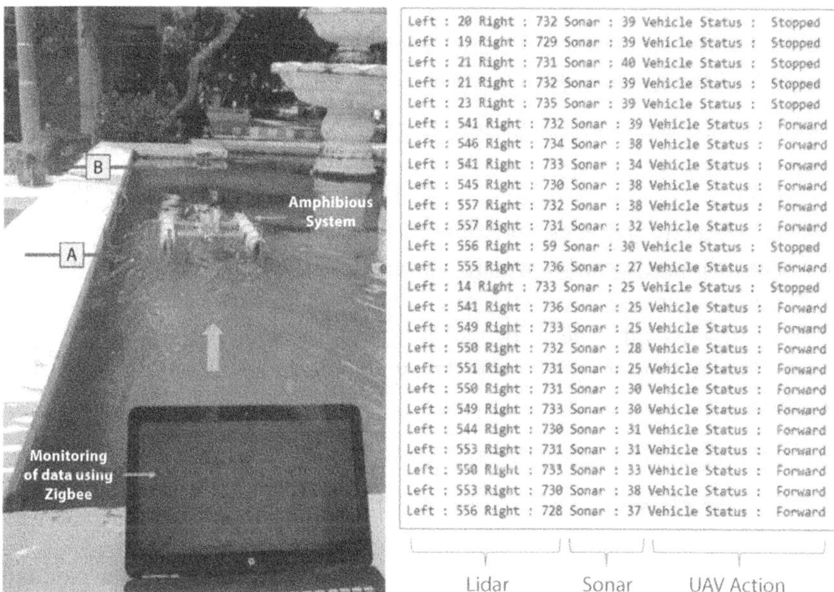

(**a**) Testing of SS module in water (**b**) Data log of Sensors

Figure 8.8 Evaluation of the surround sense module.

A typical test of the amphibious system and the data logs are shown in Figure 8.8. Data collected from the field trials were analyzed to validate the floating performance and the functioning of control systems.

The Velocity-Braking Distance Profile shown in Figure 8.9 of the amphibious system (Float) tested in still water with a depth of 40 cm reveals crucial insights into the system's performance during obstacle detection using the SSM. The variable testing component involves different obstacle-detecting ranges (D) set at 0.300m, 0.225m, and 0.150m. These distances represent the system's ability to detect obstacles at varying proximities.

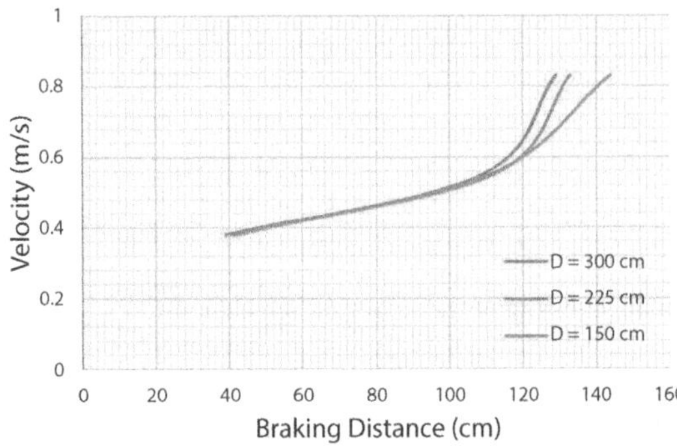

Figure 8.9 Velocity-braking distance profile.

In the absence of depth variation in the water throughout the test, the graph illustrates the velocity of the amphibious system under different obstacle-detecting ranges and operating speeds. The results depict that the responsiveness of the system varied linearly at low velocity profiles, and sudden increase in velocity due to water scattering as speed increases, which provides valuable information for optimizing the system's performance and enhancing its obstacle avoidance capabilities in still water conditions.

8.6 Conclusion

A modular and reconfigurable amphibious UAU system with a payload capacity of 4.5 kg, integrated with Sonar and LiDAR modules, performed well in tests conducted on water. These sensors enabled broad and real-time situational awareness of the surroundings of the UAV in different environmental conditions. The results of the system field test proved that the sensors detect and trigger accurately with high precision and reliability in water-based missions. From the Velocity-Braking Distance profile, it is very clear that the SSM system performance is linear at low velocity and changes abruptly due to water scattering at higher speeds. It gave valuable information for optimizing the system's performance and enhancing its obstacle avoidance capability in still water conditions. The field tests regarding the stability, buoyancy, and control aspects of developed amphibious UAVs prove they can become highly suitable for water-based missions to collect water quality parameters and depth information related to remote water bodies. Despite the system's success, certain limitations were observed in performance variations, such as extreme weather conditions, strong winds, and running waters. Additionally, limitations in sensing in the surround sense module's range and effectiveness in complex environments such as water reflections and floating vegetation were noted. These insights from testing provide valuable information for further refinement and improvement of developed modular and multifunctional UAVs with advanced dynamic situation awareness systems.

Acknowledgement

We acknowledge the funding support from the DBT project Ref. No. PR40444. We also acknowledge the technical and knowledge assistance from the Centre for Autonomous System Research and the financial aid from Vel Tech Rangarajan Dr. Sagunthala R&D Institute of Science and Technology, Avadi, Chennai under VTU SEED FY 23-24_023 are also significantly acknowledged.

References

1. Ayamga, M., Akaba, S., Nyaaba, A.A., Multifaceted applicability of drones: A review. *Technol. Forecast. Soc. Change*, 167, 120677, 2021.
2. Alzahrani, B., Oubbati, O.S., Barnawi, A., Atiquzzaman, M., Alghazzawi, D., UAV assistance paradigm: State-of-the-art in applications and challenges. *J. Netw. Comput. Appl.*, 166, 102706, 2020.
3. Esakki, B., Ganesan, S., Mathiyazhagan, S., Ramasubramanian, K., Gnanasekaran, B., Son, B., Choi, J.S., Design of amphibious vehicle for unmanned mission in water quality monitoring using internet of things. *Sensors*, 18, 10, 3318, 2018.
4. Mohsan, S.A.H., Othman, N.Q.H., Li, Y., Alsharif, M.H., Khan, M.A., Unmanned aerial vehicles (UAVs): Practical aspects, applications, open challenges, security issues, and future trends. *Intell. Serv. Rob.*, 16, 1, 109–137, 2023.
5. Huang, C.Y., Chung, P.H., Shyu, J.Z., Ho, Y.H., Wu, C.H., Lee, M.C., Wu, M.J., Evaluation and selection of materials for particulate matter MEMS sensors by using hybrid MCDM methods. *Sustainability*, 10, 10, 3451, 2018.
6. Vigna, B., Lasalandra, E., Zerbini, S., Aleo, M., Silicon Sensors, in: *Springer Handbook of Semiconductor Devices*, pp. 635–698, Springer International Publishing, Cham, 2022.

7. Mukherjee, A., Misra, S., Raghuwanshi, N.S., A survey of unmanned aerial sensing solutions in precision agriculture. *J. Netw. Comput. Appl.*, 148, 102461, 2019.
8. Kotarski, D., Piljek, P., Pranjić, M., Grlj, C.G., Kasać, J., A modular multirotor unmanned aerial vehicle design approach for development of an engineering education platform. *Sensors*, 21, 8, 2737, 2021.
9. Hang, K., Lyu, X., Song, H., Stork, J.A., Dollar, A.M., Kragic, D., Zhang, F., Perching and resting—A paradigm for UAV maneuvering with modularized landing gears. *Sci. Rob.*, 4, 28, eaau6637, 2019.
10. Atyabi, A., MahmoudZadeh, S., Nefti-Meziani, S., Current advancements on autonomous mission planning and management systems: An AUV and UAV perspective. *Annu. Rev. Control*, 46, 196–215, 2018.
11. Allen, B.D., Alexandrov, N., Puig-Navarro, J., ATTRACTOR: Toward Trustworthy and Trusted Autonomous Systems, in: *AIAA Scitech 2021 Forum*, p. 1684, 2021.
12. Ganesan, S., Esakki, B., Mathiyazhagan, S., Pandimuthu, V., Design conception and evaluation of an unmanned amphibious aerial vehicle using systematic approach. *Aviation*, 26, 1, 41–53, 2022.
13. Manoharan, D., Gajendran, C., Padmanabhan, M.K., Vignesh, S., Rajesh, S., Bhuvaneshwaran, G.D., Design and development of autonomous amphibious unmanned aerial vehicle for in situ water quality assessment and water sampling. *J. Unmanned Veh. Syst.*, 9, 1, 182–204, 2021.
14. Alzu'bi, H., Mansour, I., Rawashdeh, O., Loon copter: Implementation of a hybrid unmanned aquatic–aerial quadcopter with active buoyancy control. *J. Field Rob.*, 35, 5, 764–778, 2018.
15. Mishra, V., Avtar, R., Prathiba, A.P., Mishra, P.K., Tiwari, A., Sharma, S.K., Jain, K., Uncrewed Aerial Systems in Water Resource Management and Monitoring: A Review of Sensors, Applications, Software, and Issues. *Adv. Civ. Eng.*, 3544724, 2023.
16. Kohli, L., Saurabh, M., Bhatia, I., Sindhwani, N., Vijh, M., Design and development of modular and multifunctional UAV with amphibious landing, processing and surround sense module, in: *Unmanned aerial vehicles for internet of things (IoT) concepts, techniques, and applications*, pp. 207–230, 2021.
17. Gokul, R.P., Esakki, B., Ganesan, S., Evaluation of mechanical strength characteristics of double ducted unmanned amphibious aerial vehicle using finite element analysis. *Int. J. Math. Eng. Manage. Sci.*, 4, 2, 420, 2019.
18. Raja, V., Kumar, A., Pacheco, D.A.D.J., Multi-disciplinary engineering design of a high-speed nature-inspired unmanned aquatic vehicle. *Ocean Eng.*, 270, 113455, 2023.
19. Sheng, T.K., Esakki, B., Ganesan, S., Salunkhe, S., Finite element analysis, prototyping and field testing of amphibious UAV. *UPB Sci. Bull. Ser. D Mech. Eng.*, 81, 125–140, 2019.

20. Zufferey, R., Siddall, R., Armanini, S.F., Kovac, M., Zufferey, R., Siddall, R., Kovac, M., Sailing and Flying with a Multimodal Robot, in: *Between Sea and Sky: Aerial Aquatic Locomotion in Miniature Robots*, pp. 167–195, 2022.
21. Josipovic, N. and Viergutz, K., Smart Solutions for Municipal Flood Management: Overview of Literature, Trends, and Applications in German Cities. *Smart Cities*, 6, 2, 944–964, 2023.
22. Vikram, P., Balasubramanian, E., Manova, M., Surendar, G., Aerodynamic Interaction Studies on Amphibious Vehicle in Forward and Hovering Flight. *Int. J. Automot. Mech. Eng.*, 16, 4, 7391–7412, 2019.
23. Darvishpoor, S., Roshanian, J., Raissi, A., Hassanalian, M., Configurations, flight mechanisms, and applications of unmanned aerial systems: A review. *Prog. Aerosp. Sci.*, 121, 100694, 2020.
24. Mathaiyan, V., Murugesan, R., Madasamy, S.K., Gnanasekaran, R.K., Sivalingam, S., Jung, D.W., Conceptual design and numerical analysis of an unmanned amphibious vehicle, in: *AIAA Scitech 2021 Forum*, p. 1285, 2021.
25. Ganesan, S. and Esakki, B., Computational fluid dynamic analysis of an unmanned amphibious aerial vehicle for drag reduction. *Int. J. Intell. Unmanned Syst.*, 8, 3, 187–200, 2020.
26. Esakki, B., Mathiyazhagan, S., Moses, M., Rao, K.J., Ganesan, S., Development of 3D-printed floating Quadrotor for collection of algae in remote water bodies. *Comput. Electron. Agric.*, 164, 104891, 2019.

9

Implementing Mission Critical Public Safety Using Communication in Drones Network

V. Nithya[1], Mani Deepak Choudhry[2*], M. Parimala Devi[3] and S. Jayachitra[4]

[1]Department of CSE, KGiSL Institute of Technology, Coimbatore, Tamil Nadu, India
[2]Department of Computing Technologies, SRM Institute of Science and Technology, Chennai, Tamil Nadu, India
[3]Department of Electronics and Communication Engineering, Velalar College of Engineering and Technology, Thindal, Erode, Tamil Nadu, India
[4]Department of Electronics and Communication Engineering, PSNA College of Engineering and Technology, Dindigul, India

Abstract
Drone technology is adopted by public safety operations to assist in disaster control and emergency relief. Drones need a proficient communication network for optimal functioning. The book chapter gives an outline of how drone network communication can be adapted for public safety. The chapter takes into consideration the unique needs of conventional public safety operations such as low coverage in remote and inaccessible areas, hostile environments, and limited response time. Thereafter, the chapter delves into the design and architecture of drone network communication systems with a view to harmonious integration with existing public safety communication infrastructure. This book chapter is a comprehensive guide for policymakers, practitioners, and researchers to understand the technicalities, opportunities, and challenges in leveraging communication technology to unlock the full potential of drones to safeguard communities and save lives.

***Keywords*:** Drones, communication, public safety, disaster management, edge computing, UAV networks

**Corresponding author*: manideepakjns22@gmail.com

A. Prasanth, Rajesh Kumar Dhanaraj, Munish Sabharwal, Vandana Sharma and Seifedine Kadry (eds.) Drones for Transportation Logistics and Disaster Management, (199–220) © 2026 Scrivener Publishing LLC

9.1 Introduction

Public safety operations rely significantly on communication systems, but these are frequently heavily dependent on fixed infrastructure and therefore exposed to major disruption during natural disasters such as earthquakes, tsunamis, hurricanes, floods, and landslides. Such disasters cause catastrophic human suffering, life casualties, shift of communities, and heavy economic losses. In the event of an emergency, economically efficient and timely temporary retrieval technologies are essential. *Ad hoc* wireless networks seem like a perfect fit, which corresponds to the Federal Communications Commission's emphasis on spearheading broadband innovations in public safety communications. This allows the primary medium to effectively direct and accept critical speech, audiovisual, and information, enabling life-saving operations as well as crime prevention. The ease of establishing these networks makes recovery easier and provides critical communication support to first responders [1]. *Ad hoc* wireless networks offer advantages such as greater efficiency, economizing, flexibility, self-containment, wider reach, and resilience. These characteristics are of great help in assisting the first responders and rescue teams during disasters, helping in the smooth performance of their tasks [2].

In the modern era, wireless communications are an indispensable component of our lives. The need and significance of communication networks under various circumstances illustrate the importance of their reliability and stability. Even though interruption of communications brings trouble in our everyday routine, failure during natural calamities may result in more catastrophic effects. Failure of critical infrastructure can result in catastrophic impacts on society and the security of disaster victims. Rendering these systems resilient and robust is important in securing our interdependent world.

Base Station Systems are the underlying pillars of wireless communication networks, providing a means for the communication of nodes and User Equipment (UE) with each other. Though cellular systems based on infrastructure are successful under normal situations, they prove to be susceptible in natural disaster situations. In such incidents, one or more components of a Base Station System can get deactivated, damaging the wireless communication structure and impacting the networked devices. To counter such challenges, Wireless Sensor Networks and Device-to-Device (D2D) communication can serve as backup solutions, with the provision of substituting the traditional framework-dependent communication system temporarily to transmit very important emergency and safety information.

Drones have attracted much attention in studies and other uses due to their resourcefulness and versatility in solving numerous issues [3]. Drones are used in the security, surveillance, exploration, and management of inaccessible land areas, which is especially useful for first responders during natural disasters. Drones perform a revolutionary function in the delivery of medical aid and support for emergency situation management such as forest fire fighting, safeguarding key infrastructure, coastal surveillance, and law enforcement support. They also support increasing public safety in intelligent cities. When natural disasters occur, with normal sources of information such as television, radio, and broadband affected, it is important to assist trapped survivors with necessary geographical information to lead them to shelter or rescue sites. This involves the transmission of risk maps, comprehensive regional maps, and data on available bridges, roads, safe paths, and recent disaster information.

High data transfer rates are essential for swift and effective communication in such situations. Recent studies have revealed that in well-established nations, dedicated communication approaches for disaster response are often lacking. Instead, rescue management services rely on cellular base stations for communication with rescue teams. Public safety communication still heavily relies on steadfast technologies like Project 25 (P25) and Terrestrial Trunked Radio (TETRA), which have been utilized for over two decades. They are primarily designed for rescue teams and not for the victims in affected areas. Additionally, they are ill-suited to aid high-bandwidth use cases, as they are predominantly based on narrowband technology.

There are various aspects related to emergency communication networks and their evolution over time:

1. **Significance of Visual and Data-Driven Information:** This refers to the significance of visual information such as images and videos in interpreting emergencies, such as river levels and flood levels. It further provides arguments for the urgency index to enhance data being added to *ad hoc* networks.
2. **Standby System Classification of Communication:** This feature allows communication in standby systems to be classified under various priority levels. Classification allows communications to be optimized during emergencies.

3. **Scheduling and Priority Levels:** The chapter emphasizes the necessity of an implementation where priority levels would be assigned to different classes of emergency calls and messages. This type of system allows for scheduling and communication management.
4. **Infrastructure-Less Communication:** The authors have mentioned in the paper that infrastructure-less communication systems play an important role during disaster communication. They have declared that technologies like D2D, Machine-to-Machine (M2M) communications, Internet of Things (IoT) [4–6], intelligent networks, sensor networks [7], and Unmanned Aerial Vehicles (UAVs) are major components of emergency communication systems.
5. **Characteristics of Infrastructure-Less System:** It provides the key features of infrastructure-less communication systems, including their ability to function without traditional infrastructure support.
6. **Drawbacks of Narrowband Services:** The article identifies narrowband service drawbacks, including the absence of data broadcast capability and the use of legacy technology.
7. **Trend towards Broadband MCS:** The author notes that with technological evolution and the global deployment of Long-Term Evolution (LTE) networks, interest has been on the rise for broadband Mobile Communication Systems (MCS).
8. **Challenges with Service Platforms:** It raises concerns about the reliability of commercial service platforms, including cloud services, which can be vulnerable. Battery-powered backups are mentioned but are noted as a temporary solution.

This chapter provides a detailed view of various implementation strategies involved in public safety through UAV communication. This chapter's overview is as follows: Section 9.2 provides the assessment of numerous research works about the implementation of UAVs in public safety. Section 9.3 offers an introduction to UAV attributes, categorization, and strategies for leveraging UAVs to enhance communication and service platforms. Section 9.4 concludes the chapter with future directions.

9.2 Related Work

UAVs are becoming a focal point in the development of next-generation wireless networks [8]. These versatile vehicles are gaining attention from both industry and academia due to their on-demand capabilities [9–12].

Drone-operated base stations significantly enhance connectivity, extend coverage, and boost energy efficiency compared to conventional land-based stations [13]. They offer scalability and reduced chances of communication obstructions. Their advanced capabilities enable UAV-to-everything (U2X) communications, where they collect essential data from various sources, including vehicles, machinery, and pedestrians. This data collection is especially valuable in epidemic outbreaks and emergency scenarios as shown in Figure 9.1. Mobile operators are increasingly interested in UAVs as they offer cost-effective, reliable, flexible, and energy-efficient communication solutions. UAVs can provide additional network capacity during emergencies and temporary events [14–16]. During natural disasters that disrupt terrestrial base stations, UAVs can swiftly provide high-quality wireless connectivity [17–19] for the liberation process to carry out public safety [20].

Figure 9.1 UAV to U2X communication.

Optimized UAV placement is crucial to maximize coverage while minimizing energy consumption. A suggested three-tier disaster-resilient framework incorporates UAVs at the network's periphery for uninterrupted communication in emergencies. Amid extensive natural disasters, strategically positioning drones paves the way for a versatile and capable solution, as depicted in Figure 9.2. However, contempt the innumerable merits related with the use of UAVs in such circumstances, the design process presents significant and pressing challenges that need to be addressed.

Several surveys on UAVs have been the topic of research, and numerous research papers have explored the features, necessities, problems, and trials of UAV communication systems. Table 9.1 highlights the distinctive aspects of the survey.

In [14], the authors presented a compact UAV platform, examining potential challenges and analyzing vital components, including navigation sensors, communication modules, and the onboard processing unit. [20] conducted a survey investigating core design aspects, challenges, routing protocols, and unresolved research issues concerning Flying *Ad-Hoc* Networks (FANET). [21] provided a thorough survey on the assessment of autonomous building UAV systems (UAS). [22] offered an overview of UAV networks in civil applications, highlighting their key characteristics for communication prospects and providing insights from various project-based experimental results.

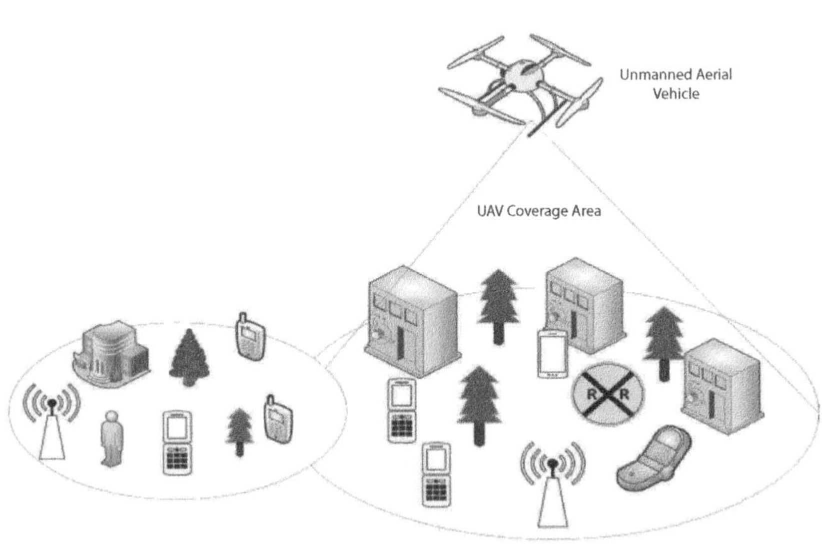

Figure 9.2 UAV societal safety application.

Table 9.1 Existing research on UAVs.

References	Description
[14]	Primary issues in the UAV Communication Platform survey
[20]	Review on UAV-based IoT interaction
[21]	UAV-based optimization review
[22]	UAV interaction survey
[23]	Automated Inspection of Structures a review
[24]	UAV interaction-based challenges
[25]	UAV network-based uses
[26]	UAV classification survey
[27]	UAVs based simulations review
[28]	Airborne network review
[29]	Collection of images by UAV for various applications a review
[30]	Small UAV-based remote sensing use cases

In [23], a review centered on UAV-facilitated wireless network guidance, covering mathematical tools and analytical frameworks applied to tackle core issues. [24] offered an in-depth comprehension of present method categorization, practical deployment, system identification context, and applications for compact, budget-friendly UAVs. [25, 26] introduced an overview of a flexible UAV simulation tool compatible with commercial and open-source uses. It covered a wide range of unmanned vehicles, including micro, subsurface, surface, and ground vehicles. In [27], an examination covered protocols and mechanisms for crafting UAVs of various natures. A brief review in [28] focused on the deployment of drone-controlling approaches of fusion UAVs. [29] presented publicly accessible automated control systems designed for minor UAVs. [30] analyzes adversity investigation and administration, with a focus on imagery acquisition using UAV systems.

Numerous surveys have been conducted on UAVs; however, a momentous opening in the literature regarding surveys that specifically address the energy efficiency of UAVs, predominantly in the background of communal protection situations. In this chapter, our primary objective is to bridge this knowledge gap. In doing so, we will not only explore the nuances of UAV energy efficiency but also emphasize the associated challenges, and viable solutions, and outline promising avenues for future research in this domain.

9.3 UAV – Characteristics and Strategies

9.3.1 What is UAV?

UAVs are equipped with multiple components envisioned to interrupt or eliminate adversary aims, and piloted autonomously, remotely, or by both. With significantly more endurance and range than comparable manned systems, UAVs can be extremely effective.

The main components of a drone are shown in Figure 9.3. It is made up of its exterior framework, turbines, and controller as seen. Each of these drone parts contains a variety of drone components, which are detailed below:

> **Drone Body:** The body contains sensors, a control system, and a battery power supply. Drones employ accelerometers and gyroscopes as sensors to help in stabilization. A camera, Wireless Fidelity (Wi-Fi) transmitter/receiver, and high gain directional antenna make up an advanced drone UAV.

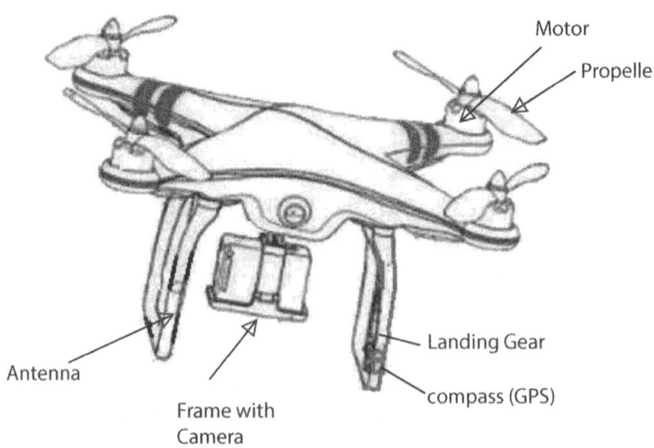

Figure 9.3 Components of drone.

Drone Frame: Carbon fiber and nylon are used to make the frame. A drone is robust and lightweight as a result. The body's shell and frame bars are held in place by the frame. It has propellers in it. The landing gear is installed at the end of the frame bars.
Propellers: These components are responsible for generating the necessary force to lift the drone into the air. Four propellers are used, with each pair of two rotating in opposite directions (clockwise and counterclockwise). These propellers are powered by electric motors, and the motors themselves are supplied with power from a battery.
Motors: Four motors, one for each propeller, provide the mechanical energy required to spin the propellers. These motors are electric and receive their power from a battery source.
Controller: The controller is a crucial component that manages various aspects of the drone's operation. It includes several functionalities:

- Wi-Fi Communication
- Propeller Speed Control
- Microcontroller
- Ground Control Communication
- Predefined Route Path
- Remote Control

9.3.2 How Does UAV Work?

The working operation of a drone UAV is exposed in Figure 9.4. It highlights three modes of motion for piloting the drone: Lift, Yaw, and Pitch. Further, two distinct frequencies are utilized in the operation of the drone: one for controlling the drone from the ground system and the other for video or image transmission. The remote controller for the drone typically includes the following buttons and features:

1. **Antennas:** Antennas are used to relay both aircraft control signals and video signals.
2. **Mobile Phone Holder:** This feature provides a secure space to mount a mobile phone or a compatible device. It is often used for viewing the stream from UAV surveillance or for running the drone control app.

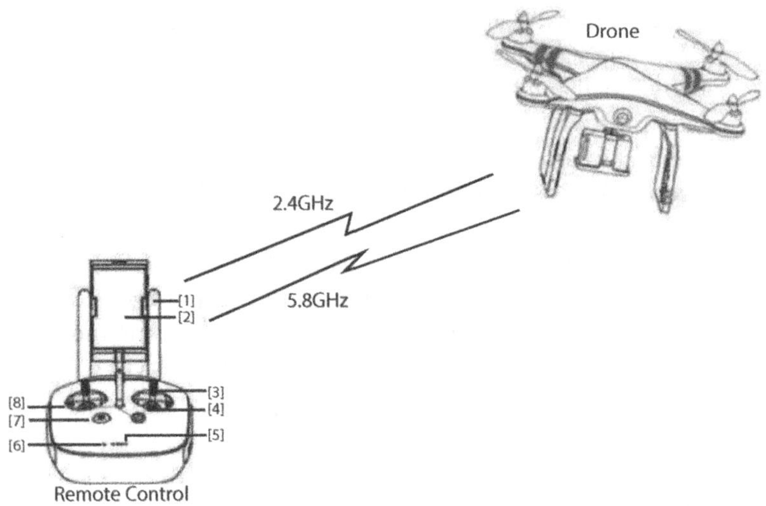

Figure 9.4 Working of drones (UAV).

3. **Control Stick:** The control sticks, or sticks, are used to manipulate the movement and orientation of the drone. They allow the operator to control aspects such as throttle (Lift), rotation (Yaw), and tilting (Pitch) of the drone.
4. **Return to Home (RTH) Button:** The Return to Home button is a crucial safety feature. When pressed, it commands the drone to automatically return to its takeoff point or a predefined home location. This function is valuable in case of emergencies or to end the flight safely.
5. **Battery Level Display:** The remote controller typically has a display or Light Emitting Diode (LED) indicator that shows the remaining battery level of the controller itself. The operator needs to monitor the controller's battery to ensure uninterrupted operation.
6. **System Status:** The system status indicator or display provides information about the drone's operational status. It may show details such as Global Positioning Systems (GPS) signal strength, drone readiness, or any error messages.
7. **Power Button:** The power button is used to turn the drone's power ON and OFF. It is a fundamental control element for initiating and ending drone operations.

8. **RTH Status LED Indicator:** This LED indicator is specifically associated with the Return to Home function. It provides information about the status of the drone's return-to-home feature, indicating whether it is actively engaged or not.

9.3.3 UAV Categorization

This section puts UAVs in three broad categories which are represented in Figure 9.5:

1. **Altitude-Based:** This type is founded on the altitude upon which UAVs fly. It provides a means of discrimination between various categories of drones that fly at different altitudes, such as low-altitude, mid-altitude, and high-altitude drones.
2. **Network-Based:** This type is centered on UAVs' network and communications ability. It classifies drones according to their connectivity and networking ability, for example,

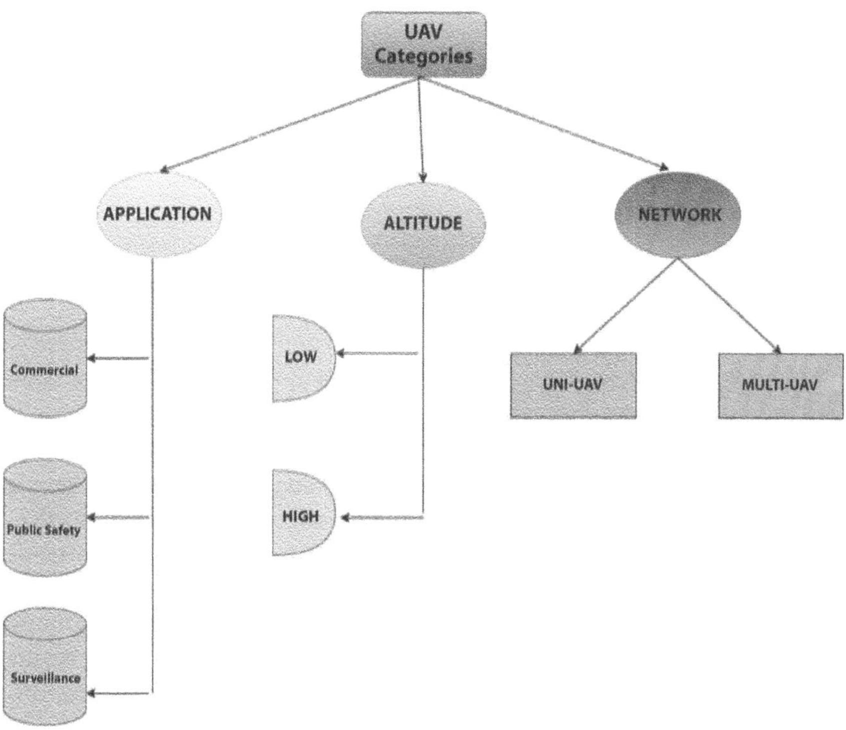

Figure 9.5 Categorization of UAV.

autonomous drones, remotely piloted drones, or drones integrated into broader communication networks.
3. **Application-Based:** This type categorizes UAVs based on their specific applications and uses. It categorizes drones into their principal uses. Such categories offer a systematic approach to comprehending the variety and diversity of UAVs concerning their operational features, networking elements, and extensive range of activities in diversified industries and departments.

9.3.4 Drones for Public Security

A public safety network, commonly known as a Public Security Communications Network (PSCN) or Public Safety Broadband Network (PSBN), is a dedicated communication infrastructure intended to serve first responders and emergency services during emergency situations. The networks are essential for facilitating the effective coordination of emergency response activities and improving public safety. Some of the main features of a public safety network include:

1. Dedicated Infrastructure
2. Priority and Preemption
3. Interoperability
4. Data and Multimedia Support
5. Geographic Coverage
6. Security
7. Local and National Networks
8. Funding

Drones play a pivotal role in various civilian applications, offering rapid and autonomous aerial capabilities for tasks like transportation, communication, agriculture, disaster mitigation, and emergency response. Their flexibility, quick deployment, and dynamic adaptability make them a promising technology, particularly in public safety networks where effective emergency communication is paramount. Drones can swiftly provide wireless communication services during and after disasters, facilitating search and rescue operations. They capture and transmit images to ground stations, aiding in the coordination of first responders and surveying affected areas, as highlighted in a multi-drone network architecture discussed in [31]. This technology enhances disaster assistance and ensures the efficient allocation of resources for maximum impact in times of crisis.

Recent studies have focused on addressing various design challenges related to drones, including 3D arrangement, energy competence, and drone take-off constraints. Notably, the ideal 3D arrangement of drones has been explored to expand coverage areas and enhance Quality of Service (QoS), with one study in [32] supporting this concept. Another study, [33], has delved into the 3D arrangement of numerous UAV base stations to exploit exposure while maintaining strong links with ground stations, employing practical swarm optimization techniques. This 3-dimensional (3D) deployment of drones holds significant potential for delivering facilities in civic care and adversity administration, as highlighted in [34], who projected a practical swarm optimization method to ensure extensive coverage, connectivity, and adherence to QoS necessities and drone size. Additionally, [35] introduced 3D deployment strategies for swarm drones, aiming to prolong drone lifetimes and optimize total throughput for all users, further contributing to the advancement of drone technologies in various applications.

9.3.5 UAV Placement Strategies

UAVs during disaster situations present a distinct challenge, differing from conventional terrestrial base stations due to unique constraints. UAV deployment offers flexibility in terms of altitude adjustments to accommodate specific scenarios and user demands, but altering the UAV altitude directly affects the coverage area. The literature has seen the development of various energy-efficient algorithms aimed at addressing issues related to drone positioning. However, most of these research papers primarily focus on optimizing UAV positioning to maximize capacity or minimize transmit power for specific UAVs in affected areas, with the primary goal of reducing interference and improving communication reliability in disaster-stricken regions. In [36], the study focuses on the strategic placement of drones to enhance coverage and performance in public safety communications, particularly in emergency or disaster scenarios. By deploying multiple drones equipped with communication technologies these drones establish connections with each other and with first responders in affected areas as shown in Figure 9.6, effectively restoring communication services in the wake of disasters that typically disrupt traditional communication and power infrastructure. This approach leverages drone technology to extend coverage and deliver crucial services in times of crisis.

In [37], the authors employed a heuristic algorithm to determine UAV locations in localities with varying utility concentrations, aiming to optimize 3D placement for meeting users' Quality of Service (QoS) requirements.

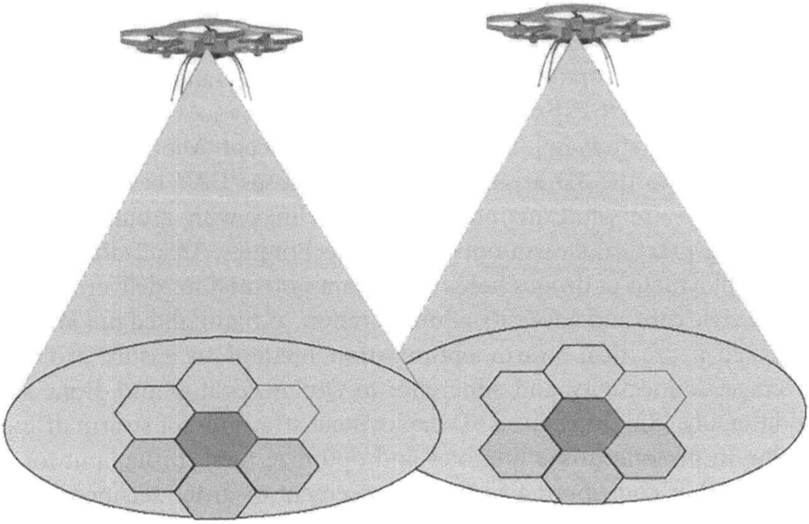

Figure 9.6 Public security by group of UAVs.

Meanwhile, [38] delved into an LP approach of varying integers to address the UAV 3D placement problem, optimizing location, altitude, and horizontal user distribution, achieving a 95% coverage rate using a bisection search method. In [39], the authors utilized ideal route efficiency and ability position techniques with four UAVs to establish an energy-efficient communication framework for IoT, resulting in lower uplink transmit power and a substantial 78% increase in energy efficiency. These efforts collectively contribute to enhancing UAV deployment strategies for improved coverage, reduced energy consumption, and optimized communication services. The mathematical tools and methodologies employed to address the UAV assignment challenge are illustrated in Figure 9.7.

The utilization of UAVs significantly enhances the communication bandwidth between UAVs and Wireless Sensor Networks (WSN), leading to reduced energy consumption during data transmission and expanded coverage areas. This synergy between WSN and UAVs yields efficient monitoring and data collection. Their collaborative efforts are pivotal in improving energy efficiency, local data processing, decision-making capabilities, and coverage extension through the application of clustering techniques, as depicted in Figure 9.8. This approach contributes to more effective and resource-efficient operations in data gathering and monitoring.

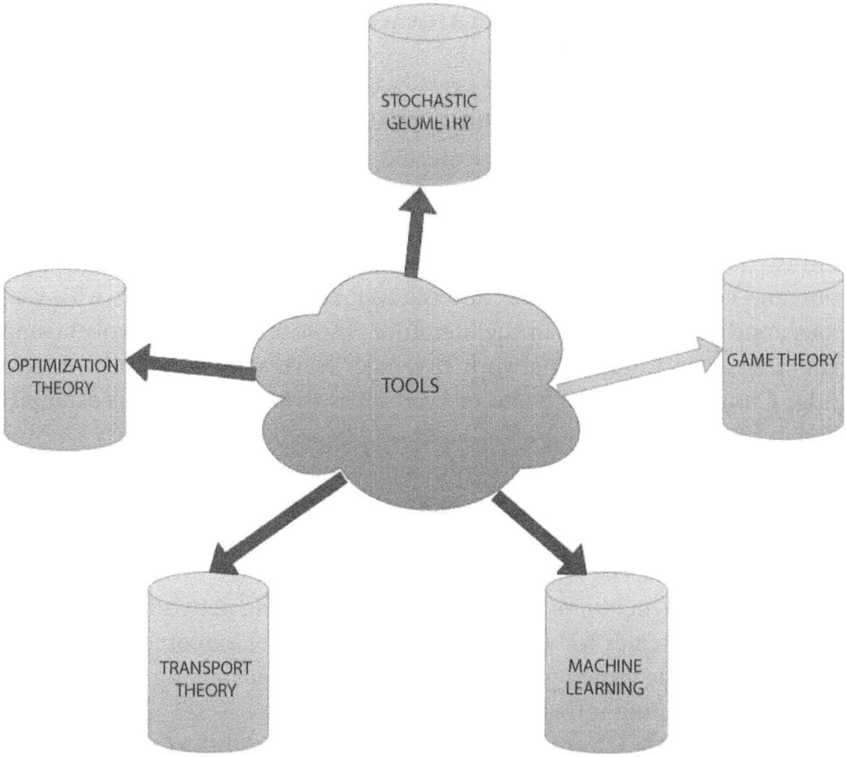

Figure 9.7 Mathematical tools for providing solutions for UAV placement.

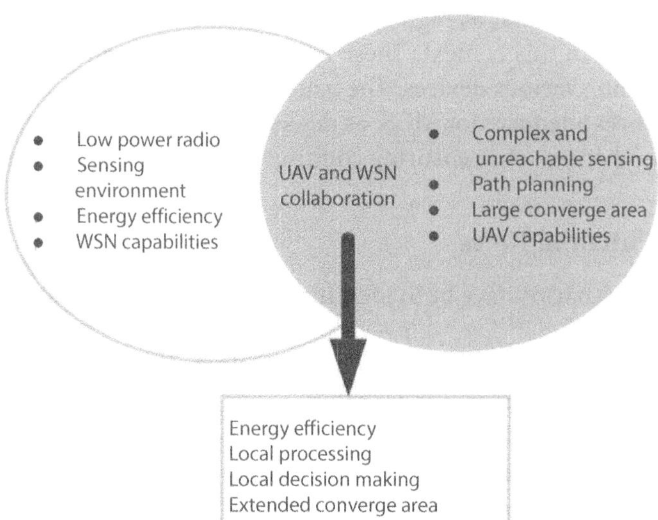

Figure 9.8 Combination of UAV and WSN.

9.4 Communication Framework Support by UAVs

UAVs have already proven their worth in disaster management, where they primarily fulfill roles in inspection, surveillance, and small parcel transportation. Their significance is expected to grow, particularly in supporting Mobile Communication Systems (MCS) during disasters or power grid failures by providing on-demand communication services when existing infrastructure is compromised. The integration of UAVs into communication and computing infrastructure is a well-established concept, with various approaches detailed in the literature. These approaches involve using UAVs to extend mobile network coverage, serve as mobile Multi-Access Edge Computing (MEC) hosts, or even act as flying 5G nodes, as exemplified in [40] and [41].

UAVs play a pivotal role in reducing latency and data transmission to central clouds, especially in time-critical IoT solutions, as seen in [42]. The concept of UAVs assisting in communication, caching, and edge computing, such as the Aerial Group of Mobile Edge Nodes (AGMEN) concept in [43], highlights their versatile potential in enhancing network capabilities. Numerous strategies for UAV-based computing are explored in [44, 45], underscoring the growing importance of UAVs in advancing communication and computing infrastructures.

Utilizing UAVs as mobile nodes in MCS is a straightforward and practical solution, especially considering the characteristics of UAVs described in the chapter. Multi-rotor UAVs are the preferred choice due to their suitability for the task, as high speed and extended range are not primary requirements in this context. These nodes typically weigh between 2 to 5 kg, combining various devices. The battery life of tiny UAVs, lasting 1 to 5 hours, proves adequate for MCS, as depleted UAVs shall readily substitute with fully charged ones, ensuring continuous operation during missions.

9.4.1 Potential Roles

Two main scenarios have been identified for the roles that UAVs can play:

1. **Enhancing Functioning Infrastructure:** When the existing communication infrastructure is operational, UAVs within the MCS framework can be organized to encompass exposure and deliver specialized facilities and stages. These UAVs may be utilized for rescue operations or dedicated tasks,

taking advantage of the existing infrastructure's connectivity and services to enhance their performance.
UAVs in these systems serve multiple roles:
- Transporters and Robotic Installers
- Deployers of Local Wireless Networks
- Deployers of Dedicated Sensors

2. ***Ad Hoc* Connectivity in Infrastructure Failure:** In situations where infrastructure is damaged, e.g., in the case of disasters or power grid collapses, drones provide seamless services. This UAV function becomes essential in maintaining communication continuity and expanding the reach of vital services in the event of non-functional infrastructure.
MCS should fulfill several critical functions:
    - Provide Connectivity
    - Basic Communication Services for Impacted Individuals
    - Backup Software Platforms

9.5 Conclusion

In the majority of cases, the use of UAVs in MCS offers a comprehensive solution to the communications issue. UAVs can carry out a variety of tasks, including deploying and gathering environmental data, as well as serving as sensors and supply carriers. Although UAVs are widely used in many wireless communications domains, numerous challenges must be resolved before their full potential can be realized in a particular scenario. The use of drones in municipal security communications was covered in this chapter. It also highlighted crucial topics like trajectory planning, communication procedures, and UAV placement. The chapter identified potential research directions for academics engaged in this field, emphasizing opportunities for further advancement.

References

1. Dabić, M., Ferreira, J., Lopes, M., Consumer Preferences and Barriers in the Adoption of Drone Delivery Services: A Comprehensive Analysis. *IEEE Trans. Eng. Manage.*, 72, 47–61, 2024.
2. Zhou, L., Silva, D., Smith, E., Locating Drone Stations for a Truck-Drone Delivery System in Continuous Space. *IEEE Trans. Evol. Comput.*, 29, 158–171, 2025.

3. Ali, K., Huan, X., Quoc, V., Review and implementation of resilient public safety networks: 5G, IoT, and emerging technologies. *IEEE Netw.*, 35, 2, 18–25, 2021.
4. Aruchamy, P., Balraj, L., Sowndarya, K., An energy-aware link fault detection and recovery scheme for QoS enhancement in Internet of Things-enabled wireless sensor network. *Comput. Electr. Eng.*, 123, 1–20, 2025.
5. Sankar, S., Ramasubbareddy, S., Dhanaraj, R.K., Balusamy, B., Cluster Head Selection for the Internet of Things Using a Sandpiper Optimization Algorithm (SOA). *J. Sens.*, 2023, 1–11, 2023.
6. Ali, M.A., Sharma, A.K., Dhanaraj, R.K., Heterogeneous features and deep learning networks fusion-based pest detection, prevention and controlling system using IoT and pest sound analytics in a vast agriculture system. *Comput. Electr. Eng.*, 116, 1–19, 2024.
7. Prasanth, A., Lakshmi, D., Dhanaraj, R.K., Balusamy, B., Technological Advancement in Internet of Medical Things and Blockchain for Personalized Healthcare: Applications and Use Cases. *CRC Press*, 1–290, 2024.
8. Kaleem, Z., Muhammad, Y., Aamir, Q., UAV-empowered disaster-resilient edge architecture for delay-sensitive communication. *IEEE Netw.* 33, 124–132, 2019.
9. Puri, A., A survey of unmanned aerial vehicles (UAV) for traffic surveillance, in: *Tech. Rep*, pp. 1–29, Dept. Comput. Sci. Eng., Univ. Tampa, FL, USA, 2005, https://www.ugpti.org/smartse/research/citations/downloads/Puri-A_Survey_of_Unmanned_Aerial_Vehicles_for_Traffic_Surveillance-2005.pdf.
10. Sitek, P. and Wikarek, J., Capacitated vehicle routing problem with pickup and alternative delivery (CVRPPAD): Model and implementation using hybrid approach. *Ann. Oper. Res.*, 273, 257–277, 2019.
11. Nielsen, I., Dang, Q.V., Bocewicz, G., Banaszak, Z., A methodology for implementation of mobile robot in adaptive manufacturing environments. *J. Intell. Manuf.*, 28, 1171–1188, 2017.
12. Guo, W., Devine, C., Wang, S., Performance analysis of micro unmanned airborne communication relays for cellular networks, in: *Proc. 9th IEEE Int. Symp. Commun. Syst., Netw. Digit. Sign (CSNDSP)*, Manchester, U.K, pp. 658–663, 2014.
13. Zhang, M., Su, C., Liu, Y., Hu, M., Zhu, Y., Unmanned aerial vehicle route planning in the presence of a threat environment based on a virtual globe platform. *ISPRS Int. J. Geo-Inf.*, 5, 1–30, 2016.
14. Cai, G., Dias, J., Seneviratne, L., A survey of small-scale unmanned aerial vehicles: Recent advances and future development trends. *Unmanned Syst.*, 2, 175–199, 2014.
15. Stansbury, R.S., Vyas, M.A., Wilson, T.A., A survey of UAS technologies for command, control, and communication (C3). *J. Intell. Robot. Syst.*, 54, 61-78, 2009.

16. Anwar, M.Z., Kaleem, Z., Jamalipour, A., Machine learning inspired sound-based amateur drone detection for public safety applications. *IEEE Trans. Veh. Technol.*, 68, 2526–2534, 2019.
17. Kaleem, Z., Hui, B., Chang, K., QoS priority-based dynamic frequency band allocation algorithm for load balancing and interference avoidance in 3GPP LTE HetNet. *EURASIP J. Wirel. Commun. Netw.*, 185, 1–18, 2009.
18. Chang, Hi, K., Kaleem, Z., Bing, H.U., II, System and method to mitigate interference of 3GPP LTE heterogeneous network according to priority of service. U.S. Patent 9,369,969, June 14, 2016.
19. Baldini, G., Karanasios, S., Allen, D., Vergari, F., Survey of wireless communication technologies for public safety. *IEEE Commun. Surv. Tutor.*, 16, 619–641, 2014.
20. Prajith, P., Yokesh, V., Aruchamy, P., Artificial Intelligence-Enabled Fully Echoed Q-Routing and Adaptive Directional Medium Access Control Protocol for Flying Ad-Hoc Networks. *Int. J. Commun. Syst.*, 38, 1–32, 2025.
21. Rakha, T. and Gorodetsky, A., Review of unmanned aerial system (UAS) applications in the built environment: Towards automated building inspection procedures using drones. *Autom. Constr.*, 93, 252–264, 2018.
22. Shakhatreh, H., Sawalmeh, A.H., Al-Fuqaha, A., Dou, Z., Almaita, E., Khalil, I., Othman, N.S., Khreishah, A., Guizani, M., Unmanned aerial vehicles (UAVs): A survey on civil applications and key research challenges. *IEEE Access*, 7, 48572–48634, 2019.
23. Mozaffari, M., Saad, W., Bennis, M., Nam, Y.H., Debbah, M., A tutorial on UAVs for wireless networks: Applications, challenges, and open problems. *IEEE Commun. Surv. Tutorials*, 21, 2334–2360, 2019.
24. Hoffer, N.V., Coopmans, C., Jensen, A.M., Chen, Y., A survey and categorization of small low-cost unmanned aerial vehicle system identification. *J. Intell. Rob. Syst.*, 74, 129–145, 2014.
25. Safdar, S., Zeb, A., Khan, A., Kaleem, Z., Android based vehicle tracking system. *EAI Endorsed Trans. Energy Web*, 5, 1–6, 2018.
26. Cao, X., Yang, P., Alzenad, M., Xi, X., Wu, D., Yanikomeroglu, H., Airborne communication networks: A survey. *IEEE J. Sel. Areas Commun.*, 36, 1907–1926, 2018.
27. Saeed, A.S., Younes, A.B., Islam, S., Dias, J., Seneviratne, L., Cai, G., A review on the platform design, dynamic modeling and control of hybrid UAVs, in: *2015 International Conference on Unmanned Aircraft Systems (ICUAS)*, Denver, CO, USA, pp. 806–815, 2015.
28. Chao, H., Cao, Y., Chen, Y., Autopilots for small unmanned aerial vehicles: A survey. *Int. J. Control Autom. Syst.*, 8, 36–44, 2010.
29. Adams, S.M. and Friedland, C.J., A survey of unmanned aerial vehicle (UAV) usage for imagery collection in disaster research and management, in: *Proc. 9th Int. Workshop Remote Sens. Disaster Response, vol. 8*, Stanford, CA, USA: Stanford Univ, pp. 1–8, 2011.

30. Whitehead, K. and Hugenholtz, C.H., Remote sensing of the environment with small unmanned aircraft systems (UASs), part 1: A review of progress and challenges. *J. Unmanned Veh. Syst.*, 2, 69–85, 2014.
31. Yanmaz, E., Yahyanejad, S., Rinner, B., Hellwagner, H., Bettstetter, C., Drone networks: Communications, coordination, and sensing. *Ad Hoc Netw.*, 68, 1–15, 2048.
32. Alzenad, M., El-Keyi, A., Yanikomeroglu, H., 3-D placement of an unmanned aerial vehicle base station for maximum coverage of users with different QoS requirements. *IEEE Wirel. Commun. Lett.*, 7, 38–41, 2017.
33. Shi, W., Li, J., Xu, W., Zhou, H., Zhang, N., Zhang, S., Shen, X., Multiple drone-cell deployment analyses and optimization in drone assisted radio access networks. *IEEE Access*, 6, 12518–12529, 2018.
34. He, X., Yu, W., Xu, H., Lin, J., Yang, X., Lu, C., Fu, X., Towards 3D deployment of UAV base stations in uneven terrain, in: *2018 27th international conference on computer communication and networks (ICCCN)*, IEEE, Hangzhou, China, pp. 1–9, 2018.
35. Chou, S.-F., Yu, Y.-J., Pang, A.-C., Lin, T.-A., Energy-aware 3D aerial small-cell deployment over next generation cellular networks, in: *2018 IEEE 87th Vehicular Technology Conference (VTC Spring)*, IEEE, pp. 1–5, 2018.
36. Košmerl, J. and Vilhar, A., Base stations placement optimization in wireless networks for emergency communications, in: *2014 IEEE international conference on communications workshops (ICC)*, IEEE, pp. 200–205, 2014.
37. Kalantari, E., Yanikomeroglu, H., Yongacoglu, A., On the number and 3D placement of drone base stations in wireless cellular networks, in: *Proc. IEEE 84th Veh. Technol. Conf. (VTC-Fall)*, Montreal, QC, Canada, Sep. 2016, pp. 1–6.
38. Bor-Yaliniz, R., II, El-Keyi, A., Yanikomeroglu, H., Efficient 3-D placement of an aerial base station in next generation cellular networks, in: *2016 IEEE international conference on communications (ICC)*, pp. 1–5, May 2016.
39. Mozaffari, M., Saad, W., Bennis, M., Debbah, M., Mobile unmanned aerial vehicles (UAVs) for energy-efficient Internet of Things communications. *IEEE Trans. Wirel. Commun.*, 16, 7574–7589, 2017.
40. Viana, J., Francisco, C., Americo, C., Rui, D., Pedro, S., MIMO relaying UAVs operating in public safety scenarios. *Drones*, 5, 1–12, 2021.
41. Bekkouche, O., Samdanis, K., Bagaa, M., Taleb, T., A service-based architecture for enabling UAV enhanced network services. *IEEE Netw.*, 34, 328–335, 2020.
42. Alsamhi, S.H., Shvetsov, A.V., Kumar, S., Shvetsova, S.V., Alhartomi, M.A., Hawbani, A., Rajput, N.S., Srivastava, S., Saif, A., Nyangaresi, V.O., UAV computing-assisted search and rescue mission framework for disaster and harsh environment mitigation. *Drones*, 6, 1–21, 2022.
43. Cheng, N., Xu, W., Shi, W., Zhou, Y., Lu, N., Zhou, H., Shen, X., Air-ground integrated mobile edge networks: Architecture, challenges, and opportunities. *IEEE Commun. Mag.*, 56, 26–32, 2018.

44. Alsamhi, S., Alexey, V., Santosh, K., Jahan, H., Computing in the sky: A survey on intelligent ubiquitous computing for UAV-assisted 6G networks and industry 4.0/5.0. *Drones*, 6, 1–29, 2022.
45. Yazid, Y., Imad, E., Antonio, G., Ahmed, M., UAV enabled mobile edge-computing for IoT based on AI: A comprehensive review. *Drones*, 5, 1–30, 2021.

10

Assessing the Impact of Drones on Students' Engagement and Learning Outcomes

M. Narendran[1]*, S. Surendiran[2], Alamgir Khan[3] and Mahender Singh[4]

[1]Department of CSE, Sona College of Technology, Salem, India
[2]Department of EEE, Bharathiyar Institute of Engineering for Women, Deviyakurichi, Salem, India
[3]Department of CSE, Xinjiang University School of Software, Urumqi, China
[4]Electronics and Communication, Guru Tegh Bahadur Institute of Technology, Delhi, India

Abstract

Drones can provide students with an entirely different, immersive learning experience. Drones stimulate students' interests and passions. Drones can capture the attention and interest of pupils through visuals and interactivity. Drone activities and projects are hands-on and this encourages students by empowering them. Drones can enable students to learn in a multi-disciplinary way. Science classes can utilize drones for experiments, data gathering, and environmental studies. Students who study geography and history can take drone-assisted virtual field trips to remote locations. Drones can enhance students' learning by creating a realistic context that blends theoretical concepts with practical application, thereby improving academic performance.

Perceiving the use of drones as a new phenomenon, this chapter aims to evaluate what impact drones have on students' engagement and learning outcomes. The chapter's drone-related goals are structured in what is referred to as the rationale portion of the chapter. It analyzes the literature on the use of drones and their impacts in education. An overview of the current research methodology, sampling, and study on students' learning outcomes related to the use of drones will be presented. The chapter examines drone-enabled learning activities designed

**Corresponding author*: Naren0812@gmail.com

A. Prasanth, Rajesh Kumar Dhanaraj, Munish Sabharwal, Vandana Sharma and Seifedine Kadry (eds.) Drones for Transportation Logistics and Disaster Management, (221–240) © 2026 Scrivener Publishing LLC

and developed within the curriculum, and later, the results and their significance are analyzed. The conclusions are directed toward practicing teachers and educational policy analysts. Finally, it summarizes the scope of the current understanding of the influence of drones on students' engagement and learning outcomes.

***Keywords*:** Drones, student learning, drone-assisted learning, learning outcomes, visual interactions

10.1 Introduction

10.1.1 Background and Context of the Study

Drones are an educational resource which have gained prominence recently because they can actively involve students in capturing and using practical learning scenarios with an immersive experience. The excitement behind their aerial view capturing, hands-on activity engagement, and linking theory to practice is what makes further integrating drones into the educational environment worth exploring. This research [1] examines students' motivation associated with participating in an online question generation activity and their learning motivation. It attempts to study the relationship between students' engagement in developing questions through an online platform and their motivation towards learning. The objective of this research is to discover students' engagement and its outcomes on their interest, commitment, and inspiration towards learning.

Education has usually pursued change besides enhancing with the purpose of responding to students' needs in a rapidly converting global utilization. The implementation of drones inside the classroom nurtures a shift in the direction of the cutting-edge technique of teaching practices which contain active participation by means of students and learning through doing. It is significant to recognize the context in which drones are being blended into the classroom with the intention to absolutely understand the effect they will have on student learning outcomes.

10.1.2 Significance of Exploring Drone-Assisted Learning

The significance of studying drone-assisted learning is its capability to alternate the way students learn. Drones can stimulate interest, imagination, as well as logical reasoning among learners at the beginning stage. The learners have the capacity to alternate the way of learning distinct subjects are taught *via* presenting a new view and real-existence applications that may assist barriers beyond the bounds of traditional teaching.

The influence of drones on student engagement and academic outcomes necessitates the similar studies to make predictions on the future of learning. Expertise on the impact of drone devices on students' academic performance, motivation to learn, and experience can provide perspective on the way to make use of drones within the educational system. This examination goes beyond merely accepting the newest technological innovations. It's about using tools that match with the needs of today's digital learners to improve the educational experience. Figure 10.1 demonstrates the benefits of drone-assisted learning steps in various aspects. The authors from [2] want to determine how the dual employment of drones for the first and last mile with trucking affects fuel energy and its environmental cost. More precisely, it examines the potential for decreased energy consumption, emissions, and negative effect on the environment. The Research incorporates drones as additional steps to the traditional trucking approach. It seeks to evaluate the advantages and disadvantages concerning the conservation of energy and protection of the environment within logistics and transportation systems.

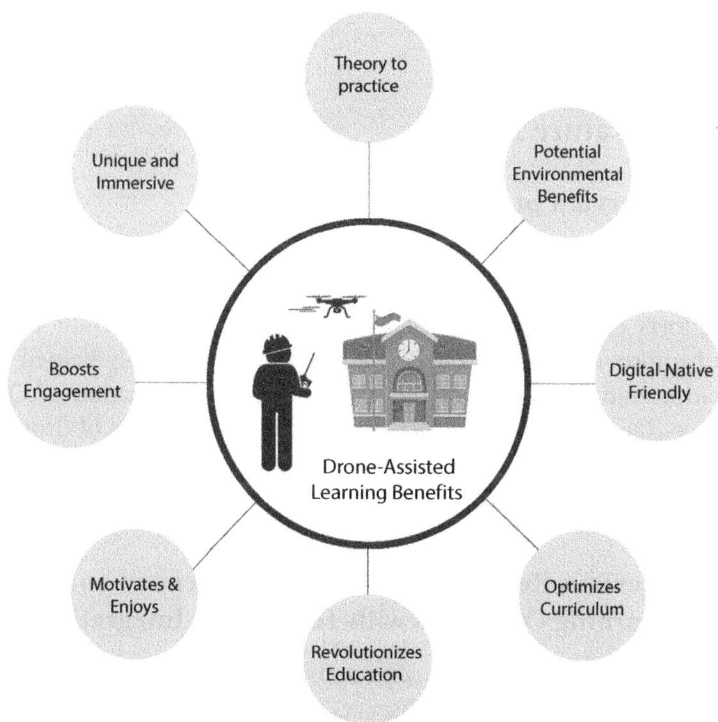

Figure 10.1 Drone-assisted learning benefits.

10.1.3 Research Objectives

This study aims to evaluate the degree of student engagement by measuring interest, motivation, and actual engagement levels before and after the implementation of drone-assisted learning tasks. Evaluation of academic achievement and knowledge gained from the pre- and post-assessments would determine the direct effect of the research on learning outcomes. Also important, is the assessment of the effectiveness of the pedagogical strategies used in drone-assisted learning, especially with regard to the goals they are intended to achieve and the curricula to be covered. The study also analyzes the perceptions of the learners through guided interviews and focus group discussions. In addition, the study intends to identify the limits and challenges of the use of drones in teaching including logistical and safety issues. This work is meant to illustrate the relevance of drone-assisted learning for a wide range of subjects by studying the cross-disciplinary relations it fosters. The article is intended to offer guidance to educators, educational authorities, and people involved in the design and implementation of curricula. The work aims to expand the information base in educational research by providing evidence on the impact of drones.

10.2 Literature Review

10.2.1 Overview of Drones in Education

In [3], the author's review and comments on the video and the research components helped refine and enrich the quality and scope of this work. Their efforts ensuring relevance and accuracy was important in refining and improving the work. Their effort and input highly helped in the elaboration and deepening of this study on the optimization of Quality of Service (QoS) and network performance in the interface of drones and the Internet of Public Safety Things in smart city settings. Their valuable comments are highly appreciated.

10.2.2 Previous Research on Student Engagement and Learning Outcomes with Technology Integration

In [4], the author analyzes the aspects of cognitive psychology, particularly focusing on how a person observes and identifies other people from a bird view perspective, including the drones. The work attempts to analyze cognitive actions that include optical recognition, recalling information, and paying attention to show the bounds and difficulties related to recognizing a person from a drone. Knowing these concepts is fundamental in understanding the identification of a person through the use of drones and the challenges that come with it.

10.2.3 Theoretical Frameworks Supporting the Use of Drones in Education

In [5], the scope of powerful, multi-functional Coleopter drones will most likely look into their architecture, working principles, and further possibilities in application. In attempting to understand these specific drones, the research may aid in understanding new engineering designs, the mechanics of these drones, and how these Coleopter models can be modified. This effort is probably aimed toward helping understand the designs of such drones which remodel in form and can be used for any reason, such as surveillance, transport, or aerial work.

10.3 Methodology

10.3.1 Research Design and Approach

The research plan to decide the impact of drones on the academic engagement and consequences of students is designed to use a blended-approach method. This method integrates both qualitative and quantitative strategies for data evaluation and interpretation. The quantitative method helps measurement and statistical scrutiny of the effect, at the same time as the qualitative approach captures the experiential size. The numerous researches that make use of drones in those fields for the functions of monitoring, data collection, and surveillance simply display their usefulness in wildlife safety, habitat monitoring, and conservation. This research appears into the developments, issues, and possibilities of using drones as essential tools for conservation projects that inform a big tale to ecologists, conservationists, and government decision makers [6].

The study makes use of a group quasi-experimental design to measure the variations in effects of students taking part in drone-assisted getting to know activities and traditional teaching methodologies. Similarly, the research will try and integrate elements of action research, so one can create an interdisciplinary partnership between instructors and researchers to enhance pedagogical strategies using drones for education.

10.3.2 Participant Selection and Sampling Strategy

The participants will consist of students across diverse age groups and educational levels. A purposive sampling strategy will be employed to select schools or educational institutions willing to participate in the Research. The sample will encompass students from varying academic backgrounds to ensure a comprehensive representation of the impact of drones on learning outcomes.

The investigation [7] focused on aspects like how to avoid an obstacle, navigate through dynamic environments, optimizing an activity's efficiency, and any decision-making techniques related to planning a route. Solving these problems, the research makes, probably, new claims and marked out the possibilities to alter autonomous drone capabilities. This document stands as a reference that is important for researchers, engineers, and practitioners working in the field of autonomous drones because it provides innovative concepts that shape the development of algorithms on problem and route determination along with the navigation of drones systems.

In order to randomize participant assignment, they will be assigned either experimental (drone-assisted learning) or control (traditional methods) groups separated by the teaching method to preserve the integrity of the Research. Selection of participants will also take into account academic performance, age, and drone technology experience exposure demographically.

10.3.3 Data Collection Methods

1. Pre- and Post-Assessments
Before the introduction of the drone-integration into lessons, pre-assessment tests will measure the level of knowledge as well as identify the level of engagement a student has. Post-assessments conducted after the intervention will evaluate the learning progress/achievement alongside critical and problem-solving thinking skills.

2. Surveys and Questionnaires
Both experimental and controlled groups will receive surveys and questionnaires. With these tools, information on the students' views, experiences, and their levels of participation in the learning process involving drones will be collected for statistical analysis.

3. Interviews
An in-depth interview approach will be employed among the sampled students, educators, and administrators. These qualitative interviews will

focus on alienated aspects of the participants' experiences, including challenges and benefits associated with drone-assisted learning.

4. Focus Group Discussions

Focus groups will serve as a forum for students to discuss their views and recommendations regarding drone-assisted learning. They will provide significant qualitative information about group interactions and their collective understanding of the learning with drones.

10.3.4 Data Analysis Plan

Assessment tests, surveys, and questionnaires will provide quantitative information that will be analyzed statistically with appropriate software such as SPSS or R. Descriptive and inferential statistical analyses will be performed to evaluate the before and after intervention outcomes of the experimental and control groups.

The qualitative data collected from the interviews and focus group discussions will be subject to qualitative analysis based on predetermined and emerging themes. Engagement, learning experience, challenge, and perception toward drone-assisted learning will be the main focus of the analysis.

Combining quantitative and qualitative facts gives a holistic assessment of the effect of drones in learning and engagement, thus examining their performance in education from distinctive angles.

10.4 Implementation of Drone-Assisted Learning Activities

10.4.1 Design of Curriculum-Aligned Drone Activities

To integrate drones within an educational framework, relevant and engaging courses must be developed. The courses must be crafted considering certain learning outcomes and the distinctive features that drones offer. For example:

Thematic Science Projects: Designing research projects that use drones to capture images, scout environments, or illustrate concepts such as motion, force, or ecosystems.

Geographic and Historical Study: Students can "travel" to previously unreachable historical or geographic locations *via* drone imagery. These videos can supplement lessons and allow for more enhanced learning opportunities.

Mathematics and Technical Problems: Students can be challenged to solve geometric issues related to the angle of the drone's flying, its horizontal distance from a certain waypoint, and the weight of the cargo it is carrying.

Creative Writing and Drones: Drones can be captured conducting creative writing or storytelling projects that require students to narrate stories or describe settings in which aerial footage can be incorporated, making students use their descriptive skills and imagination.

Every task will correspond with certain learning outputs and make certain that drones add to the level of comprehension and engagement while fulfilling educational goals from a variety of subjects. The research work [8] is probably affecting the implications of the rapid actuation of control surfaces on the aerodynamic performance of drones. This research is anticipated to advance the optimization of control mechanisms with their effects on drone flight, which may result in innovations and aerodynamics with the control systems.

10.4.2 Integration Across Different Subjects and Disciplines

Drones can be effortlessly incorporated into virtually any field, which makes them particularly useful in education. Drone-based learning will be coordinated with other subject teachers. Here are a few examples:

Science classes: Drones can be used in the collection and analysis of data for biology, physics, chemistry and even environmental science.

Geography and History: Drones can be used for geographical mapping, historical documentation, topographical understanding, and landscape appreciation.

Mathematics: Drones may be used in geometry related training sessions like visualization of shapes, area computation, and spatial relationships.

Arts and Humanities: Drones can be integrated into creative expression through using them in art images, and cinematography.

The aim of the integrated technique is to expose that the usage of drones can serve many purposes in many fields of research and to provide a significant interdisciplinary learning experience.

10.4.3 Logistical Considerations and Safety Measures

Integrating drone-assisted learning activities-based strategies that implementation must be considered logistical and protection measures. Schools and teachers ought to comply with safety regulations and local regulations on drones.

Training and Certification: An educator can make utilization of the drones for teaching, must completely require training and acquire the certification to make sure the drone is operated accurately and efficiently.

Risk Assessment: The complete thorough risk assessments before any drone activity to decide if safe flight operations may be performed. This consists of thinking about weather, airspace regulations, and hazards.

Student Briefing and Supervision: Providing a brief explanation of the relevant safety strategies and the rules guiding the usage of drones. It is essential that the teacher supervises every drone activity to ensure safety measures are adhered to.

Equipment Maintenance: The drone's mechanical components need to be checked often to assure they function properly and do no longer cause accidents due to mechanical failure.

Teachers can make certain safety measures for the students along with the team of workers simultaneously for maximizing the educational benefits. Drone-assisted learning activities will aid in effectively addressing safety logistics, protocols, and instructions.

10.5 Results and Analysis

10.5.1 Quantitative Analysis of Pre- and Post-Assessment Data

Measuring the effects that drone-assisted learning activities have on students' academic performance and learning outcomes is the goal of the quantitative analysis of pre- and post- assessment data. Statistical methods such as paired t-tests or ANOVA will be employed to compare the results between the experimental (drone-integrated) and control groups.

Control Group: This group typically represents the standard or existing condition against which changes are compared. In a Research, the control group usually needs to receive the experimental treatment or intervention being tested. They might follow the conventional teaching methods or standard curriculum without the inclusion of drone-assisted learning activities.

Experimental Group: This group undergoes experimental treatment or intervention—in this case, drone-assisted learning activities. The experimental group experiences the new teaching method or intervention being tested, which, in this scenario, involves integrating drones into the learning process to enhance engagement or academic performance.

In Table 10.1, each row with "Control" or "Experimental" in the "Group" column corresponds to individual students' data within their respective groups. The data collected for each student includes pre-assessment and post-assessment scores, participation rates, task completion percentages, and interaction levels, allowing for a comparison between the control group (without drone-assisted learning) and the experimental group (with drone-assisted learning) regarding academic performance and engagement metrics.

Assessment Scores: Analysis will focus on changes in assessment scores pre and post-drone-assisted learning interventions. This comparison will ascertain whether there is a statistically significant difference in academic achievement between the two groups. This table contains fictitious data for the "Control" group and the "Experimental" group. The dataset includes columns for each group's student ID, pre-assessment scores, post-assessment scores, participation rates, task completion percentages, and interaction levels during drone activities.

Engagement Metrics: To ascertain the degree of interest and involvement drone-assisted learning activities generated, participation rates, task completion, and level of interaction were analyzed in relation to student engagement.

Figure 10.2 depicts how drone-assisted learning activities influenced students' academic performance or engagement metrics qualitatively and quantitatively over time, or through different cycles of the learning sessions. It measures the shifts in scoring assessments or engagement levels before and after the drone-integrated learning approach was utilized. Each point on the line represents a particular engagement metric or an assessment score at a specific stage of the drone-assisted learning process, which

Table 10.1 Comparison of assessment scores and engagement metrics in control and experimental groups.

Labels	Student ID	Pre-assessment score	Post-assessment score	Participation rate (%)	Task completion (%)	Interaction level (1–10)
Control	1	78	82	75	85	6
Control	2	85	88	80	90	7
Control	3	72	75	70	75	5
Control	4	90	92	85	95	8
Control	5	79	81	72	80	6
Experimental	6	76	84	78	88	7
Experimental	7	82	87	75	85	6
Experimental	8	70	76	68	78	5
Experimental	9	88	91	80	90	8
Experimental	10	77	83	74	82	7

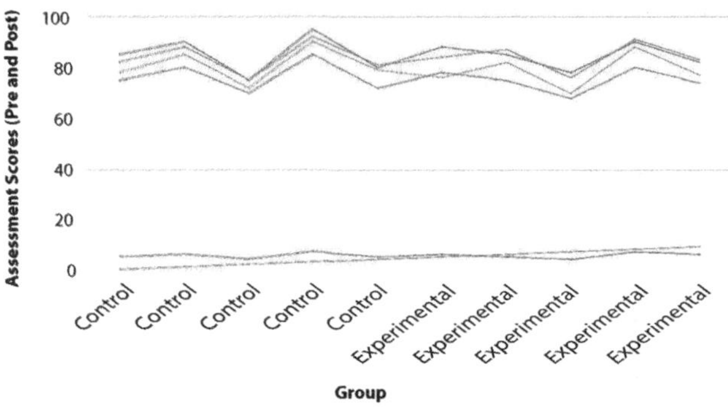

Figure 10.2 Drone-assisted learning impact: academic performance and engagement over time.

shows the trend or progress noted throughout the Research duration. The chart helps visualize how the use of drones changes students' academic performance or engagement during the learning activities.

The work in the reference [9] carried out a wind tunnel study with an emphasis on the factors that might contribute to spray drift in the application of pesticides using a UAV. The study emphasizes in particular wind velocity, type of nozzle, spray pressure, speed of travel of the drone, and other related factors which influence the UAV's pesticide spray pattern. This study intends to evaluate the mechanisms of spray drift utilizing a wind tunnel so that UAV pesticide application methods can be refined. Maximizing accuracy and UAV aerial application efficiency while reducing negative environmental effects and unintended herbicide spray from the UAV is the desired outcome. The statistical analysis will verify the hypothesis of the effectiveness of drone usage in educational settings and its influence on the student's achievement.

10.5.2 Examination of Survey and Questionnaire Responses

Analysis of survey and questionnaire responses will offer insights into students' perceptions, experiences, and attitudes toward drone-assisted learning. Quantitative data from these instruments will be analyzed using statistical software to derive meaningful patterns and trends.

Descriptive Analysis: Frequency distributions, means, and standard deviations will be calculated to understand students' general sentiments and opinions regarding their engagement and learning experiences with drones.

Correlation Analysis: Exploring correlations between different survey items or questionnaire responses to identify relationships between drone-based activities and aspects such as motivation, comprehension, or interest in specific subjects.

The authors in [10] explored the effects of action civics service-learning on the civic outcomes of eighth-grade students. The research investigates how engaging in action-oriented civic education and service-learning experiences influences civic development among students at this grade level. The focus likely includes examining changes in civic knowledge, skills, attitudes, and behaviors resulting from participating in hands-on projects that address community issues. This analysis will provide quantitative evidence of students' perspectives and preferences regarding drone-assisted learning, complementing the assessment of their academic performance.

10.5.3 Thematic Analysis of Interview and Focus Group Data

The set of qualitative data obtained from the interviews and group discussions will be thematically coded to provide details on the participants' and the educators' interactions with the drone-assisted learning system within the classroom environment.

Cognitive and Categorization: The transcripts from the interviews and discussion will be analyzed step by step to extract all the themes, perceptions, issues, and advantages related to the drone-assisted education activities.
Conceptual Pattern Matching: The pattern matching within the qualitative data will be done in an iterative manner to discern the overarching sentiments, worries, or positive viewpoints that were solicited from the participants.
Conceptual Data Interpretation: Making sense of the qualitative data relative to relevant educational concepts, teaching methods, and learner activities to ascribe meaning and significance around what was analyzed for possible practical use in the future.

Thematic coding analysis will achieve qualitative richness to be integrated within the quantitative outcomes, understanding of the effects resonates with drone assisted learning in terms of engagement, perception, and educational achievement of the students from different schools.

10.6 Discussion

10.6.1 Interpretation of Findings in Relation to Research Objectives

The discussion section starts with the analysis of the Research findings focusing on the previously set research objectives. This mapping analysis describes in detail the extent to which the outcomes respond to the initial expectations of the research, which in this case is the influence of drones on learners' active participation and achievement.

> **Objective 1:** This will be evaluated using the survey results, test scores, and the qualitative comments for the students' participation in various subjects before and after the use of the drones.
> **Objective 2:** Analyzing the comparative data of the two assessments will elucidate the impact drone-assisted learning activities have on the students' academic performance and overall educational achievements.
> **Objective 3:** In order to determine how drones might be meaningfully incorporated into the curriculum, conversations about the planning and execution of drone projects associated with the curriculum will be held.
> **Objective 4:** Overcoming obstacles and constraints found doing the Research will improve the next iterations of drone-based educational strategy implementations.

The Research [11] aims at finding out how teachers' knowledge of their professional and pedagogical disciplines impacts their use of cooperative learning strategies within the science curriculum. It investigates the implementation and outcomes of teachers using cooperative learning strategies, and tries to show the relationship between these teaching competences and students' performance and achievements in science.

10.6.2 Comparison with Existing Literature and Theoretical Frameworks

The focus of this section is to merge the Research's results with literature, theories, and frameworks within the field of educational technology, pedagogy, and student engagement. These research outcomes will be integrated with the existing theoretical models or previous studies on technology integration in education.

Integration of Literature Review: The outcome of the Research was analyzed in relation to established theories, in this case, constructivism or the theory of technology acceptance in relation to the educational processes.

Theoretical Framework Consistency: Analyzing the results for any connection to the theory provided, looking at if the outcomes that were identified matched what the theory said would happen or if there was a need to change some of the existing theories.

This Research sets out to understand the effects of teacher-scaffolding integration in a mobile-assisted collaborative learning environment on student learning outcomes, engagement, and attitudes in English reading classes. The research looks into the implementation of teacher-directed support strategies within collaborative frameworks enabled by mobile technology, and how these strategies affect students' academic performance, participation, and attitude towards learning [12].

10.6.3 Implications for Student Engagement and Learning Outcomes

This part will examine the more profound consequences of the Research's findings on student's engagement and learning outcomes from their drones' utilization. It will focus on analyzing the benefits and challenges of drones in education.

Increased Engagement: Describe how drones positively influence children's engagement through stimulating their curiosity, imagination, and application of theoretical concepts.

Better Learning Outcomes: Discussing the impacts of drones on the students' academic performance and the results in critical thinking, problem analysis, and comprehension of the subjects taught in class.

10.6.4 Addressing Limitations and Potential Challenges

Being mindful of the challenges and limitations of the Research helps in forming a complete understanding of the results. It aids in remaining objective.

Limitations: Discuss boundaries like sample size, duration of the Research and observable contextual limitations that would affect the outcomes of the study is of most challenge to focus on.

Challenges: Offer implementation or logistical challenges related to the usage of drones in educational practices and suggest how to deal with these challenges for later usages.

In conclusion, the components listed are integrated as an explanation to broaden the reasoning part of the research so as to explain the findings, their significance, and how they fit to previous studies and general theories and concepts in educational technology and didactics.

10.7 Recommendations

10.7.1 Practical Implications for Educators and Policymakers

This section will outline practical suggestions based on the researcher's results which will assist drones' users, educators, and policy makers in utilizing them in education for drone feature enhancement.

Professional Development: Actively motivates teachers and educators to enroll in courses aimed at drones as well as drone pedagogy and safety so that they can assimilate drones into their teaching practices.
Curriculum Development: Promoting these new instructional strategies should include the development of drone-assisted learning courses matched with appropriate grade and subject learning goals.
Resource Allocation: Recommend the establishment of funds to provide the necessary equipment and drone infrastructure so that their use can be implemented in classrooms.
Policy Considerations: Advise educational decision makers to review policies and programs while paying particular attention to safety issues, ethical policies, and pedagogical guidelines concerning the use of drones in teaching.

10.7.2 Guidance for Successful Integration of Drones in Education

The research work in [13] presumably analyzes how different sociological variables, features of the course design, and levels of participation in community service activities impact students' civic identity, skills, and self-development. The research has the overarching goal of analyzing the impact of service-learning on university students' civic engagement and development in Egypt.

Pedagogical Strategies: Focusing on creating curriculum-centered lessons that involve students in interdisciplinary learning through the distinctive aspects of drone technology.
Safety Protocols: Teaching the importance of following safety protocols and training both teachers and students on the responsible use of drones within the context of local regulations and guidelines.
Collaborative Approach: Encouraging collaboration among teachers, school leaders, and technology specialists to create engaging and innovative drone-centered instruction.
Evaluation and Assessment: Outlining the necessity for drone integration evaluation and assessment to evaluate how these tools affect learner participation and educational achievement.

In [14], the authors have conducted a comparative analysis focusing on students' engagement levels and academic achievements in live online learning at the university level. This research was designed to explore the effectiveness of active online learning by evaluating the engagement measurements and learning results of participants in active online classes. The study analyzes the degree of student participation, interaction, academic achievement, and success rate in courses taught through live online teaching.

"Combat engagement of drones in USA counter-terrorist operations" is studied in [15] which covers the use of drones in the Counter Terrorist Operations of the USA. It examines the roles, strategies, and implications of deploying drones as a crucial tool in combating terrorist activities, shedding light on their effectiveness and challenges in such operations. This research compares the efficacy of classical and quantum machine learning methods by investigating the identification of drone swarm activities. It explores the potential of quantum computing in enhancing the recognition and analysis of complex drone swarm behaviors compared to traditional machine-learning approaches [16].

In forestry education [17], the research examines integrating drone technology with GPS data collection to enrich interactive field experiences for forestry students. It highlights how this integration enhances hands-on learning and data collection methods in forestry education programs.

This research [18] concentrates on online instructional design education. It explores how incorporating authentic, real-world activities within online learning environments enhances practical knowledge and skills among students Researching instructional design in online education settings.

Focusing on using swarming drones in real-world scenarios, the authors in [19] introduce Drone Swarming for Real-world Missions. It aims to demonstrate the practical applications and potential advantages of swarming drone technology in various real-world missions and operations.

Investigating the role of theoretical risk models was projected in [20], which explores their potential to aid the comprehension and management of hazardous situations. It examines the applicability and effectiveness of theoretical risk models in enhancing the understanding and strategic handling of difficult circumstances.

10.7.3 Future Research Directions

In this study, the researchers outline the scope of the respondents' future research activities based on the clouds and opportunities identified during the current Research.

Long-term Impact Studies: Longitudinal studies are recommended to ascertain the enduring effects that drone-assisted learning has on students' academic achievements as well as their overall career preparedness.
Diverse Student Populations: This further suggests that research should be directed towards other demographic groups and their educational settings to examine how drones can be designed to cater to specific learning needs.
Advanced Educational Applications: Encouraging further research for specialized use of drones in STEM education, vocational education, and even special education is bound to advance the position of drones in education.
Ethical and Social Implications: Further, there is a need to examine drone use in education from an ethical, social, and cultural perspective, including issues of privacy and how society views these concepts.

This section aims to ensure that drones are successfully and meaningfully integrated into educational systems by assisting the relevant stakeholders in formulating implementation and policy strategies as well as additional research undertakings using these broad recommendations.

10.8 Conclusion

The scope of this study resulted in relevant information as it pertains to the use of drones and their effect on students' participation and their overall learning achievements. The results show that students from different academic subjects dramatically increased their participation as a result of learning with the aid of drones. The drone activities were so absorbing and practical that those drones aroused curiosity, fostered creativity and advanced students' participation.

The use of drones in education has no doubt great potential and is revolutionary. The findings highlight the immense value drones offer in education by adding new dimensions, encouraging learning by doing, and providing ways to relate abstracted ideas to actual phenomena. With the advancement of technology, drones have captured students' attention and can serve as an amazing resource that helps develop skills needed in the future. Drones allow for the creation of stimulating and interactive multi-disciplinary lessons which challenges the conventional approaches to teaching and offers the students a wider learning scope.

To summarize, this chapter exposed that drones positively influence students' level of participation, comprehension, and learner-centric practices. The ongoing use and implementation of drone technology in educational sectors may redefine learning paradigms while equipping learners with the needed skills for the digital age.

References

1. Sabri, M., Khalid, F.K., F, K., Li, L, K., Assessing Students Engagement in an Online Student Question-Generation Activity towards Their Learning Motivation. *Int. J. Lang. Lit. Ling.*, 2, 23–31, 2016.
2. Bányai, T., Impact of the Integration of First-Mile and Last-Mile Drone-Based Operations from Trucks on Energy Efficiency and the Environment. *Drones*, 6, 1–12, 2022.
3. Alsamhi, S., Ma, O., Ansari, M., Gupta, S., Collaboration of Drone and Internet of Public Safety Things in Smart Cities: An Overview of QoS and Network Performance Optimization. *Drones*, 3, 1–21, 2019.
4. Fysh, M. and Bindemann, M., Person Identification from Drones by Humans: Insights from Cognitive Psychology. *Drones*, 2, 1–19, 2018.
5. Barrett-Gonzalez, R., High Performance Convertible Coleopter Drones. *Drones*, 6, 1–25, 2022.
6. Jiménez López, J. and Mulero-Pázmány, M., Drones for Conservation in Protected Areas: Present and Future. *Drones*, 3, 1–18, 2019.
7. Gugan, G. and Haque, A., Path Planning for Autonomous Drones: Challenges and Future Directions. *Drones*, 7, 1–32, 2023.

8. Panta, A., Marino, M., Fisher, A., Mohamed, A., Exploring the Impact of Rapidly Actuated Control Surfaces on Drone Aerodynamics. *Drones*, 7, 1–19, 2023.
9. Grant, S., Perine, J., Abi-Akar, F., Lane, T., A Wind-Tunnel Assessment of Parameters That May Impact Spray Drift during UAV Pesticide Application. *Drones*, 6, 1–15, 2022.
10. Hart, S. and Wandeler, C., The Impact of Action Civics Service-Learning on Eighth-Grade Students' Civic Outcomes. *Int. J. Res. Serv.-Learn. Commun. Engage.*, 6, 1–30, 2018.
11. Geletu, G.M., The effects of teachers' professional and pedagogical competencies on implementing cooperative learning and enhancing students' learning engagement and outcomes in science: Practices and changes. *Cogent Educ.*, 9, 1–19, 2022.
12. Xu, Q., Sun, D., Zhan, Y., Embedding teacher scaffolding in a mobile technology supported collaborative learning environment in English reading class: students' learning outcomes, engagement, and attitudes. *Int. J. Mob. Learn. Organ.*, 17, 1–24, 2023.
13. Shalabi, N., Service-Learning in Egypt: Effects of Demographics, Course Features, and Community Engagement on Civic and Developmental Outcomes for University Students. *Comp. Int. Educ.*, 46, 1–16, 2017.
14. Kobicheva, A., Comparative Research on Students' Engagement and Academic Outcomes in Live Online Learning at University. *Educ. Sci.*, 12, 1–33, 2022.
15. Jovic, Z., Combat engagement of drones in USA counterterrorist operations. *Bezbednost Beograd*, 58, 171–190, 2016.
16. Barbeau, M., Recognizing Drone Swarm Activities: Classical versus Quantum Machine Learning. *Digitale Welt*, 3, 45–50, 2019.
17. Unger, D., Hung, I., K., Zhang, Y., Kulhavy, D., Integrating Drone Technology with GPS Data Collection to Enhance Forestry Students Interactive Hands-On Field Experiences. *Higher Educ. Stud.*, 8, 49–62, 2018.
18. Lowell V. L. and Moore, R.L., Developing Practical Knowledge and Skills of Online Instructional Design Students through Authentic Learning and Real-World Activities. *Tech. Trends*, 64, 581–590, 2020.
19. Sauter, J., DSOARS: Swarming Drones for Real-world Missions. *Scientia*, 5, 1–20, 2021.
20. Bui, P., Can Theoretical Models of Risk Facilitate the Management and Understanding of Dangerous Situations. *SSRN Electron. J.*, 5, 1–22, 2015.

11

Implementation of Delivery Drones in the Logistics Business Process

Jeevitha D.[1]*, S. Venkatesh[2], R. Sathya[3] and Raffaele Mascella[4]

[1]*Department of Computer Science and Engineering, Jeppiaar Engineering College, Chennai, India*
[2]*Department of Data Science and Business Systems, Faculty of Engineering and Technology, SRM Institute of Science and Technology, Chennai, India*
[3]*Department of Information Technology, Kongunadu College of Engineering and Technology, Trichy, India*
[4]*Faculty of Communication Sciences, University of Teramo, Teramo, Italy*

Abstract
The logistics industry is currently experiencing changes as it adopts delivery drones, also known as piloted aerial systems. In recent times these drones have emerged as a solution to improve and streamline logistics operations. In this chapter, we will delve into the integration of delivery drones into the business process of logistics, focusing on the advantages, challenges, and important considerations associated with this technology, by examining aspects such as technology, operations, and regulations. This chapter aims to provide an overview of how delivery drones can bring about a revolutionary transformation in the logistics industry.

***Keywords*:** Delivery drones, logistics industry, aerial systems, drone integration, regulatory challenges, operational efficiency

11.1 Introduction

In recent years the logistics industry has been actively exploring methods to improve their delivery operations and overall efficiency. One technology that has garnered attention is the utilization of delivery drones. These Unmanned

**Corresponding author*: jeevithadamotharanjpr@gmail.com

A. Prasanth, Rajesh Kumar Dhanaraj, Munish Sabharwal, Vandana Sharma and Seifedine Kadry (eds.) Drones for Transportation Logistics and Disaster Management, (241–258) © 2026 Scrivener Publishing LLC

Aerial Vehicles (UAVs) have the potential to bring about a transformation in the logistics sector by providing more economical and environmentally friendly delivery solutions. A study conducted by PwC reveals that integrating delivery drones into the logistics industry could result in cost savings, shorter delivery times, and opportunities for last-mile deliveries.

11.1.1 Logistics Industry Background

The logistics agency is an important part of exchange and trade around the world, responsible for transporting and controlling the movement of goods. This includes many services such as transportation, warehousing, inventory management, packaging, and distribution [1]. The origins of the logistics industry can be traced back to historical times when civilizations developed systems to move goods efficiently. For example, the historical Egyptians built problematic canals, roads, and canals to facilitate exchange along the Nile. Similarly, the Romans built major streets such as the famous "Appian Way" to add to their vast empire and to allow the passage of goods and naval supplies. The modern export industry flourished with the Industrial Revolution in the late 18th and 19th centuries. The advent of steam-powered machinery and mass production created a need for new green methods of transporting raw materials and finished goods.

Railways and steamships emerged as major modes of transportation, facilitating the transportation of goods over long distances. During the twentieth century, advances in transportation, and technology led to dramatic changes in the logistics industry. The advent of mechanized vehicles, especially passenger cars, has made transportation much faster and more convenient. The development of air transport and freight transport further expanded global options by reducing road congestion and simplifying the transportation of goods. With increased globalization, the logistics industry has become even more important. Companies now operate globally, sourcing from unique international sources and selling in a variety of markets.

This has resulted in complex distribution chains, with goods moving through more than one channel and through various intermediaries such as manufacturers, distributors, and retail outlets. In addition, the virtual revolution has had a significant impact on the logistics industry. Widespread adoption of computer systems, communications, and statistical technology systems has enabled real-time monitoring, supply chain visibility, selection, and record-keeping. Technologies such as GPS, RFID, and cloud computing have practical business efficiencies in size, bank management, and customer support in the logistics process. Today, the logistics industry is rapidly catching up [2].

The expertise lies in sustainability, operations, and the integration of e-commerce and omnichannel retail requirements. Innovations such as autonomous motors, drone shipping, robotics, and synthetic intelligence are reshaping the logistics industry, aimed at creating faster, greener, and more cost-effective distribution chains. Overall, the management team occupies a key position in connecting suppliers, manufacturers, stores, and customers around the world. It ensures the continuity of clean resources and data, drives economic growth, and drives global change.

11.1.2 Emergence of Drones

Drone conveyors have emerged as a revolutionary technology in the logistics industry, offering efficient and fast deliveries. These UAVs are capable of changing the way goods are transported and delivered. The rapid growth of e-commerce has led to a demand for faster and easier delivery options. Delivery drones offer the potential to speed up the delivery process, reduce transit time, and increase customer satisfaction.

Governments around the world are working to implement laws and regulations that will enable the safe operation of drone deliveries. This regulation is intended to address climate control, privacy, and public safety issues. Organizations such as the Federal Aviation Administration in the United States are actively involved in developing these regulations. Drone delivery offers a variety of benefits, including reduced logistics costs, a reduced carbon footprint, and access to remote locations, but for collection, atomic ubiquity requires addressing challenges such as limited logistics, climate, and public acceptance.

11.2 Logistics Advantages of Delivery Drones

Drones can bypass traffic congestion and deliver packages directly to the destination, significantly reducing delivery time. Capable of flying at high autonomous speeds, drones can travel long distances in fractions of the time that traditional delivery methods would. The last miles of the logistics chain are often the most time-consuming and expensive [3]. Drones excel in last-mile deliveries, reaching local areas efficiently and directly to customers' doorsteps. This reduces reliance on traditional delivery vehicles and improves overall efficiency.

Many drone applications include real-time tracking systems, allowing customers to monitor the progress of their deliveries. This transparency increases customer satisfaction and allows for better planning and control of the distribution process. Drone technology continues to evolve, with steady improvements in battery life, flight paths, and obstacle avoidance systems. These upgrades improve drone performance, safety, and reliability. While the deployment of drones depends on regulations and weather controls, the industry is working closely with aviation authorities to establish guidelines for safe and responsible drone operations.

Drone technology provides significant cost reductions for businesses across industries. Drones automate tasks that would otherwise require manpower, reducing labor costs associated with manual tasks such as inspection, inspection, and delivery. By replacing human workers with drones, companies can save on wages, profits, and training costs. Drones excel at last-mile deliveries, eliminating the need for expensive ground transportation services. You can also save on the fuel, maintenance, and labor costs associated with traditional delivery vehicles. In addition, drones can reach remote or harsh environments without the need for special infrastructure, significantly reducing transportation costs.

Drones can improve safety measures by performing hazardous tasks. For example, using drones to monitor hazardous environments or critical infrastructure reduces the need for human personnel to engage in potential hazards. This not only reduces the risk of an accident but also reduces liability and insurance costs. The adoption of drone technologies is in line with sustainability goals because, compared to conventional vehicles, machines have passive control for low carbon emissions. By incorporating environmentally friendly practices, firms can reduce their environmental impact and achieve cost-related benefits through incentives or reduced taxes. It's important to note that while drones offer the opportunity for cost savings, there may be an initial investment in drone hardware, software, training, and compliance, but with the long-term benefit of operating costs, the discount typically exceeds this original cost. The workings of drones are represented in Figure 11.1.

11.3 Challenges and Considerations

11.3.1 Technical Limitations and Reliability

While drone technology has advanced significantly in recent years, there are still some technical limitations and reliability considerations to be aware of.

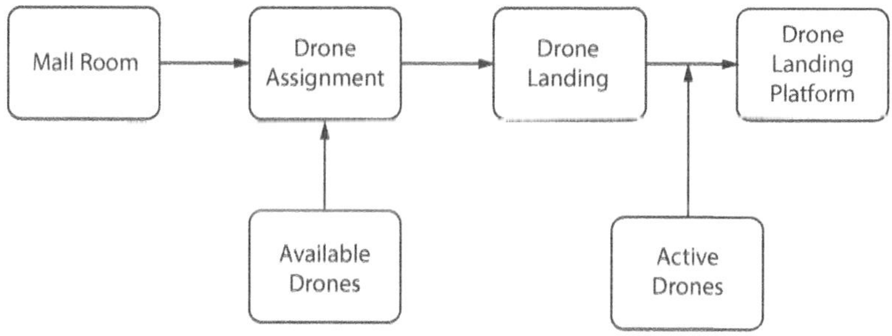

Figure 11.1 Working of drones.

Runtime is usually less thanks to battery life. Most drones have a flight range of 15 to 30 minutes, while commercial drones can be longer but still constrained. This restriction depends on where they are used and for how long they are used, requiring careful planning and monitoring of drone operations [4]. Adverse weather conditions, such as strong winds, heavy rain, or poor visibility, can significantly affect drone operations. High winds can make it difficult for drones to maintain stability and maneuver safely. Rain or snow can block sensors and compromise the reliability of data collection. Pilots must consider weather conditions and take precautions to ensure safe and reliable flights.

Drone communications rely on wireless signals, which can be susceptible to interference. In crowded urban areas or areas with high levels of electronic devices, interfering signals may deteriorate or weaken the drone's communication with the remote controller or ground station. This may result in a loss of control or degradation of data communication and affect reliability and security. Like any industrial device, drones do not respond to system malfunctions.

Technical failures, software errors, or sensor errors may occur, resulting in corruption or incorrect data collection. Routine maintenance, thorough preflight inspections, and contingency planning help reduce the risk of system failure. Despite these limitations and reliability considerations, the drone industry continues to innovate and overcome these challenges. Ongoing research and development is aimed at increasing flight time, payloads, navigation systems, and overall reliability [5]. Prioritize regulatory compliance, investment in quality equipment, proper training, and maintenance to ensure safe and reliable drone operations.

11.3.2 Weather Regulations and Safety Issues

Weather regulations and safety concerns are important aspects of drone technology to ensure safe operation and reduce potential hazards. Most countries have specific regulations for the use of drones to maintain weather safety. These rules specify flight altitude, restrictions on restricted areas (such as airports or critical areas), drone registration, pilot licensing, and due requirements if local aviation authorities comply with them. These rules are essential to ensuring the safe and legal operation of drones. Drones are equipped with sensors, cameras, and other collision avoidance systems to detect and avoid obstacles in flight. These systems help prevent mid-air collisions, especially in narrow spaces. Pilots must understand the capabilities and limitations of these systems and operate the drone within their assigned limitations.

A comprehensive risk assessment should be conducted before operating a drone. This includes considering factors such as weather conditions, nearby obstacles, and potential risks to people, property, and other aircraft. Mitigation measures such as maintaining safe distances between people and buildings, pre-evacuation surveys, and post-evacuation contingency plans help reduce risks. Drones equipped with cameras or sensors raise privacy and data security concerns. Practitioners must adhere to privacy laws and respect individual rights [6]. Appropriate data storage, encryption, and deletion protocols must be established to protect the data collected and ensure compliance with privacy laws. These proposals aim to ensure that drones are safely integrated into airborne systems to minimize risks to people, property, and other aircraft.

11.3.3 Environmental Factors

Weather and environmental factors play an important role in drone technology as they can affect the overall safety, efficiency, and effectiveness of drone operations. Strong winds can also affect the stability and maneuverability of drones. High winds can cause drones to veer off course, struggle to maintain altitude, or even crash. Pilots should check weather reports and ensure wind conditions are within safe operating limits specified by the drone manufacturer. Excessive temperatures can affect battery performance, flight time, and overall drone operation. Cold temperatures can drain batteries over time and cause fluctuating power, while excess heat can remain, causing overheating and possibly system failure [7].

Pilots should consider the temperature limits specified by the drone manufacturer and take appropriate precautions. Poor visibility due to darkness, fog, or low light can make it difficult to control and hold a visible drone. Reduced visibility increases the risk of collisions with obstacles and other aircraft. Pilots should prioritize safety and avoid flying in low visibility. Environmental factors such as metallurgy or electromagnetic fields can interfere with drone compasses and navigation systems. Pilots should exercise caution when operating drones near power lines, radio towers, or strong magnetic fields to reduce the risk of navigation errors. Sky understands and manages these conditions and environments to enable safe drone operations. Drone pilots must regularly check weather forecasts, be aware of local conditions, and take precautions to ensure the safety of people, property, aircraft, and drones.

11.4 Integration Strategies

11.4.1 Last-Mile Deliveries

Last-mile delivery refers to the last step in the delivery process, where goods are delivered from a delivery center or local business to the end customer's location. Drones can bypass road traffic and fly directly to the delivery point, which allows for faster and more efficient delivery. This is particularly useful in congested urban areas or situations where rapid delivery of supplies is necessary, such as medical supplies or emergency response. Drones can reach remote or difficult-to-reach areas that can be difficult for traditional delivery vehicles, such as rural areas, islands, mountains, or obstacles. When surface propulsion is used, the use of a UAV can offer logistics, allowing customers to move up.

Drone technology enables real-time delivery tracking and monitoring. Customers can receive updates on the status of their packages, including estimated delivery times and live demand. This increases transparency and provides greater convenience and peace of mind. With advanced navigation systems, obstacle avoidance technology, and geo-fencing capabilities, drones can operate safely and avoid collisions with aircraft or other objects. Furthermore, encrypted communication channels and onboard sensors are for delivery protection, stealth, or protection from interruption [8]. While drone technology enables last-mile transportation, there are still challenges to overcome. These include regulatory framework, climate control, public acceptance, and resource considerations. Collaborative efforts from industry stakeholders, policymakers, and the public are needed to responsibly integrate drones into the distribution ecosystem.

11.4.2 Warehouse-to-Warehouse Transport

Warehouse-to-warehouse transport refers to the movement of goods between warehouses or distribution centers. Drone technology has the potential to transform this segment of logistics by providing an efficient and cost-effective solution. Drones can dramatically reduce the time it takes to move goods from warehouse to warehouse. They fly directly between warehouses in order to bypass congestion and other obstacles. This allows for faster shipments, allowing for faster and more efficient deliveries.

Drones offer flexibility and switch from warehouse to warehouse transport. It can be deployed as needed, allowing for quick response and flexibility to changing requirements. Drones can handle a variety of sizes, from small to large, making them suitable for a wide variety of transport applications [9]. Drones excel in remote or inaccessible areas where traditional modes of transportation can face challenges. This is especially useful for warehouses located in remote areas, islands, or areas with difficult terrain. Drones can establish reliable communications between such locations, ensuring timely and efficient transportation.

Transporting drones from warehouse to warehouse requires compliance with local aviation regulations. Companies must comply with airspace restrictions, obtain necessary permits, and ensure safety measures are in place. Cooperation between drone operators, carriers, and regulators is essential to establishing a safe and effective regulatory framework. While drone technology has great potential for warehouse-to-warehouse transportation, there are still challenges to overcome. These include regulatory restrictions, technical restrictions, payload capacity, and flight planning. Continued improvements in drone technology and collaboration between stakeholders in warehouse logistics will be paramount.

11.4.3 Remote Area Accessibility

Drone technology has revolutionized access to remote areas, bringing several benefits to businesses. Drones provide infrastructure-free access to remote areas, eliminating the need for roads, bridges, or other transportation infrastructure. This is particularly valuable in rural or undeveloped areas where traditional transportation systems may be limited or nonexistent. Drones play an important role in emergencies and disaster mitigation efforts. They are able to reach remote areas affected by natural disasters quickly, providing vital supplies, medical assistance, search services, and rescue services.

Drones allow for faster deployment, making emergency response teams more efficient and effective. Equipped with sensors and cameras, drones enable close environmental monitoring in remote locations. They can assess biodiversity, monitor wildlife populations, analyze forest fires, monitor deforestation, and assess the effects of climate change. Drones provide valuable data for conservation efforts and scientific research.

Drones can monitor infrastructure in remote locations, including power lines, pipelines, bridges, and dams [10]. They can analyze the condition of these buildings, identify potential issues, and facilitate preventive maintenance. Compared to traditional manual inspections, drone scanning improves safety and reduces costs. Drones help document and preserve remote cultural heritage sites. They support wildlife conservation efforts with drones that can take high-resolution aerial photographs, 3D scan, construct archaeological sites, historic monuments, and reconstructed aboriginal lands, view endangered species animals, and fight poaching in remote areas. They provide valuable information to conservationists and help protect biodiversity. The infrastructure delivered by drone technology has the potential to transform remote areas, bridging the gap between isolated communities and the resources, services, and support they need.

11.4.4 Emergency and Medical Supply Delivery

Emergency medical supplies using drone technology have emerged as a revolutionary solution, providing quick and effective assistance in critical situations. Drones enable rapid response in demanding situations by delivering essential supplies to the affected areas. They can easily get from congested traffic or challenging terrain to remote or disaster-stricken areas where traditional modes of transportation may face restrictions. Drones play a vital role in delivering life-saving medicines, such as medicines and blood products, vaccines, and diagnostic tools. In emergencies, the timely delivery of these supplies can be critical to saving lives and reducing the impact of health hazards. Drones increase access to remote areas, enabling them to deliver medicine to underserved communities. Especially in areas with limited infrastructure or complex geographical resources, drones provide a valuable tool to overcome infrastructure barriers, and access to essential health facilities is guaranteed.

Equipped with a remote control device, it facilitates remote consultations between healthcare providers and patients in remote areas. Vital signs, medical images, and other information can be transmitted, enabling real-time diagnosis and treatment guidelines even when physical access is limited. If drones are used along with emergency medical supplies, it can reduce transportation costs, especially in remote or hard-to-reach areas. Drones eliminate the need for expensive manufacturing and can operate automatically, reducing material costs associated with traditional modes of transportation.

In emergencies and public health crises, drones provide valuable surveillance and capabilities. They can monitor crowds, enforce quarantine measures, and collect data on the spread of diseases to aid in public health interventions [10]. The integration of drones into emergency medical supplies requires compliance with local aviation regulations and cooperation with relevant authorities. Establishing clear guidelines and policies ensures the safe and responsible use of drones in emergency and health care delivery. Using drones in emergency medicine can save lives, improve access to health care, enhance disaster response capabilities, and support flexible and effective emergency management systems.

11.5 Technical Components of Delivery Drones

The major technical components of drones are as follows:

11.5.1 Communication Systems

Communication systems play an important role in drone technology, enabling data transfer and management between drones and ground control stations. Drones generally use radio-frequency communications for remote control. The pilot or operator uses a hand transmitter that sends control signals to the drone's receiver. These signals control throttle, pitch, and roll airspeed. Drones equipped with built-in cameras typically use a video link to transmit live video captures to a dedicated ground station or projector. This gives the pilot or operator a first-person view of the drone, which helps navigate and capture the aerial footage. The key hardware and components of the drone is enumerated in Table 11.1.

Advanced drones may incorporate cellular modules to connect to the Internet. This allows for remote control, autonomous flight *via* cloud-based platforms, or data transfer to online services for mapping, surveying, or live-streaming purposes. Drones can use satellite communication systems, which have been used in remote or inaccessible areas where traditional communication systems are limited. Satellites can provide global context and enable communication with drones, even in areas without cellular or Wi-Fi connectivity. It is important to note that the specific

Table 11.1 Hardware and components.

S. no.	Hardware and components	Uses
1	Frame	The frame acts as the structural backbone of the drone, providing support and room for all other components.
2	Motors	Drones typically use brushless electric motors to provide input power. The number and configuration of devices depend on the design of the drone, such as quadcopter (four devices), hexacopter (six devices), or octocopter (eight devices).
3	Propellers	A rotating blade attached to the motor provides lift and elevation. Propellers come in a variety of sizes, shapes, and materials depending on the needs of the drone.
4	Flight Controller	The flight controller is the brain of the drone, responsible for stability, control, speed, and processing output from sensors and commands issued by the operator.
5	Electronic Speed Controllers	The ESCs individually control the speed of each motor based on the signals received from the flight controller. Flight controller commands are converted to specific motor RPM.
6	Batteries	The drones are powered by rechargeable lithium polymer (LiPo) batteries. The battery supplies the electricity needed for instrumentation, flight control, and other on-board systems.
7	Transmitter and Receiver	These features facilitate communication between the drone and the remote control unit. The transmitter sends commands to the drone, while the receiver receives and interprets these commands.
8	Telemetry Systems	The telemetry system collects real-time data and transmits it from the drone to the ground control station. This data includes flight parameters such as altitude, speed, battery status, and GPS coordinates.

communications protocol used in a drone depends on factors such as the purpose of the drone, remoteness requirements, legal restrictions, and manufacturer choice.

11.5.2 Sensors and Navigation

Sensors and navigation systems are essential components of drone technology, enabling autonomous flight, positioning, and obstacle avoidance. GPS receivers receive signals from a number of satellites to provide accurate positioning information. Drones use GPS data to determine their location, navigate to checkpoints, return home, and enable features such as geo-fencing.

Barometric pressure sensors measure changes in atmospheric pressure, allowing drones to calculate altitude. This data helps monitor a specific height above the ground and allows for precise height control. Optical flow sensors use a camera and image processing algorithms to track objects visible on the ground. By analyzing changes in these parameters, drones can calculate their speed, adjust airflow, and even continue running at low speeds even without a GPS signal. Drones typically use a compass sensor to measure the Earth's magnetic field and navigate. A compass provides vital navigational information, and navigation aids, and maintains a specific course.

Drones can be equipped with optical and thermal cameras to capture visible and infrared images. These cameras offer applications such as aerial photography, surveillance, search, and rescue operations, and agricultural monitoring. These sensors and navigation systems work together to give drones situational awareness, position precision, and the ability to maneuver in challenging environments.

11.6 Data Management and Analytics

Data Collection

Collecting telemetry data is an important part of drone technology, allowing real-time monitoring and analysis of various aspects of the aircraft. Telemetry systems collect information about the drone's flight parameters, including altitude, speed, heading, GPS coordinates, battery voltage, motor rpm, and behavior (roll, pitch, and yaw). These parameters provide important information about the performance of the drone, as operators are also capable of monitoring and controlling its aircraft in real-time.

The telemetry dataset contains information about the internal structure and components of the drone. This can include data on motor health, propeller rotational speed, communication link characteristics, GPS signal strength, and any error messages or system analysis. Analysis of this data helps to ensure that the drone is operating properly and enables proper maintenance and troubleshooting. Radiofrequency (RF) communication is typically used to wirelessly transmit telemetry data from the drone to the ground control station. Some drones can also use other wireless technologies, such as Wi-Fi or cellular networks, to transmit telemetry data remotely or for remote sensing.

Telemetry data can be sent to the drone's internal storage or transmitted in real-time to a ground control station, where it is recorded on tape for further analysis [11]. It can be performed to assess aircraft performance, detect trends, detect abnormalities, improve operational efficiency, and improve safety measures. Telemetry data collection plays a key role in ensuring safe and effective drone operations by providing real-time insights into drone operations and environmental conditions for operators.

Performance Monitoring

Performance monitoring is an important aspect of drone technology related to drone technology where various metrics need to be tracked and analyzed to assess drone efficiency, flight characteristics, and overall performance. Performance monitoring involves collecting flight metrics such as speed, altitude, distance traveled, flight duration, and battery limits are analyzed. These metrics provide insight into the drone's flight efficiency, power consumption, and endurance.

Drones are equipped with sensors that provide valuable performance monitoring data. Sensor data, including GPS readings, accelerometer data, gyroscope data, and magnetometer data, can be analyzed to verify drone positioning accuracy, performance, and stability during flight. Monitoring of environmental conditions is needed to check the drone's versatility [12]. Factors such as airspeed, temperature, humidity, and air pressure can affect aircraft characteristics and performance. Performance monitoring systems collect data on these environmental factors to assess their impact on the drone's performance.

Using renewable energy sources such as solar and wind energy to charge drone batteries can significantly reduce carbon footprints. By combining charging stations with renewable energy infrastructure, drones can be further equipped with cleaner and more sustainable components, further reducing reliance on fossil fuels, improving the quality of roads and logistics, and reducing the carbon footprint of effective fleet management systems.

By strategically deploying drones in industry and optimizing routing, users can reduce overall distance and associated emissions. Ongoing research and development efforts in drone technology focus on increasing energy efficiency, battery technology, and sustainability. These developments aim to significantly reduce the emissions of drones and encourage environmentally friendly operations. By implementing these techniques and embracing technological advances, the drone industry can play an important role in reducing carbon footprint emissions and promoting sustainable practices.

11.7 Environmental and Social Impact

11.7.1 Carbon Footprint Reduction

Reducing the carbon footprint of drone technology has become a priority as the industry recognizes the environmental impact of drone operations. One of the most important steps to reduce carbon emissions is to switch to non-electric drones. Unlike conventional gasoline-powered jets, electric drones produce no direct emissions during flight.

The use of lithium-ion batteries and other electrical systems reduces the carbon footprint associated with drone operations. Drones made of lightweight materials such as carbon fiber reduce energy consumption during flight. The smaller the drone, the less energy it needs to stay in the air, thus reducing carbon emissions. In addition, an air management system can improve aircraft performance and further reduce energy consumption. Effective fleet management systems can improve routing and logistics and reduce carbon emissions.

Ongoing research and development efforts in drone technology focus on increasing energy efficiency, battery technology, and sustainability [13]. These developments aim to significantly reduce the emissions of drones and encourage environmentally friendly operations. By implementing these techniques and embracing technological advances, the drone industry can play an important role in reducing carbon footprint emissions and promoting sustainable practices.

11.7.2 Employment Opportunities

The use of drone technology has opened up a wide range of career opportunities in various industries. Experienced cockpit pilots are in high demand. Companies and organizations need licensed and experienced pilots to operate drones for use in aerial photography, videography, mapping, surveying, and other purposes. In most cases, it is important to obtain a remote pilot certificate from the appropriate aviation authority.

While drones capture a lot of information, such as images, videos, and sensor readings. It needs data analysts and geographic information systems experts to process, analyze, interpret, and visualize the data [14]. Drones play an important role in gaining meaningful insights and generating usable information for various industries, such as agriculture, construction, environmental monitoring, etc. Most carry special payloads, such as thermal cameras, lidar scanners, multispectral sensors, gas detectors, etc., and are in demand by professionals equipped with sensors that can integrate and measure these useful features to improve efficiency.

Sensors are ensured to capture and process data accurately for specific applications, such as infrastructure monitoring, crop monitoring, search and rescue operations, etc. As drone technology evolves, the rules and regulations that govern its use safely and responsibly continue to improve.

Professionals with expertise in drone regulations and compliance play a vital role in ensuring that operating units are in compliance with local aviation regulations [15]. They may have worked with government agencies, industry associations, or consulting firms to provide guidance and assistance in the legal process.

11.7.3 Public Perception and Acceptance

Public opinion and acceptance of drone technology have evolved over the years, influenced by a variety of factors including media representation, legal frameworks, privacy concerns, and the actual use of drones [16]. In commercial settings, drones have gained wide acceptance. Industries, including real estate, agriculture, manufacturing, and media, have embraced drone technology to gather data, improve efficiency, and improve productivity.

Privacy is a major factor that has changed the public perception of drones. The use of drones equipped with cameras and sensors has raised concerns about potential invasions of privacy [17]. Generally, public acceptance of drones is based on clear guiding privacy rules, such as limits on surveillance powers and guidelines for data collection and storage. Communities will communicate and engage in discussions about drone operations, which have become increasingly important.

Open discussion, transparency, and addressing community concerns can help increase public acceptance. This participation may include public meetings, informational meetings, and workshops with local authorities to establish guidelines for drone use in specific areas where drones are used. Balancing the benefits with addressing safety, security, and privacy considerations is essential for widespread public acceptance.

11.8 Case Study

Amazon Prime Air

Amazon Prime Air, a delivery company, will use UAVs or drones to deliver packages to customers in 30 minutes or less. Amazon CEO Jeff Bezos first proposed the idea in December 2013. The case study focuses on the creation and launch of the Prime Air service, which intends to revolutionize logistics through faster and more efficient delivery of goods using drones. Amazon Prime Air uses sophisticated drone technology, such as specially designed sensor drones, with cameras and new navigation systems. While delivering supplies, these drones are designed to fly on their own and safely maneuver around obstacles.

A key focus of the case study is the challenges Amazon faces in obtaining necessary regulations and approvals from aviation authorities. The case study examines how well Amazon worked with law enforcement agencies to address safety concerns and develop guidelines for the use of commercial drones. The case study delves into the development of the distribution system required for Prime Air to operate at scale. This includes establishing assembly lines near customer locations, optimizing drone delivery routes, and creating safe and efficient landing zones for takeoff and parking. Amazon Prime Air aims to provide customers with fast and convenient delivery options. The case study discusses how drones can be used to enhance the overall customer experience, enabling faster delivery of small items to designated residential destinations.

The case studies also examine the potential uses of Prime Air beyond logistics, such as disaster relief, medical supplies to remote locations, and environmental monitoring and research support. It is important to note that the specifics and conclusions of the Amazon Prime Air Case Study may vary depending on where and when it was published. For an in-depth analysis of the issue, it is recommended to consult academic journals, research papers, or official publications related to the Amazon Prime Air project.

11.9 Conclusion

In summary, drone technology has the potential to fundamentally revolutionize logistics. In the logistics industry, drones have the potential to increase overall operational efficiency, reduce costs, and increase productivity. Drones can help with last-mile transportation due to their ability to operate automatically and deliver packages, especially in isolated or hard-to-reach areas. This can save delivery time and has become faster and more efficient, reducing the need for traditional overland methods. Drones with sophisticated sensors and imaging capabilities can help manage the supply chain and inventory. They can track inventory levels, inspect air warehouses, and identify potential problems or challenges in the logistics process, but there are hurdles to overcome before drones can be widely used in the export of goods. Establishing a legal framework is essential.

References

1. Wang, W.F., Lin, L.Y., Petros, A., Ioannou, L., Vlacic, W., Liu, X., Transportation 5.0 The DAO to Safe, Secure, and Sustainable Intelligent Transportation Systems. *IEEE Trans. Intell. Transp. Syst.*, 7, 10262–10278, 2023.
2. Li, L.J., Qin, S.R., Olaverri-Monreal, M.C., Prodan, R., Wang, W.F., Logistics 5.0: From Intelligent Networks to Sustainable Ecosystems. *IEEE Trans. Intell. Veh.*, 7, 3771–3774, 2023.
3. Ajakwe, A.S.O., Saviour, S.I.I., Kim, K.J.H., Kim, S.D.S., Lee, L.J.M., BANDA: A Novel Blockchain-Assisted Network for Drone Authentication, in: *Fourteenth International Conference on Ubiquitous and Future Networks (ICUFN)*, pp. 120–125, 2023.
4. Attenni G., Arrigoni V., Bartolini N., Maselli G., Drone-Based Delivery Systems: A Survey on Route Planning, in: *IEEE Access*, vol. 11, pp. 123476–123504, 2023.
5. Prabhakar, P., Yokesh, V., Aruchamy, P., Nanthakumar, S., Artificial Intelligence-Enabled Fully Echoed Q-Routing and Adaptive Directional Medium Access Control Protocol for Flying Ad-Hoc Networks. *Int. J. Commun. Syst.*, 38, 1–22, 2025.

6. Madani, M.B. and Ndiaye, N.M., Drone Delivery Systems for Logistics Operations: Identification of Regulatory Challenges, in: *IEEE International Conference on Engineering, Technology and Innovation (ICE/ITMC)*, pp. 1–7, 2023.
7. Elsaid, E.F., Sanchez, S.E.T., Li, C.Y., Khamis, K.A., Optimal Placement of Drone Delivery Stations and Demand Allocation using Bio-inspired Algorithms, in: *IEEE International Conference on Smart Mobility (SM)*, pp. 39–44, 2023.
8. Rao, N. and Sundaram S., Integrated Decision Control Approach for Cooperative Safety-Critical Payload Transport in a Cluttered Environment, in: *IEEE Transactions on Aerospace and Electronic Systems,* vol. 59, no. 6, pp. 8800–8811, Dec. 2023.
9. Liu B., Ni W., Liu R. P., Guo Y. J., Zhu H., Optimal Routing of Unmanned Aerial Vehicle for Joint Goods Delivery and in-Situ Sensing, in: *IEEE Transactions on Intelligent Transportation Systems,* vol. 24, no. 3, pp. 3594–3599, March 2023.
10. Park Y., Lee S., Sung I., Nielsen P., Moon I., Facility Location-Allocation Problem for Emergency Medical Service With Unmanned Aerial Vehicle, in: *IEEE Transactions on Intelligent Transportation Systems,* vol. 24, no. 2, pp. 1465–1479, Feb. 2023.
11. Podsędkowski, M., Konopiński, R., Lipian, M., Sound noise properties of variable pitch propeller for small UAV, in: *International Conference on Unmanned Aircraft Systems (ICUAS)*, pp. 1025–1029, 2023.
12. Arafat M. Y. and Moh S., JRCS: Joint Routing and Charging Strategy for Logistics Drones, in: *IEEE Internet of Things Journal,* vol. 9, no. 21, pp. 21751–21764, 1 Nov., 2022.
13. Lee, S., Hong, D., Kim, J., Baek, D., Chang, N., Congestion-Aware Multi-Drone Delivery Routing Framework. *IEEE Trans. Veh. Technol.*, 9, 9384–9396, 2022.
14. Michelberger, B., Mutschler, B., Reichert, M., Process-oriented Information Logistics: Aligning Enterprise Information with Business Processes, in: *IEEE 16th International Enterprise Distributed Object Computing Conference*, pp. 21–30, 2012.
15. Shmeleva, A.G., Ladynin, A.I., Smirnova, E.E., Ryabchik, T.A., Transport Logistics Management Information System, in: *IEEE Conference of Russian Young Researchers in Electrical and Electronic Engineering (EIConRus)*, pp. 1471–1473, 2019.
16. Pan, L., Analysis of Logistics Management Mode and Operation Process in The Era of Internet of Things, in: *International Conference on Education, Information Management and Service Science (EIMSS)*, 2021, 417-420.
17. Zhang Z., Ruan J., Liu X., Wang J., Wang P., Wang X., Logistics information systems model designing. *World Autom. Congr.* 2012, 1–4, Puerto Vallarta, Mexico, 2012.

12

Social and Ethical Issues of Using Drones in Logistics and Disaster Management

M. Nalini[1]*, N. Prashithaa[2], Rajesh Kumar Dhanaraj[3] and F.H.A. Shibly[4]

[1]Electronics and Instrumentation Engineering, Sri Sairam Engineering College, Chennai, India
[2]HCL Tech, Chennai, India
[3]Symbiosis Institute of Computer Studies and Research (SICSR), Symbiosis International (Deemed University), Pune, India
[4]Department of Information Technology, South Eastern University of Sri Lanka, Oluvil, Sri Lanka

Abstract
An unmanned flying aircraft without a pilot on board will be referred to as a drone. Drones can be operated remotely or equipped with onboard processors for a certain level of autonomy. The extent to which a drone can be deemed autonomous is a topic of discussion. Numerous drone types exist in a wide range of sizes and shapes, and each type of drone has unique ethical concerns. The premise that what a drone plays is an action, but that this action is an extension of what someone does, is the inspiration of moral issues surrounding drones. This study is a radical analysis of the ethical troubles raised by approximately five thematic classes: community, era, surroundings, fitness, and safety.

***Keywords*:** Ethics, social issues, disaster management, emergency management, geographic facts systems

12.1 Introduction

Unmanned aerial vehicles, or drones, have grown to be greater well-known in recent years as adaptable gadgets with a huge variety of viable uses of.

**Corresponding author*: nalini.ei@sairam.edu.in

A. Prasanth, Rajesh Kumar Dhanaraj, Munish Sabharwal, Vandana Sharma and Seifedine Kadry (eds.) Drones for Transportation Logistics and Disaster Management, (259–278) © 2026 Scrivener Publishing LLC

The fields of catastrophe help and logistics are two sectors wherein drones have had a large effect. Drones are beneficial in various fields for a wide range of functions, from helping disaster reaction and rehabilitation packages to streamlining delivery chains [1]. However, incorporating drones into emergency assist and logistics poses several hard social and ethical problems that require careful attention and sizable resolution.

12.1.1 An Overview of Drones for Disaster Relief and Logistics

Drones are self-sustaining or remotely managed aerial cars. They go by way of the use of the names unmanned aerial automobiles (UAVs) or unmanned aircraft structures (UAS). Because in their versions in size, talents, and usage, they may be flexible gear with a large number of packages. Drones can trade logistics by lowering fees associated with product transport and transportation and boosting productivity. They may be used for the methods related to delivery chain monitoring, inventory management, and final-mile transport, offering answers to some of the chronic problems facing the logistics industry.

For many important reasons, drones have been verified to be quite beneficial in disaster management. Modern drones ready with cameras and sensors have the capability to take actual-time photos and facts of locations devastated *via* failures. These facts may be used to plan comfort efforts, assess damage, and behavior search and rescue missions. They can get too risky or difficult-to-attain places and give emergency responders essential data. In the aftermath of herbal catastrophes like hurricanes, wildfires, and earthquakes, drones have proved essential in improving disaster reaction and restoration.

12.1.2 Importance of Addressing Social and Ethical Issues

Although there are many blessings of employing drones for logistics and disaster relief, there are also many moral and societal troubles to consider. These issues are crucial for us to ignore because they may have a massive effect on human lives, groups, and the environment [2].

Figure 12.1 emphasizes the increasing importance of social and moral awareness cultivation within the scope of drones. It acts as a compelling call, which urges practices responsible for the harmonious integration of technology in society. It is important to address these issues for many important reasons:

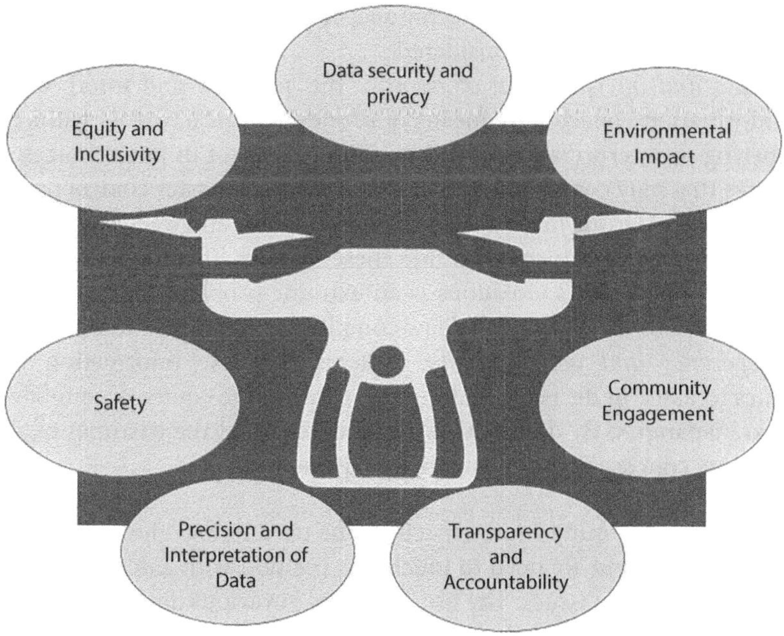

Figure 12.1 Social and ethical awareness: a call to action.

Data security and privacy: drones, especially with cameras, have the ability to record private and sensitive data. The affected population, privacy and data protection rights and dignity are essential in disaster-stricken places where people are already at risk.

Equity and inclusion: Drone technology cannot be widely available or used in all communities or places. It is necessary to guarantee the same access to drone-assessed services for a deprived or marginalized population to prevent the growth of inequalities already existing.

Environmental effects: Drone may have adverse effects on the ecosystem and may cause noise pollution between other environmental issues. To reduce their ecological impression, responsible drone operations should take into account these effects and take steps to reduce them.

Security: To avoid accidents and human injuries, it is important to guarantee safe functioning of drones. An important element of protection is prevention of injury, adherence to airspace restrictions, and provision of relevant training for users.

Community engagement: Including the affected groups in choice-making strategies at some stage in disaster management, not only makes sense morally but also enhances the efficiency of reaction operations. To assure a

humble and culturally touchy technique, cultural characteristics and community values need to be considered.
Accurate and interpretation of records: The accuracy and interpretation of information collected by means of drones is critical. Errors within the gathering or interpretation of information can result in poor choices and features that may compromise the effectiveness of disaster control projects as a whole [3]. Drone may be used for each military and tracking purposes, making their employment ethical in these regions. The possibility of misuse and human rights violations is an extreme problem that needs to be addressed with the aid of global law compliance and regulation.
Transparency and accountability: Systems that use information from drones to assist in decision-making should preserve equity, accountability and transparency. By doing this, it's guaranteed that the pastimes of all of the events concerned are substantially represented.

The drone has quite a few capacity to be used in logistics and catastrophe alleviation, but we need to touch this method with a deep know-how of moral and social issues. The use of drones in various domains can benefit the society, while all those issues preserve the rights, welfare and dignity if these issues are accepted and responsible, moral solutions are pursued. This book chapter will detect these subjects in detail, suggest the moral and social results of using drones for case studies, in-depth analysis, and logistics and disaster management.

12.2 The Role of Drones in Logistics and Disaster Management

Drone has become an invaluable tool in disaster relief and logistics, changing the operational processes of these areas. This chapter examines the important role that drones play in these domains, seeing their benefits, uses and current developments that are affecting their use.

12.2.1 Utilizations and Benefits

Logistics Using Drones
Last-Meal Shipping: The use of drones to distribute things quickly and effectively at the doors of customers is increasing. They are particularly useful to distribute high-primaries, small products quickly and economically.

Inventory management: Drones with sophisticated sensors and cameras are used to manage inventions in real time in warehouses. They track item locations, monitor stock levels, and improve the overall effectiveness of supply chain processes. Drone is an effective tool for monitoring and safety because they can be used to track cargo, guard warehouse flow, and are looking for any safety violations. They are necessary to maintain the security of the logistic network.
Environment Monitoring: Logistics business appoints drones to monitor the environment and make intelligent options about routes and transport modes. This reduces the effect that is on the logistics environment. Drone helps in traffic management by helping logistics companies, monitoring traffic and congestion on transit routes. This helps them to improve delivery routes and schedule, which is ultimately the result of low cost and delay.

Disaster Management Using Drones
Search and rescue: An essential component of search and rescue operations is a drone fitted with cameras and infrared imaging technologies. They can find the remaining people, enter dangerous or remote areas, and provide access to the rescue respondents up to real-time data.
Damage evaluation: Drone is used to evaluate degrees of damage to buildings, infrastructure and important features after disasters. For efficient destruction response and resource allocation, this information is necessary.
Communication infrastructure: Drone connection equipped with communication relay equipment can restore connectivity in disaster-affected areas where traditional infrastructure has been disrupted. They enable the impacted communities to coordinate relief efforts and maintain communication with rescuers.
Medical supply distribution: Drones make it simple to deliver medical supplies, including medications, immunizations, and first aid kits, to isolated or damaged areas. In an emergency, they offer a vital resource for medical assistance [13].

Drones are used for mapping applications and data collection [4]. They aid in decision-making, resource allocation, and damage assessment by creating maps of high-resolution disaster-affected areas. They offer crucial information needed for effective control of destruction.

Benefits of Using Drones
The drones' quick deployment makes them ideal for disaster response and time-sensitive logistics. When something happens and the conventional reaction team turns up, they behave like an event.

Cost-effectiveness: Drone utilization is cost-effective in a variety of scenarios. They lower the cost of infrastructure, transportation, and physical labor, which lowers the cost of logistics and disaster management.
Hard-to-wheel sites can be reached by drones: drones can approach areas that are inaccessible to people or conventional vehicles. Since access to hazardous or remote places is frequently required, it is particularly beneficial in disaster management.
Real-time data: By offering real-time data and imagery, drones help people make more informed decisions and improve their status awareness. This crisis is required for management as well as logistics.
Security: By offering an initial assessment of potentially dangerous conditions, drones can help disaster managers to reduce the risk to human respondents. They can help in monitoring security compliance and complete accidents in logistics.
Environmental aspects: Drone corresponds to stability objectives because they have lower carbon footprints than traditional automobiles, especially powered by electric engines.

12.2.2 Current Drone Usage Trends

Advanced payload: To collect data and increase its capacity for processing, the drone is being prepared with some time-upheld payloads, such as gas sensors, lidar and multispectral cameras.
Drone herd: Using multiple drones in a planned herd is becoming more and more popular. This method is collected and improves the effectiveness and access and rescue activities of discovery and rescue activities.
Urban air mobility: drone distribution and transportation in urban areas are being investigated; this can revolutionize the distribution of products and services in crowded cities.
Artificial intelligence and machine learning: Drone is becoming more functional and adaptive by integrating AI and machine learning algorithms for autonomous flight, image recognition and data processing. Governments and international organizations are trying to standardize and regulate drone operations to guarantee their responsible and safe use.
Public-commercial participation: This type of involvement is growing in popularity and combines the expertise of several stakeholders to enhance drone capabilities and deployment. These include collaboration between governmental organizations, private companies, and nonprofit organizations.

12.3 Privacy Concerns

Drones are currently very useful tools in many fields, including logistics and disaster relief. However, the growing usage of drones raises significant privacy concerns. This chapter discusses privacy concerns related to drone use, with a particular emphasis on data collecting and monitoring, as well as legal privacy safety precautions.

12.3.1 Observation and Data Collection

Illegal infiltration: Without permission, drones with cameras and other sensors can record and take pictures of people, houses, and other private property. The right to privacy is called into doubt by this unwelcome interference into people's private lives.
Data collection: The drone gets a lot of data, including images, geolocation information, and other private information. Data protection issues might surface if this information is gathered without the individuals' consent or knowledge. Because of their constant monitoring capabilities, the drones can continually observe people or resources. Secondly, the potential for inappropriate surveillance breaches individuals' privacy. Drones may covertly and covertly record chores or occurrences without the subject's awareness. It raises questions about the ability to misuse and secret nature of drone monitoring [5].
Data safety: There are security problems while storing and sending data that the drone has captured. Unauthorized access to drone data can result in misuse or privacy violation of sensitive information.

12.3.2 Legal Frameworks for Privacy Protection

Regulations by Federal Aviation Administration (FAA): Drone operations are subject to FAA rules in the United States. Some privacy issues are addressed by subtlety by these rules, including restrictions where drones can fly and how close they are for people.
Data safety law: Acceptance, storage and use of personal information is controlled by data security laws in many different countries. These rules often apply to how drone operators handle the data collected in flight [6].
Warrant requirements: Similar to legal requirements for traditional discovery and seizure operations, law enforcement agencies may be required to obtain a warrant before deploying a drone for monitoring.

Notice and consent: Under some circumstances, people will have to make them aware that the drone is being seen, and it may be necessary to allow them to collect data. For owners of both private and commercial drones, this is very important. Drones with geo-fencing technology are limited to their ability to fly in areas that have been declared as a no-fly zone, which may include personal properties. This aggressive drone reduces the possibility of espionage.

European Union (EU) General Data Protection Regulation (GDPR): Personal data security laws of GDPR apply to drones used to collect data within the European Union. Operators who want to protect user privacy, they have to follow GDPR.

Case law and examples: The way privacy rights are interpreted and the use of drones is regulated, can be affected by examples related to court decisions and secrecy cases related to drones.

Ethical guidelines: Manufacturers and Business Association for drones often provide moral guidelines for proper use of drones, some of which may include suggestions for Pro.

12.4 Challenges and Considerations

Finding the right balance between privacy and utility: It can be difficult to create a proper balance between privacy concerns and useful applications of drones in disaster relief and logistics. Laws should be prepared to protect privacy without disrupting creativity.

Border–border operations: Because drones can operate internationally, it is difficult to harmonize privacy laws and their enforcement worldwide.

Emerging technologies: As drone technology develops, new issues and privacy concerns may occur on the surface, the ongoing legal structure requires modifications.

Public education and knowledge: It is important to inform people about their rights and safety measures and to increase public knowledge of privacy concerns related to drones.

12.4.1 Safety and Security

In the moral and social context of drone technology, it is highly important to ensure the safety and safety of drone operations. This involves the possibility of accidents and accidents, as well as risk that may compromise unattractive drone usage to national security.

12.4.2 Risk of Accidents and Collisions

Increasing number of drones in the sky increases the risk of accidents and conflicts. Man-made aircraft and drones share airspace, so any event can have disastrous results. Strict adherence to laws and regulations, such as flying heights, no-fly zones and collision-revival technologies, is required to ensure safe drone operations [7]. To reduce the possibility of accidents and conflicts, it is mandatory to encourage honest and moral drone operations. In addition, the implementation of technological innovations such as remote identification systems and geo-fighting is important in improving drone safety.

12.4.3 Unauthorized Drone Use and Security Risks

Using drones for illegal activities, such as trespasses, espionage, and safety threats, poses risk for public safety. To preserve public safety and safety, unpublished drone activity is required to be identified and addressed. Law enforcement and regulatory organizations are developing counter-drain technology and procedures to detect and address security risks brought by crook drone operators [8]. To maintain a safe and safe operating environment, moral drone users use the importance of following the rules and regulations and reporting any suspicious or unauthorized drone activities.

12.4.4 Environmental Impact

1. *Carbon footprint of drone operations*: The drones, like many other technological innovations, are carbon footprints mostly due to their energy use. Greenhouse gas emissions are a result of electricity required for drone operation and battery charging. Although drones are considered more environmentally friendly options than traditional techniques in many applications, such as air photography and survey, there are still concerns about their carbon footprint. More energy-efficient drones are being developed, renewable energy sources are being used to charge, and customized flight tracts are considered as a solution to this problem. Maintaining moral and ecological criteria requires drone operation to reduce your carbon effect.
2. *Special concerns and interruption in habitat*: research, protection, and environmental monitoring all use drones.

On the other hand, their presence in delicate ecosystems can lead to disturbances of wildlife and residence destruction. Concerns about the environment include migrant patterns, potential effects on nesting birds, and stress that is caused by drone appearance wildlife. In ecological contexts morally the use of drones requires following rules that reduce disturbances and protect the natural environment. Drone should be used morally and responsible by finding a balance between the benefits of collecting data and environmental protection.

3. *Accessibility and equity*: The difficulties around the socio-economic difference in access and the need to guarantee inclusion in disaster response are brought to light by the access and equity of drone technology. Since drones become unavoidable tools in many areas, solving these issues is essential for an equitable and moral application of this technique.
4. *Socio economic diligence in drone access*: Uneven access to various socio-economic classes for drone abilities is one of the main moral concerns in the field of drone technology. Because they can be expensive to buy and maintain, drones are not equally accessible to low-income people or communities. The unfairness of this digital difference is called into question because rich people can benefit from drone technology in areas such as agriculture, survey and disaster relief, while the disadvantaged population may be damaged. Efforts to increase the strength and access to drone technology are necessary to address these inequalities, encouraging them in its use.
5. *Ensuring inclusion in disaster response*: An area where responsible deployment of drones is important, is disaster response. After a natural disaster, drones can be a major help in finding people, evaluating damage and supplying people. However, underprivileged groups do not have the same access to drone technology and are often affected by natural disasters. Regardless of the financial level, it is necessary to ensure inclusion, accessible to all, accessible to drone-assisted disaster response operations. It acknowledges efforts to guarantee that there is proper access to drone-based services and resources in the disadvantaged population to achieve the benefits of this technique [9].

Finally, it is important to overcome the concerns of justice and access to drone technology to guarantee justice and moral applications of these powerful equipment. Reducing the socio-economic gaps in goals and use for inclusion would be important to achieve the full potential of drones for the benefits of all communities, regardless of their financial situation, because they are significant in disaster response and other important areas.

12.4.5 Ethical Dilemmas

Drone perfection often creates complex moral quandaries, especially when important bets are involved and it is necessary to consider all the possibilities carefully. These moral dilemmas attract attention to the difficulties involved in using drones and emphasize how important moral issues are in decision making.

Making Choices in Dangerous Situations

Drone operators and decision makers often have to walk carefully if it comes to morality in areas of military operations and disaster response. These condoms are the centers on the fine line that are different from collecting important data from maintaining dignity and privacy. For example, the operators in the disaster response will assemble which photos and data keeping in mind the sensitivity of the situation. The use of drones for air strikes, monitoring and reconnaissance presents moral challenges in military settings. To avoid injury to innocent citizens, decision-makers must balance military purposes against the possibility of collateral damage and the number of civil casualties. This requires complete moral analysis [10].

Ethics in Drone Deployments

In order to direct responsible and morally righteous activities, moral concepts must be considered completely for drone purposes. In particular, when drones are fitted with cameras and sensors, privacy and data security become extremely important moral ideas. To protect individuals' privacy, regulations and rights of standards, they should specify areas in which drones can fly, they can collect data, and manage and store that data.

Environmental morality is also very important, especially when using drones in ecosystems that are delicate. Careful preparation and attention on moral norms is required to reduce ecological damage in light of potential disruption for wildlife and ecosystems [11]. In the end, using drones will essentially lead to morality, especially when there are complex options at stake. It is necessary to take deliberate and prudent decisions to maintain a constant struggle to balance the benefits of drone technology with moral ideas. Drone operators and decision-producers can manage the complications of drone deployment by honoring moral principles and limiting these moral problems and ideas.

12.4.6 Community Engagement

Including the community in the drone initiative is an essential way to address moral and social concerns associated with the use of drones. To enhance the establishment and public perception of the trust, there is a need to consult the stakeholders and involve them in the decision-making process.

1. **Involving Stakeholders**
   Stakeholders including local companies, communities, governmental organizations, and environmental groups must be involved in the design phases of drone projects. By minimizing the negative effects of drones, this approach ensures that their positive effects are maximized. It also makes it possible to identify and resolve moral dilemmas. Engagement techniques that may be employed to gather feedback and foster a feeling of shared accountability include open dialogues, public ceremonies, and surveys.
2. **Building Trust and Public Perception**
   Confidence is a key component of responsible drone operation. A devotion is essential for openness, accountability and moral behavior to earn the trust of the public and the community. This makes the objectives of drone operations, data security procedures, and privacy security transparent. Education and awareness campaigns may dispel misconceptions, promote ethical and responsible drone use, and help people comprehend the advantages of drone technology [12].

In conclusion, community involvement is a crucial tactic to address ethical and societal issues surrounding drone use. By involving stakeholders and building trust *via* openness and education, drone programs may successfully break down ethical obstacles and actively foster public support for the ethical and responsible use of drones.

12.5 Regulatory Frameworks

Drone laws are necessary to determine the moral and social context of the use of drones. In this section, we give a general review of drone laws, which face difficulties, and changing policies that are being created to deal with the complexity of drone operations.

12.5.1 A Brief Overview of the Drone Regulations

To guarantee moral, safe and responsible use of unmanned aircraft in various situations, drone rules are important. Government organizations and aviation officers usually implement these laws to determine rules for drone operators. They address things such as licenses, registration, flight boundaries, security procedures and privacy issues. The main goal of these policies is to achieve balance between using drone technology for public safety, privacy and environmental protection.

12.5.2 Issues and Changing Policies

Many applications of rapid progress and drone technology have given rise to constantly developed regulatory structures. In this frequently changing regulatory environment, some major issues and concerns are as follows. Figure 12.2 displays issues and changing policies in drone technology.

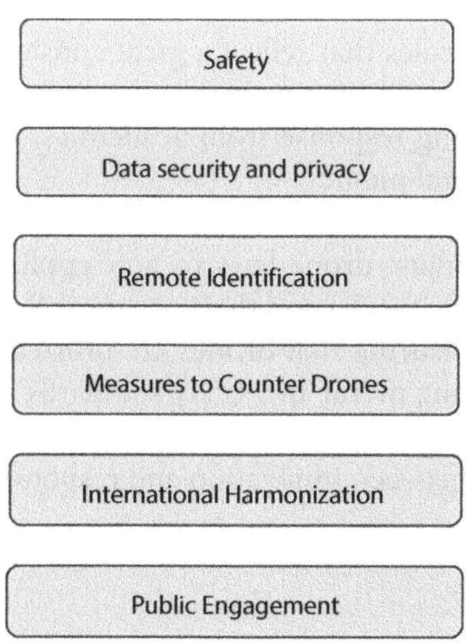

Figure 12.2 Social and ethical awareness: issues and changing policies.

i. Safety
It is paramount to ensure proper functioning of drones in an area shared with manned aircraft. To reduce the risk of conflict and promote safe flight, regulatory organizations are drafting rules for drone operators.

ii. Data Security and Privacy
Concerns about privacy invasion are raised by drones that fit with cameras and other sensors. These concerns are being addressed by rules that ban areas in which drones are allowed to fly and protect data along with pictures.

iii. Remote Identification
To facilitate rapid identification and trekking of drones while in flight, authorities are examining ways to install distance identity systems for drones. Both airspace management and law enforcement depend on this.

iv. Measures to Counter Drones
Since drones can be used in a malicious manner, laws are being drafted which allows the deployment of counter-drain techniques in specific conditions to reduce security risks.

v. International Harmonization
Due to drone activities worldwide, international cooperation is necessary. Efforts are being made to standardize laws and standards, in an attempt to promote equal moral and safety standards and enable cross-border drone operations.

vi. Public Engagement
To establish rules that reflect a great consensus and handle moral concerns, policy makers and regulatory agencies are rapidly seeking response from academics, industry players and the general public.

The necessity to adjust drone laws to new applications, technologies, and moral issues is what is driving their progress. Regulatory frameworks will be essential in ensuring that drones are utilized in ways that benefit society while respecting moral and safety standards as the drone industry grows and matures [14]. An ongoing theme in the creation of these policies is striking a balance between innovation and responsibility.

12.6 Case Studies

Analyzing case studies from real-world situations offers insightful information about the moral and societal concerns related to drone use in many settings. Here, we highlight a few of these case examples to highlight the difficulties and ramifications associated with drone operations.

1. Residential Area Privacy Concerns
   Neighbors expressed worries when a hobbyist drone operator often buzzed a camera drone on homes in a suburban neighborhood. Even though the drone operator claimed to be using it for fun, the locals felt that their privacy had been breached. The moral conundrum of balancing freedom is shown in this case study.
2. Environmental Impact of Commercial Drones
   The environmental impact of commercial drone use was observed when a business used one to monitor and study the Remote Nature Reserve. However, because they unintentionally disturbed wildlife behavior and nesting locations, their drone activities created ecological problems. This illustration emphasizes how crucial it is to consider environmental concerns and ecological ethics before deploying drones in fragile habitats.
3. Use of Drones in Human Response
   During a natural catastrophe, a human team utilized drones to determine the number of individuals still alive and assess the extent of the damage. Local organizations, meanwhile, are concerned about drones hovering over culturally significant locations. The case study has revealed the ethical difficulties in striking a balance between the benefits of using drones for disaster response and cultural sensitivity and respect for community consent.
4. Military Drone Strikes and Civilian Casualties
   Serious ethical concerns have been raised by reports of military drone strikes in conflict zones, particularly in relation to the deaths of civilians. These occurrences draw attention to the necessity to uphold international human law and the ethical quandaries surrounding the use of military drones in combat.

5. Drone Delivery Services
   A large online store introduced drone delivery services in an effort to expedite order supply. However, increased drone use in urban areas also increased the potential of security breaches, noise pollution, and blocked airways. This particular scenario illustrates the ethical issues surrounding the requirement for legislation to regulate the use and handling of drones in urban areas. These case studies effectively illustrate the ethical and social concerns surrounding the use of drones in commercial, military, entertainment, and human domains. They stress how crucial it is to strike a balance between the advantages of aerial vehicles and the moral standards that dictate their appropriate usage. As they engage with complicated ethical problems surrounding the drone era, policy officials, industry stakeholders, and the general public may all gain from looking at these incidents.
6. Mitigation Strategies
   In order to ensure responsible and moral drones, effective mitigation mechanisms must be developed in light of the complex moral concepts that are emerging with the growing usage of drones in numerous domains. These projects must incorporate best practices, technological advancements, and international collaboration in order to completely address these concerns.

12.6.1 Best Practices for Addressing Ethical Concerns

One of the primary mitigating strategies is the establishment and promotion of best practices to address ethical concerns around drone use. These protocols ought to provide guidelines for drone operators, stressing the importance of abiding by moral principles in all circumstances, whether they are business, military, or recreational. Accountability, openness, and privacy are the three core tenets that must be incorporated into these procedures. Regular training and instruction for drone operators may help foster a culture of responsibility and ethical conduct. By exchanging ideas and experiences, Straits, including government agencies, corporate leaders, and specialists, may further enhance these best practices to suit evolving moral standards.

12.6.2 Technological Solutions for Safer and More Ethical Operations

Technical solutions for safe and more moral operation: Development of new technologies is necessary to address moral issues. It is mandatory to develop and implement technologies that improve privacy security, drone safety and moral decision making. The dangers and concerns of the drone can be greatly reduced, for example, to avoid strong conflicts, by incorporating the encryption protocol for data protection and sophisticated AI algorithm for moral decisions. Programming drones must be at the forefront of technical solutions to emphasize safety precautions and moral values as they are more autonomous. In addition, development in no-fly zones can help protect sensitive places and maintain moral boundaries.

To navigate the complex web of moral issues raised by increasing use of drones, it is necessary to create, create and create efficient mitigation techniques. If the best practices, technology progression and international cooperation are combined, the world will see safe, more moral and more responsible drone operations. This strategy guarantees that drones are used with the highest relationship for moral standards and social welfare, in addition to protecting the potential benefits of drone technology.

12.6.3 Future Directions

There are some important issues in navigating the constraints of drone technology and its moral implications that demand further investigations and considerations. Above these, new techniques will continue to shape the moral structure around the use of drones. New moral issues will be resulted in innovations such as enlarged sensors, sophisticated AIs and long-term flight capabilities. For example, if machines are more capable of making complex decisions, the autonomy of the drone may be more sophisticated, requiring maximum evaluation of moral responsibility of their operators. In addition, the theme of drone morality is constantly changing, requiring adaptability. The deployment of drones will generate more and more moral discussion as they are included in more areas of our lives, such as environmental monitoring and distribution services. As a result of this development, we should be aware of how the moral paradigms are changing to keep our structures and guiding principles present and useful while dealing with novels and unexpected problems.

Future interaction on drone morality will probably focus on most privacy, data security and social effects of drones. In addition, it is expected that the rules and international cooperation that control the use of drones will become more important. In areas such as growing drones worldwide, especially in areas such as logistics and disaster relief, uniform moral standards and procedures will be required to ensure that the border operations are coordinated. More international cooperation for creating a drone code of conduct worldwide is estimated in the future, encouraging moral and responsible drone use everywhere. By addressing these potential future paths, we can ensure that drones respect the highest moral standard.

12.7 Conclusion

In this chapter, moral and social implications of using drones for disaster relief and logistics were discussed. These field drones have changed as a result, providing precious air visual and data collection equipment. However, their integration presents complex ethical issues through which there is a need to consider carefully. Community, technology, environment, health and safety ideas are all involved in moral issues, representing the interests of stakeholders and cultural variables. Drone will always be necessary for these industries, so it is important to find a happy medium between technological advancement and morality. To do this, there should be a continuous discussion of stakeholders, legislators and the general public. Only then moral standards can be met and the full potential of these amazing equipment used in crisis management and logistics can be realized. By encouraging this balance, we can successfully interact with moral complications and feel the benefits of drones perfectly in these important areas, eventually more efficient and moral disaster relief and positive effects on logistics operations.

References

1. Sandbrook, C., The social implications of using drones for biodiversity conservation. *Ambio*, 44, Suppl 4, 636–647, 2015.
2. Wang, N., Christen, M., Hunt, M., Ethical considerations associated with humanitarian drones: A scoping literature review. *Sci. Eng. Ethics*, 27, 4, 51–55, 2021.
3. Tatsidou, E., Tsiamis, C., Karamagioli, E., Boudouris, G., Pikoulis, A., Kakalou, E., Pikoulis, E., Reflecting upon the humanitarian use of unmanned aerial vehicles (drones). *Swiss Med. Wkly.*, 149, 13–14, w20065–w20065, 2019.

4. Wang, N., We Live on Hope..: Ethical Considerations of Humanitarian Use of Drones in Post-Disaster Nepal. *IEEE Technol. Soc. Mag.*, 39, 3, 76–85, 2020.
5. Hodgson, J.C. and Koh, L.P., Best practice for minimizing unmanned aerial vehicle disturbance to wildlife in biological field research. *Curr. Biol.*, 26, 10, R404–R405, 2016.
6. Alsumayt, A., El-Haggar, N., Amouri, L., Alfawaer, Z.M., Aljameel, S.S., Smart flood detection with AI and blockchain integration in Saudi Arabia using drones. *Sensors*, 23, 11, 5148–5151, 2023.
7. Robakowska, M., Slezak, D., Zuraty, P., Tyranska-Fobke, A., Robakowski, P., Predkiewicz, P., Zorena, K., Possibilities of using UAVs in pre-hospital security for medical emergencies. *Int. J. Environ. Res. Public Health*, 19, 17, 10754–10758, 2022.
8. Siddiqi, M.A., Iwendi, C., Kniezova, J., Anumbe, N., Analysis on security-related concerns of unmanned aerial vehicle: Attacks, limitations, and recommendations. *Math. Biosci. Eng.*, 19, 3, 2641–2670, 2022.
9. Nadeem, A.B. and Chandna, Y., Remotely Piloted Life-Saving Effort vehicles and emergency management: An analysis on revolutionizing humanitarian assistance in Pakistan. *J. Emergency Manage.*, 16, 1, 7–14, 2018.
10. Chuang, C.-C., Rau, J.-Y., Lai, M.-K., Shih, C.-L., Combining unmanned aerial vehicles and internet protocol cameras to reconstruct 3-D disaster scenes during rescue operations. *Prehosp. Emerg. Care*, 22, 4, 479–484, 2018, https://doi.org/10.1080/10903127.2017.139266.
11. Johnson, A.M., Cunningham, C.J., Arnold, E., Rosamond, W.D., Zègre-Hemsey, J.K., Impact of using drones in emergency medicine: What does the future hold? *Open Access Emerg. Med.*, 13, 487–498, 2021, https://doi.org/10.2147/OAEM.S307123.
12. Poljak, M. and Sterbenc, A., Use of drones in clinical microbiology and infectious diseases: current status, challenges and barriers. *Clin. Microbiol. Infect.*, 26, 4, 425–430, 2020.
13. Merei, A., Mcheick, H., Ghaddar, A., Survey on Path Planning for UAVs in Healthcare Missions. *J. Med. Syst.*, 47, 1, 79–82, 2023.
14. Rejeb, A., Rejeb, K., Simske, S.J., Treiblmaier, H., Drones for supply chain management and logistics: a review and research agenda. *Int. J. Logist. Res. Appl.*, 26, 6, 708–731, 2023.

13

Regulatory Overview and Challenges of Drone Technology in Safety Regulation

S. Vijayanand[1]*, T.S. Pradeepkumar[2], M. Annalakshmi[3] and Azween Bin Abdullah[4]

[1]*Department of ECE, Sri Venkateswara College of Engineering, Chennai, India*
[2]*School of Computer Science and Engineering, Vellore Institute of Technology, Chennai Campus, India*
[3]*Department of EE (VLSI), Velammal College of Engineering and Technology, Madurai, India*
[4]*Computer Science, Perdana University, Kuala Lumpur, Malaysia*

Abstract

With the explosive growth of the commercial and recreational drone industry, making a robust but flexible regulatory framework is an incredibly challenging task for policymakers in seeking to safely integrate drones into airspace. This chapter provides an overview of the regulatory landscape of operations that are being governed at the present moment in the United States, specifically concerning regulations from the Federal Aviation Administration. It first articulates the overall main FAA rules to govern drones generally, including Part 107 for commercial operators, Certificate of Authorizations for public entities, and guidelines for hobbyists. It then emphasizes significant limitations under current rules that restrict sophisticated operations such as beyond visual line of sight flights, expanding drones over populated areas, remote identification, and autonomous operations. It outlines the patchwork of state and local privacy, data protection, and land use laws that intersect with FAA authority over airspace safety.

Keywords: Federal aviation administration (FAA), Part 107 regulations, certificates of authorization (COA), drone safety framework, beyond visual line of sight (BVLOS), state and local drone laws, privacy and data collection, airspace integration challenges

**Corresponding author*: vijayanand.s@svce.ac.in

A. Prasanth, Rajesh Kumar Dhanaraj, Munish Sabharwal, Vandana Sharma and Seifedine Kadry (eds.) Drones for Transportation Logistics and Disaster Management, (279–298) © 2026 Scrivener Publishing LLC

13.1 Introduction

The past decade has seen an improvement in the development and use of drone technology across a diverse range of commercial, recreational, and governmental applications. This rapid pace of technological advancement in drones outstripped the creation of a robust regulatory framework which would ensure that drones are integrated into the national airspace system in a safe and responsible manner. Drones integration presents new and complex challenges requiring focused attention of regulators in order to harness the benefits of the technology while minimizing risks to manned aircraft and people on the ground.

This chapter provides a detailed analysis of the current regulatory environment for drone technology, with special emphasis on safety regulations and challenges. Starting from an overview of the main regulatory bodies and frameworks in the United States under the Federal Aviation Administration (FAA), the chapter will analyze some of the major debates and open challenges. These range from beyond line-of-sight operation, flight over people, and remote identification issues, to some form of privacy issues. Consensus standards, industry partnerships, and best practices may be necessary in order to help make these types of drone operations safer. It will present views on what the future holds regarding regulation, with recommendations to actively address the need for safety in the operations as drone use becomes more common and automated.

13.1.1 Need for Regulation to Ensure Safe and Responsible Use

Effective safety regulation of drones is essential to protect human life and property as their use grows [1]. In the absence of proactive regulatory action, the dangers of collisions, interference with emergency response efforts, and privacy violations could proliferate as drone activities expand, especially for autonomous and beyond-visual line-of-sight operations. Regulators play a critical role in balancing safety with the ability to innovate through appropriate performance-based regulations [2].

13.1.2 Objectives and Scope

This chapter has two main objectives: (i) to provide a comprehensive overview of the existing drone safety regulations and challenges, and (ii) to offer perspectives on the future regulatory outlook and recommendations to enable safe integration.

The scope of this chapter is specifically focused on drone safety regulations and challenges under the FAA framework in the United States. It examines the interplay between federal, state, and local laws related to drone operations. Acknowledging broader regulatory issues surrounding privacy, data protection, and security, the chapter focuses its analysis primarily on regulations and standards aimed at mitigating safety risks and hazards associated with increased drone use in the airspace.

13.1.3 Regulatory Bodies and Framework

In the United States, drone operations are controlled by a web of regulatory bodies at the federal, state, and local levels. Operational regulations are determined by the federal authority, known as the Federal Aviation Administration, acting under the federal code [3]. Meanwhile, state and municipal laws often govern other essential dimensions, such as privacy and data collection when a drone flies over an area. There is a need for a landscape analysis of these various regulations and how they intersect to identify gaps and overlaps.

In addition to international regulations and bodies such as the European Union Aviation Safety Agency (EASA), the International Civil Aviation Organization (ICAO), and individual country rules around the world. As drones quickly diffuse across borders, a unified, consistent global approach to regulation will be crucial to ensure that the technology can be integrated in a safe and sustainable manner [4]. This chapter provides some data points and recommendations to bring together otherwise conflicting national policies toward drones by augmenting international coordination and sharing knowledge.

13.2 FAA Oversight and Authority Over US Airspace

The FAA, the first regulatory body established to oversee the national airspace system in the United States, has broad responsibility for regulating drones and also ensuring safety. It has the authority to create rules for operating, approval of waivers of certain applicable rules, its approval of pilot requirements and enforcement of regulations. The scope and limits of FAA authority are essential to provide insight into current regulatory gaps and challenges.

13.2.1 FAA Drone Regulations – Part 107, COAs, Recreational Use Rules

The rate of growth of drones has surpassed the development of regulations that govern their safe use. In the United States, the FAA oversees and authorizes all aircraft operations in the national airspace system. This broad authority allows the FAA to set rules for drones, issue operator certifications, approve airspace authorizations, and enforce regulations. Understanding current FAA drone regulations provides critical insight into existing gaps and challenges.

The primary FAA rules for drone operation include Part 107 for commercial purposes, Certificates of Authorization for public use, and guidelines for hobbyist operations under Section 336. The Part 107 sets up some operational boundaries: daylight only, line of sight, altitudes of up to 400 feet, and a test of remote pilots. Drone registration and certifications are required for commercial operation. COAs enable government agencies, universities, and other public operators to apply for case-by-case approvals for drone activities such as night flights or beyond visual line of sight. Section 336 provides recreational guidelines for hobbyists, such as registering drones and notifying airports before flying in controlled airspace.

Analysis of the limits and permissions under these FAA rule sets stresses areas without rules on passenger transport, operation beyond line of sight, and expansion over people. Other issues focused on include enforcement issues, integration with airspace, and remote identification. This regulatory analysis serves as a basis for developing policies and procedures necessary to ensure safe advancement into complex drone operations such as autonomous flights and urban air mobility and expanded commercial services. It empowers stakeholders to work cooperatively with the FAA in building a robust, scalable, and performance-based regulatory paradigm [5]. While the FAA has centralized oversight of airspace safety, state legislatures and city councils have enacted drone laws concerning privacy, data collection, use by law enforcement, land use, and nuisance concerns. Such laws create a patchwork of regulations that the drone operator must navigate [6].

13.2.2 Role of State and Local Laws Related to Privacy and Data Collection

For instance, 43 states have enacted laws regarding drone privacy issues. Laws range from requiring warrants for law enforcement surveillance, to restrictions on flying near critical infrastructure, to prohibiting voyeuristic drone use. Some states like California and Nevada have comprehensive drone privacy laws, while others have minimal protections. On data retention, states have taken varying approaches. Oregon law mandates destroying drone data within 30 days absent a warrant or exception. Comparatively, North Carolina has no data retention limits. Developing minimum federal data privacy standards could help align state policy variations.

Take off/landing sites, noise limits, flight altitudes, and distances from landmarks such as stadiums or parks are regulated by local municipal laws. Sometimes, this conflicts with the FAA's operational authorization, leading to confusion among drone pilots. Federal, state, and city officials need to coordinate more with each other as drones proliferate across jurisdictions, smoothing out the regulatory complexity.

On balance, although safety in airspace falls within the exclusive domain of federal authority, questions of privacy and land use pose substantial state and local oversight responsibilities. This presents a complexity to navigate but permits tailoring regulation to community values. Further consideration of where and how FAA's authority intersects with other regulatory needs and possible legislation can help create more uniform regulations at all levels of government.

13.2.3 International Regulations and Bodies Like EASA, ICAO, and Individual Country Rules

Drone regulations are very different between countries and across international bodies. The EASA sets requirements for EU nations, including operational authorizations, pilot licensing, technical requirements, and UAS classes. Other countries, like the UK, Switzerland, and Norway, base their policies on EASA rules. The ICAO sets worldwide standards and recommended practices for drones to harmonize policies among its member states. Countries such as China, Canada, Australia, and Singapore have developed national regulations for drones regarding pilot certification, registration, airspace restrictions, and flight rules.

Some of the key differences are permissions for BVLOS operations, privacy rules, requirements for identification from a distance, limits for flights over people, and operator qualifications. It is imperative, as the world continues to adopt drones in wider areas, for ICAO and bilateral engagements to heighten coordination toward balancing the varied laws of different countries and allow inter-border services to be carried out. Benchmarking regulations around the world provides perspective toward the promotion of international harmonization. Cross-border services: worldwide regulations benchmarking for international harmonization efforts give insight [7].

13.3 Key Regulatory Challenges

Expanding operations create significant regulatory challenges for ensuring safety of drones. The main areas of concern include collision avoidance, especially for BVLOS flights, developing adequate detect-and-avoid technologies, and establishing certifications and standards for safe flight over populated areas. With increasing drone use, regulators have to balance issues of privacy and data protection. Data collection, use, and retention from flights are governed differently by states' laws. The issues developing uniform security rules cause include mandating geo-fencing and anti-tampering safeguards, remediation of cyber threats to the systems used by the drones, and encryption of command links.

13.3.1 Safety – Collision Avoidance, Flight Over People, Certification

Safety considerations will pose one of the most severe regulatory barriers to integrating drones. A leading safety issue is that of designing suitable collision avoidance technologies and systems for safety beyond visual line of sight drone operation. In the absence of dependable detect-and-avoid features, the threats of mid-air collisions increase drastically as drone traffic is scaled up. Moreover, development of performance standards and certification for routine operation over populated territories poses challenges. The regulators will need to identify how to ensure safety for allowing operations such as drone delivery services and inspectors operating over urban skies teeming with people and buildings. This is through evaluation of minimum safety thresholds for events like loss of control or power outages. Even certification of each unique drone model comes with time and cost.

Remote pilot testing and training requirements will also be challenging for regulators when autonomous capabilities progress. More standardized, scalable approaches to pilot certifications and exemptions to certain autonomous operations will be necessary. Cybersecurity threats also present new vulnerabilities that could compromise safety. Regulatory measures for such issues as GPS spoofing, malicious control of drones, and tampering with navigation will be inevitable. It is on such complex safety concerns that proactive yet innovation-friendly solutions continue to feature in the leading thoughts of regulators as well as their industry partners as drones continue their proliferation into ever-increasing space within the air [8–13].

13.3.2 Privacy and Data Protection Laws

Drone use for more commercial activity and law enforcement applications poses major privacy regulatory issues. Some key questions include uniforming data retention requirements for drone-collected data by state. To date, states and locals are very patchy in how much protection exists to prevent improper intrusion. States such as California have enacted what will be among the strictest privacy laws of all. Federalization of data retention policy could unify some of that action. Another issue is that of requiring warrants for the police to deploy drones for surveillance activities on private land. There are conflicting rulings in courts. The other problem is how to make disclosure requirements on commercial operators regarding what information they collect and for what purpose.

Transparency regulation can follow in the lines of policies like GDPR in Europe. It would be crucial to allow citizens to request deletion of their information from commercial drone databases as services like delivery expand. Drone data collection for insurance purposes also raises concerns. Clear regulations on exemptions like emergency services versus permitted commercial data gathering uses help address the associated privacy risks. As drones proliferate, significant coordination between federal, state, and industry stakeholders will be needed to develop comprehensive privacy and data protection regulations that balance safety innovation and personal liberties.

13.3.3 Security – Geo-Fencing, Tampering, Cybersecurity

With the increase in drone usage, security presents various challenges. An important one is geo-fencing, altitude limits, and airspace restrictions around sensitive sites such as airports, government facilities, prisons, and critical infrastructure to prevent intentional or unintentional misuse. However, establishing these restrictions needs to be balanced with the goal of allowing legitimate drone operations.

Tampering and anti-tampering standards are also other concerns. The regulations should include protecting against dangerous uses of tampered with commercial or consumer drones while allowing owners to modify or repair them. Cybersecurity threats must be addressed too. Risks include GPS spoofing, taking over by hacking command links and onboard systems. Imposition of data encryption, multi-factor authentication, and other software protection will be important but cost-adding.

Development of enforcement and rapid response protocols for security breaches is also necessary. Tracking and identification of hijacked or tampered drones also present challenges. Regulators have to finalize remote identification rules in a manner that balances security requirements with privacy. It is not easy to have consistent security regulation at the federal, state, and local level but is absolutely necessary for the approach to be uniform. Collaborating with the technology companies would be key in enabling security innovations that keep up with the dynamic cyber threats.

13.3.4 Enforcement and Identification – Registration, Remote ID

Drones pose multiple challenges for regulators in effective enforcement of regulations. The issue of simplifying and harmonizing registration of recreational and commercial drones across borders is one of them. Gaps exist at the moment, with no federal requirement for the registration of recreational drones. Remote identification capabilities, as part of efforts to enhance accountability, also create problems on the cost and technology fronts.

On enforcement, detection of such violations as illegal night flights or flying in restricted airspaces will be challenging without remote ID and tracking. Development of systems to give air traffic control and law enforcement real-time monitoring capabilities for drones will be crucial. This will include developing remote ID rules that balance the safety needs with privacy concerns over constant surveillance. There are also challenges around identification and penalization of unauthorized drone operators as drones increasingly fly autonomously.

Finally, regulators need to address topics, such as registering drones that are manufactured from kits or 3D printed. In addition, updating registration databases rapidly as the consumers buy new drones is also a priority. Collaboration with emerging digital solutions such as blockchain would help answer these challenges at low cost. Consortial work with the technology companies will be critical in enabling secure but flexible drone identification and tracking.

13.3.5 Urban Air Mobility and Autonomous Operations

With urban air mobility and increasingly autonomous drones, altogether new regulatory models will be called for. Concerns around safe operation of fully autonomous passenger- carrying vehicles will be vast-issues around redundancies, dependability, security, and procedures for emergency event handling. Challenges in traffic management will arise for coordination of air traffic, including right-of-way rules, disruptive event planning for recovery, as well as intersection with manned-aircraft traffic.

Challenges also exist around autonomous flight rules and restrictions. Determining permitted operations, geo-fences, altitude limits, and advanced pilot certifications or exemptions will be complex. Enabling beyond visual line of sight operations over populated areas adds risks. Furthermore, autonomous inspections and deliveries create challenges around security, privacy, and public acceptance. Updating existing pilot-centric regulations to accommodate autonomous operations will be difficult.

Overall, regulators need a forward-thinking, data-intensive approach to appreciating that drones cannot be strictly regulated under currently applicable manned aviation rules. Creation of performance-based and risk-minimizing rules in partnership with industry and local governments is thus essential. Comprehensive coordination between key stakeholders, maybe even establishing regulatory zones of difference, will expedite the advancement of these groundbreaking technologies in safety.

13.3.6 Environmental Impact – Noise, Emissions, Wildlife

Noise pollution could be addressed by maximum sound level standards set by the type of drone and the intended use in an urban environment. For example, requirements may vary from 55–65 dB for small recreational drones to 75–85 dB for larger cargo delivery drones based on take-off weight. Another possible restriction is that of night flights over residential neighborhoods.

Air quality and emissions are also issues that must be overcome. Smaller drones do not have the opportunity to create a lot of pollution, but bigger delivery drones and air taxis will have a much greater impact over time. Emissions standards will have to consider different engine types, fuel mixes, size classes, and typical flight durations. Policymakers can work together with the industry to encourage the development of electric and low-emission drones.

Wildlife impact mitigation is also essential. Geo-fencing, altitude restrictions near protected wildlife areas, seasonal restrictions during migration or nesting seasons, and minimum separation distances from rare animal habitats can avoid causing disturbances. Tagged endangered species could be tracked in real time to dynamically reroute drones. Drone lighting standards should also be developed to minimize disruptions.

Other concerns range from visual blight and light pollution as drones populate the skies, to disposal and recycling standards for retired drones and batteries. Continual environmental impact monitoring and data-driven policy updates will be essential. Overall, a collaborative regulatory approach is required to enable drone innovation while proactively addressing sustainability [14, 15].

13.4 Industry Standards and Best Practices

Industry knowledge-sharing initiatives and voluntary consensus standards help push drone technology forward safely. ASTM International and ISO, for example, have created broad performance-based standards referenced in rulemaking by regulators in many areas including safety, design, operations, training, and maintenance. The Joint Authorities for Rulemaking of Unmanned Systems (JARUS) creates technical guidance and best practices in rulemaking between regulators around the world. Knowledge sharing is encouraged by industry organizations, such as the Drone Safety Alliance.

Accountability is being promoted through some industry programs created around pilot certification, operator standards, and requirements for insurance. The FAA along with industry, created the Trusted Operator Program for guidance on safe voluntary practices outside of regulation. The Universal Drone Insurance Framework provides tools to calculate risks of flight. Programs like these make it possible to spread operational information, learnings from safety failures, and technical know-how. This enables industry actors to cooperate in ways that can help inform the standard-setting process to create balanced, scalable regulations.

13.4.1 Consensus Safety Standards – ASTM, ISO

Industry groups such as ASTM International and ISO have developed extensive voluntary consensus standards focused on drone safety aspects such as design, operations, maintenance, training, and qualifications. These are developed collaboratively by experts from companies, academia, and government. For instance, the ASTM F38 committee has promulgated standards for remote ID, detect-and-avoid, BVLOS, night operations, and lightning protections. These stand as technical guidance to the mitigation of risks involved. ISO-published standards deal with topics such as airworthiness, traffic management, inertial navigation, and pilot competency.

Such standards are often referenced by regulators when developing official regulations. They provide granular technical detail that is collaboratively vetted to achieve a balance between safety and innovation. By incorporating input from all stakeholders, consensus standards gain broad acceptance and help accelerate adoption. Such agencies also have in place processes to update standards in real-time. The lessons learnt from mishaps and incidents are readily assimilated. Regulators collaborate with the standards bodies in the identification of the gaps that exist between voluntary standards and obligatory regulations that ought to be aligned. This facilitates a practical approach, performance-based rulemaking [16].

13.4.2 Industry Knowledge Sharing Programs and Partnerships

The drone industry has encouraged initiatives and the sharing of operational data, lessons learned, and best practices to enhance safety and inform regulations. Lead groups involved are the Commercial Drone Alliance, Drone Safety Alliance, and Association for Unmanned Vehicle Systems International (AUVSI). For example, the Drone Safety Alliance organized the UAS Integration Pilot Program alongside the FAA, enabling drone operators and local governments to provide insights on advanced operations. AUVSI's Trusted Operator Program helps certify commercial drone operators who exceed baseline regulations.

Other partnerships include the FAA's Drone Advisory Committee and JARUS (Joint Authorities for Rulemaking on Unmanned Systems), which brings industry, government, and academia together to collaborate on regulatory challenges. Companies have also formed data-sharing consortiums to assess reliability and safety issues across diverse operational environments. Such collaborative programs allow rapid transfer of field data, safety lessons, and expertise between sectors to benefit the entire industry. They facilitate coordination on issues like remote ID, BVLOS flights, and Unmanned Aircraft System Traffic Management (UTM) systems that companies independently cannot solve. Regulators also gain access to operational insights to develop performance-based regulations.

13.4.3 Operator Certification Programs and Training Requirements

Various industry organizations have developed operating certification programs that go beyond FAA requirements in terms of safety and accountability. Some examples include the AUVSI Trusted Operator Program, the operator certification by the UAS Pilot Institute, and the Remote Pilot Council's standards. These programs establish training standards and knowledge requirements in areas such as weather analysis, airspace restrictions, maintenance, emergency procedures, and privacy/security considerations. Certifications require passing practical and written exams assessed by industry experts. Some programs have tiered certifications based on operational complexity.

Maintenance of certification may mandate completing a minimum number of flights annually, undergoing recurrent training, reporting safety incidents, adhering to a code of conduct, and demonstrating current knowledge. Certified remote pilots are listed in databases for verification by regulators and insurance providers. While voluntary at this point in the US, standardized certification programs based on consensus-based standards provide a model for future FAA testing requirements as the technology of drones becomes increasingly sophisticated. These ensure professionalism, a culture of safety, and constant improvement of skill over and above base FAA requirements through collaboration of industry [17].

13.4.4 Insurance Requirements and Risk Management

Insurance is a key enabler of responsible drone operations. Industry initiatives include efforts to make coverage more accessible and standardized. The Global Unmanned Aircraft Systems Traffic Management Association developed the UAS Service Supplier and UAS Traffic Management Service Provider insurance frameworks. These define minimum coverage levels based on operational risks.

The Drone Insurance Group is also educating insurers and developing standardized policies. Insurance can incentivize proper licensing, pilot training, aircraft maintenance, operational safety procedures, and manufacturers adopting safety-by-design approaches. Usage-based insurance models using flight data are also emerging to dynamically assess and price risks. Pre-flight risk assessment tools from Allianz Center for Technology or MIT's Haystack app enables operators to look at environmental hazards, airspace restrictions, and local risks that an underwriter might want to consider. Such a framework enhances underwriting rigor with reduced costs. Overall industry risk management guidelines, insurance pools, standardized policies, tools to validate safety procedures, and usage-based models can accelerate the development of actuarial data to enable wider drone coverage. This makes insurance both more affordable and an incentive for continuous safety improvement [18].

13.5 Regulatory Outlook and Future Needs

Regulation of drone technology in different parts of the world varies, and the majority of the regions have developed or are in the process of developing regulations to ensure safe and responsible use of drones. The main purposes of drone regulations are:

- Ensure Safety: The safety of people, property, and other aircraft is an important concern. Regulations may be related to altitude, no-fly zones, and collision avoidance systems.
- Privacy and Data Protection: Regulations must achieve a balance between the use of drones for many purposes, including aerial photography, and the privacy and data of individuals.
- Commercial Use: Most regulations distinguish between recreational and commercial drone use. Commercial drone operators are often required to obtain special licenses and permits.
- Airspace Management: Regulatory bodies aim to safely integrate drones into the existing airspace system. This involves managing airspace for various types of drones and establishing operating zones.
- Remote ID and Tracking: Several countries are looking into or already implementing Remote Identification (Remote ID) systems, which will help track and identify drones in real-time and hold users accountable.
- Licensing and Certification: Licensing and certification requirements for drone pilots are common, including written exams and practical tests.

For regulations, this fast-paced rise in drone technology necessarily needs them to evolve in parallel with new uses, risks, and operational models. Some of the key areas regulators are focusing upon are plans for beyond visual line of sight flights, expanding operations over people, and enabling urban air mobility. Updating airspace integration rules to allow more complex operations would be a priority. The future requires increased automation and performance-based regulations by leveraging industry data versus prescriptive limits. This includes predictive analytics and risk modeling to allow for flexible approvals. As advanced capabilities emerge, transitioning to more scalable, data-driven oversight models will be important.

As the final implementation steps, there are other priorities such as remote ID implementation, digital drone traffic management infrastructure, and new pilot testing approaches. Cybersecurity rules and infrastructure protections will be critical too as systems become increasingly interconnected. In addition, noise standards, emissions limits, and other environmental regulations will have to be updated. Strong collaboration between regulators worldwide, industry partners, technology innovators, and local governments will be necessary to balance safety with rapid advancement. Taking a proactive, forward-looking approach to crafting regulations will maximize the benefits of this transformative technology [19].

13.5.1 Projected Regulatory Changes and Updates

The FAA and other regulators have several rulemaking initiatives in place to expand drone operations as the technology advances. Some of the key projected regulatory changes include:

- Finalizing standards for remote ID for tracking of drones in flight, which would permit expanded operations, such as over people and at night.
- Allowing routine beyond visual line of sight operations under a performance-based regulatory framework, rather than relying on a waiver-based system of approvals- a critical accommodation for many commercial applications.
- Expanded authorization for nighttime operation of qualifying drones equipped with anti-collision lighting visible from a distance of 3 miles to provide for increased operational flexibility.
- Increase the drone weight limits from the existing 55 lb to facilitate more cargo deliveries, passenger-carrying drones, and infrastructure inspection drones.
- Formulate a regulatory framework for urban air mobility through passenger-carrying autonomous drones and air taxis; thereby encouraging testing and obtaining certifications.
- Modify the integration rules of airspace and allow increased access by drones, including overpopulated areas, maintaining safety standards.

- Cybersecurity regulations would be established to develop rules to reduce risks related to hacking, spoofing, and tampering as drones become more connected.
- Creating more performance-based pilot certification under simulation, knowledge testing and assessment of competence over flight

13.5.1.1 Speed of Regulatory Innovation

Speed and flexibility for emergence of technology as well as commercial use cases are needed in an accelerated mode ensuring safety and security concerns together with the interest of the communities concerned. New but strict control structures for higher degrees of autonomy [20, 21].

13.5.2 Streamlining and Harmonizing Global Regulations

As the drone industry scales worldwide, aligning regulations across major aviation markets is important. Organizations like the Joint Authorities for Rulemaking on Unmanned Systems and the International Civil Aviation Organization facilitate this process. Key goals are to harmonize categories and operational limitations for drones, pilot certifications, airspace restrictions, and technologies like remote ID. The establishment of standards that are recognized in multiple countries helps reduce compliance burdens and enable transnational services.

For instance, JARUS issued recommendations on technical airworthiness requirements and operational categories, which many regulators such as the FAA and EASA have embraced. ICAO develops model rules and guidance documents to aid harmonization. Bilateral arrangements between countries such as the U.S.-EU arrangement also pursue mutual recognition of regulations and the sharing of oversight duties to minimize duplication. Information-sharing networks allow regulators to cooperate on UTM systems and autonomous flights.

While differences may persist due to local issues, greater international collaboration can eliminate those inconsistencies that don't make a difference. Evolution of standards and recommended practices among international working groups of experts, representing all involved sectors, should foster greater alignment. This is especially important given the advent of advanced operations, such as UAM [22, 23].

13.5.3 Performance and Risk-Based Regulations Using Industry Data

Prescriptive drone regulations based on aircraft weight or operational restrictions can hamper innovation. An emerging regulatory approach is employing performance and risk-based rules leveraging data from the drone industry. This data-driven framework focuses on desired safety outcomes rather than mandating specific technologies or procedures. For example, regulators establish detect-and-avoid reliability benchmarks that operators can meet through various technical solutions based on their operational needs.

Industrial relationships with safety data collection are most crucial. Under programs such as the FAA UAS Integration Pilot Program and LAANC, access to performance statistics, incident reporting, and real operational insights assists the regulators to set performance-based rules. Analytics and modeling identify critical risk factors and forecast scenarios as drone usage scales. This will allow for custom approvals optimized to each operator and use case rather than one-size-fits-all limits. Through time, prescriptive regulations mature into flexible, scalable ones. Even though consensus standards are still required, performance-based frameworks enable faster assimilation of emerging technologies such as autonomous features. The regulators will maintain oversight by audits and inspections to ensure desired outcomes. This data-driven approach balances the advancement of safety with rapid innovation.

13.5.4 Automated Enforcement and Monitoring Technology

As drone operations grow, the regulators are searching for technologies on automated enforcement and real-time monitoring that will support the field inspection. These shall help digitize parts of oversight to ensure efficiency and consistency. Many automated enforcement applications will be enabled by remote identification systems to track and identify drones in flight. When coupled with geospatial data about airspace restrictions, compliant drone identifications can automatically be cleared while violations trigger warnings and penalties. Machine vision technologies also promise a means for detecting unauthorized flights.

UTMs can feed the automated dashboards of localized activity of drones that regulators can give priority to the oversight. There is a way to flag machine learning models to high-risk operations requiring intervention because they are trained on past violation and incident data. However, automated systems give rise to a number of fears regarding privacy intrusion, algorithmic bias, and excessive surveillance. Therefore, this calls for maximum transparency, accountability, and human oversight of enforcement tools. Overall, measured regulation with automation and AI but not at the expense of human oversight can lead to more responsive, consistent enforcement and data-driven safety oversight as drone operations grow exponentially [24–27].

13.5.5 Regulatory Gaps and Challenges for Advanced Operations Like UAM

UAM refers to passenger-carrying air taxi drones and autonomous cargo services in urban environments. It is a quantum leap in complexity compared to current regulations centered around small hobbyists and commercial drones.

Major regulatory gaps include:

- No certification process or airworthiness standards exist yet for larger passenger-carrying drones. New standards for reliability, redundancy, and control systems are needed.
- No operational framework to manage dense, low-altitude air traffic and coordinate with conventional aircraft in urban airspace. UTM systems will be critical.
- No regulations governing vertiports, passenger loading/unloading processes, and community noise impacts.
- Unclear liabilities and insurance requirements for autonomous passenger flights and delivery services over populated areas.

The sum of factors-namely, safety issues, community acceptance, infrastructure needs, and the radically different operational models-places UAM atop a steep regulatory mountain. It will require new partnerships between regulators, cities, and industry to cooperatively develop specialized regulations and oversight frameworks tailored to this emerging application. A phased approach beginning with pilots in select locations will help collect data to inform nationwide UAM rules [28, 29].

13.6 Conclusions and Recommendations

13.6.1 Conclusions

Drone integration was proceeding at an accelerated pace, providing tremendous benefits but also new risks that regulations are trying to mitigate. Current regulations are not suitable for enabling complex operations such as autonomous deliveries, urban air mobility, and extensive flights over people. A patchwork of state and local laws combined with gaps in federal rules has created an inconsistent regulatory system that hampers innovation and scalability. Development of pragmatic policies by regulators in cooperation with the industry, academia, and local communities is the key to staying in step with the technology. Forward-looking and data-driven, such an approach needs to be outcome-based and oriented toward safety objectives rather than on prescriptive limits.

13.6.2 Recommendations

Ensure federal, state, and local laws relating to drones are increasingly harmonized across such key issues as safety, privacy, security, and environmental impact. Develop international cooperation through bodies such as ICAO and JARUS to allow worldwide interoperability and recognition of acceptances across nations. Establish industry collaboration to obtain the data needed for performance-based rulemaking, guided by consensus standards. Implement regulatory sandboxes and pilot projects to allow for safe testing and validation of new operations. Enhance community outreach and noise/privacy mitigation efforts as drone operations expand, especially for urban regions.

References

1. Federal Aviation Administration, FAA aerospace forecast: Fiscal years, pp. 2019–2039, 2019.
2. Grigonis, H., The sky's the limit for drones—if they can overcome these challenges, Digital Trends, 29 Jan. 2020.
3. FAA Reauthorization Act of 2018, H.R. 302, 115th Cong, 2018.
4. Rupprecht, J., Current drone laws in the states, JD Supra, 11 Feb. 2021.
5. Federal Aviation Administration, 14 C.F.R. § 107 (Electronic Code of Federal Regulations, Title 14, Chapter 1, Subchapter F, Part 107), 2016.

6. Federal Aviation Administration, FAA order 1100.161 - Certificates of waiver or authorization, 2021.
7. Public Law 112-95, § 336 - Special rule for model aircraft, 2012.
8. European Union Aviation Safety Agency, AMC RPAS.SPEC.020 Basic Regulation.
9. International Civil Aviation Organization, Unmanned aircraft systems traffic management (UTM) manual (Doc 10019), 2021.
10. Gettinger, D. and Michel, A.H., A balancing act: Privacy, security, and drone use in Canada, Center for the Study of the Drone, 2020.
11. Kopardekar, P., Safely enabling UAS operations: The FAA's progress and path forward, NASA, 2021.
12. Clothier, R., Silva, C.R.C., Walker, R., Williams, B.P., Challenges of safety assurance in the commercial UAS industry. *Aeronaut. J.*, 121, 1245, 1659–1680, 2017.
13. European Union Aviation Safety Agency (EASA), EASA publishes world's first rules for operation of air taxis in cities.
14. Kaminski, M.E., Drone federalism: Civilian drones and the things they carry. *Calif. Law Rev.*, 101, 1, 57–112, 2013.
15. California Civil Code § 1708.8.
16. ASTM International Committee F38 on Unmanned Aircraft Systems.
17. Joint Authorities for Rulemaking on Unmanned Systems (JARUS).
18. Federal Aviation Administration, UAS integration pilot program report, 2021.
19. Ravich, T.M., The integration of drones into the national airspace. *North Dakota Law Rev.*, 85, 4, 597–619, 2009.
20. Brumfield, D., *et al.*, Federal, state, and local regulation of unmanned aircraft systems in the low-altitude airspace. *J. Air Law Commerce*, 86, 3, 447–485, 2021.
21. Nasr, J., Drone laws: FAA drone regulations, state drone laws, local drone ordinances, JD Supra, 29 Nov. 2021.
22. Yeoh, P., A global snapshot of drone regulations, Tractable, 29 Nov. 2019.
23. Khaykin, I., The regulation of civilian drones' use in the European Union and their future. *J. Air Law Commerce*, 84, 1, 51–72, 2019.
24. Luppicini, R. and So, R., Drones and privacy. *Int. J. Inform. Secur. Syst. Manage.*, 5, 1, 51–59, 2016.
25. PwC, Skies without limits: Drones—taking the UK's economy to new heights, 2021.
26. Clothier, R.A., Williams, B.P., Washington, A., Fulton, N.L., Structuring the safety case for unmanned aircraft system operations in non-segregated airspace. *Saf. Sci.*, 79, 213–228, 2015.
27. Suau-Sanchez, P., Voltes-Dorta, A., Rodríguez-Déniz, H., Regulating drones: Desperately seeking solutions or solutions desperately seeking problems? *Transp. Policy*, 92, 1–11, 2020.

28. Hayat, S., Yanmaz, E., Muzaffar, R., Survey on unmanned aerial vehicle networks for civil applications: A communications viewpoint. *IEEE Commun. Surv. Tutor.*, 18, 4, 2624–2661, 2016.
29. Rao, B., *et al.*, Environmental impact of civil subsonic jet aircraft: Noise pollution, emission impact on air quality, and health effects. *Environ. Technol. Innov.*, 23, 101530, 2021.

14

Drone-Based Application of Warehouse Logistics – A Case Study

Sangeetha Radhakrishnan[1]*, A. Prasanth[2], K.K. Devi Sowndarya[3] and Ahmed A. Elngar[4]

[1]Department of Computer Applications UG, School of Computing Sciences, Vels Institute of Science, Technology &Advanced Studies (VISTAS), Chennai, India
[2]Department of Computer Science and Engineering, Vel Tech Rangarajan Dr. Sagunthala R&D Institute of Science and Technology, Chennai, India
[3]Department of Electronics and Communication Engineering, DMI College of Engineering, Chennai, India
[4]Faculty of Computers and Artificial Intelligence, Beni-Suef University, Beni-Suef City, Egypt

Abstract
The logistics sector has seen great promise for drone technology. By 2027, this market is expected to grow by $29 billion, averaging 20% yearly growth. Particularly in logistics and warehousing, drones have applications. Thanks to recent technological advancements like optical navigation and sensors, drones can now be used indoors. Apart from the advancements in drone technology, another important factor contributing to the growth of warehouses is the worldwide e-commerce industry. Drones can be used, for example, in intra-logistics and automated inventory checks.

The chapter offers a case study on drone-based applications of warehouse logistics. Three distinct areas of drone-based applications will be explored such as indoor intra-logistics, inventory management, and inspection & surveillance. Among all of these applications, the logistics sector can benefit greatly from drone utilization. For instance, they can move components from storage facilities to factory workshops. The ability of drones to fly along predefined courses and carry

**Corresponding author*: sangeetha0988@gmail.com

A. Prasanth, Rajesh Kumar Dhanaraj, Munish Sabharwal, Vandana Sharma and Seifedine Kadry (eds.) Drones for Transportation Logistics and Disaster Management, (299–328) © 2026 Scrivener Publishing LLC

cargo holds good capability for applications indoors, including prompt tool delivery and lubricant to an area. However, there are significant intra-logistics limitations with regard to payload, gripping/placing motions, and navigation.

Keywords: Drones, warehouse logistics, inventory management, machine learning techniques

14.1 Introduction

Sustainable logistics using various modes to deliver packages from one place to another are necessary for shipping commodities by air, sea, and land/ground. Sea transportation, for example, uses boats and ships, land/ground transportation, trucks, vehicles, two-wheelers, drones, and airplanes. Most last-mile delivery services get help from trucks. Actually, the most common instrument used in urban logistics is the truck. However using trucks for last-mile delivery has drawbacks as well [1]. For example, trucks can't deliver to some remote areas that are difficult for them to access, they emit pollution into the air when they run out of gasoline, they can't deliver packages on time because of traffic congestion, etc. Bicycles are one possible eco-friendly means of transport for last-mile deliveries. Bicycles are used by retail businesses like DoorDash and Uber Eats to deliver packages to customers in cities. Nevertheless, there are disadvantages to bicycle delivery, including the riders' exposure to traffic risks, the need for longer delivery times, and the vehicle's restricted capacity for freight. Bicycle delivery works well in neighborhoods generally, but it might not be as effective in cities.

Drones can now be used for last-mile deliveries in addition to the existing modes of transportation, thanks to advancements in drone technology. Drone technology is the invention of Abraham Karem. He began developing prototype aircraft as a youth because he had a strong interest in aviation. Following his immigration to the US in the 1970s, he established Leading Systems Inc. Leading Systems Inc. in the garage of his house. Hence began producing the Albatross, his first drone, and then the sophisticated Amber, which was the forerunner of the renowned rapacious drone which earned the title of "Drone father."

A drone's thrust system, which enables it to fly and make aerial maneuvers, is made up of speed-specific electric motors and impelling force. Drone kinds were classified by Hassanalian and Abdelkefi according to their weights, attributes, and capacities. Drones can be divided into four main categories: multirotor, single-rotor, fixed-wing hybrid, and Vertical Take-Off. These categories are determined by the propulsion system and structural design. Global industries gain from drone use in a number of ways, including increased productivity and job efficiency, reduced effort and expense of production, enhanced precision and reproducibility, improved customer interactions, service, and extensive protection issue resolution.

More retail businesses are able to embrace electronic commerce or e-commerce, thanks to the Internet and digital transformation. This is the main cause of the increase in the amount of parcels sent worldwide, which climbed from 89 billion in 2019 to over 132 trillion in 2019 and reached 250 trillion by 2028 [2]. There were only roughly 11.2 billion packages delivered in the USA by UPS, FedEx, and Amazon in 2019.

The main mode of transportation for last-mile deliveries is trucks. In addition to causing air pollution and noise pollution, fossil fuel trucks also lead to traffic jams, crowded parking lots, accidents, increased delivery costs, and other unanticipated ground vehicle problems. Approximately 4% of all vehicles on American roads are delivery trucks, which also account for 50% of the country's emissions of nitrogen oxide, 54 % of the particulates, and 9 % of the greenhouse gasses overall. Thus, a lot of retail businesses, such as Wal-Mart, amazon, UPS, and others, are beginning to use drones as a legitimate way of transport because they provide speedy, environment-amiable deliveries and lower running costs along with fewer human involvements. Other businesses, like Mercedes Benz, started cooperating with Matter Net, a manufacturer of drones, to introduce electric vans and drones which were used for the high delivery demand of the e-commerce commodities. Drones are used not only to deliver regular parcels but also to deliver food, vaccines, and other life-saving supplies during disaster relief efforts [3]. Drone delivery presents a number of challenges, though, including lowering environmental damage, lowering delivery costs, tracking parcel deliveries, preventing occasional delivery delays, keeping up with evolving technologies, etc.

The key motive of this case study is to offer a helpful overview of recently released research findings of drone application in warehouse logistics. The study's conclusions are assessed in light of major technological issues like data transmission, battery management, cargo distribution optimization, routing, and environmental preservation and the application areas of drones along with the regression algorithms.

14.1.1 Warehouse Logistics

Logistics can be defined as the careful planning, organizing, managing, and carrying out of complex tasks. A number of industries, like logistics warehousing, deal with the transportation of data and physical goods.

Thus, the many complex facets of warehousing, including management, organization, and transportation, are included in logistics warehousing. This includes the movement of tangible inventory, such shipping and receiving, as well as the flow of more intangible assets, including time and information [4]. Furthermore, warehouse logistics may involve safety rules and client returns, bug extermination, maneuver-impaired commodities, and HR management. To put it another way, warehouse logistics refers to all of the guidelines, practices, and planning instruments required to maintain the efficiency of your warehouse operations.

14.1.2 Challenges of Warehouse Logistics

It is essential to understand the exact location of every inventory item. These controls are crucial for efficient operations and strong revenue growth, but they are practically hard to implement without the right tools. Beyond these immediate issues, inventory control, management of the supply chain, cost restraints, HR, management of risk, and protection are a few of the many aspects of warehouse challenges. Thus, the queries to get answers are how to maintain sufficient offerings to satisfy customers, exercise enough control to safeguard your revenues, and still maintain enough flexibility to remain competitive.

14.1.3 Improving Warehouse Logistics

Warehouse logistic managers have to explore with a humongous amount of work. Luckily, there is an increasing amount of extremely useful and powerful tools at your disposal to help meet the needs of the customers, their demands, and the warehouse's needs. One can utilize advanced Warehouse Management Systems (WMS) to keep an eye on their warehouse in real time and give all of their employees the tools they need to operate it successfully, financially, and efficiently [5]. Actually, a WMS is a definite method to enhance warehouse logistics in addition to being necessary in today's warehousing.

Beyond the capabilities of an inventory control system, warehouse management systems manage every facet of warehouse logistics, including order fulfillment and inventory control and administration. The mobile tools that many WMSs now include allow your employees and warehouse managers to access the system *via* a smartphone while they are on the go. In order to improve warehouse logistics from inventory receipt and shipping, WMS also combines authentic management tools and a Warehouse Control System. This creates a synergy throughout the entire warehouse.

14.1.4 Advantages of Warehouse Logistics

The advantage of well-managed logistics warehousing is straightforward: higher profits. The warehouse runs more efficiently when everything goes according to plan—people, procedures, and systems all fall into place—in inventory, at the right time, the right item should be sent, when stock is added back in as needed, and when finding error is reduced. Maximized revenue is the result of fewer mistakes and issues. A robust WMS that oversees your warehouse logistics allows you to:

- ✓ Ensure accurate, current inventory counts: Know how much inventory you have and where it is located in your warehouse.
- ✓ Cut down on returns: A precise inventory check is necessary to ensure that the right item is sent the first time.
- ✓ Auto-replenish stock: Instead of waiting until you run out of stock before placing another order, let your WMS automatically restock inventory as stock levels fall.
- ✓ Make the most of warehousing space: Few WMS automatic warehouse tasks like rotation of goods and pick to cut down on the amount of floor space needed for staff. As a result, your warehouse can hold more merchandise.

Improved demand forecasting, more brightness, and transparency are further benefits of WMS, tracing commodity, decreased picking errors, optimized procedures, effective labor allotting, and enhanced user support. These elements ultimately result in lower operating costs and increased revenue.

14.1.5 Cogitation of Warehouse Influencing the Warehouse Logistics

Unquestionably, a strong WMS can optimize logistics which changes operations. However, a strong software tool can address only human error. Real-world warehouse considerations have a big impact on your logistics. The use of your WMS and equipment requires extensive training for your warehouse managers as well as, to some extent, all participating employees. Your logistics efforts must be coordinated with those of your transportation and associated contractors and suppliers.

Any modifications to your storage, such as new racking or reorganized shelving, need to be recorded in the system in order to prevent issues from arising prematurely. The logistics of a warehouse are dynamic [6].

They alter in accordance with your personnel, inventory, physical needs, and any other variables under their control. In the end, efficient operations depend on alertness, appropriate training, and changing policies even though tools and processes can be employed in place to help manage and control warehouse logistics.

14.1.6 Importance of Warehousing in Logistics

Transporting goods from the manufacturer to the end consumer requires logistic warehousing, which is vital for supply chain management. According to the benefits of logistical warehousing, an inefficient warehouse operation can paralyze the complete supply chain by causing delays, blockages, or a shortage of goods. Every logistical task carried out in a warehouse prevents it from becoming a supply chain bottleneck. In order to support manufacturing processes, warehouses must keep enough supplies, raw materials, and packaging on hand to support production and ensure continuous output. The methodical receipt of completed goods and delivery to the particular receiver can be done by permanent ingress to the goods. It's crucial to keep in mind that Such warehouses facilitate marketing activities by giving continuous access to the materials needed to finish a campaign or create promotional sets.

14.1.7 Aspects of Warehouse in a Logistics System

Transport: The important aspect of logistics that involves moving goods from the supplier to the customer is transportation. That's how commodities are traveled to the customer from the supplier. They don't deliver commodities to the buyer directly. From the location of the production to the intended destination, it should be transported.

Carry out quality controls: Accurate records of every product's size, weight, quantity, and configuration kept in a central warehouse are necessary to ensure that the right part is reached at the right location at the proper time. Visual verification and code reading from the container or commodity are instances of authentic quality control methods [7]. Verification checks which are more intricate and costly for a commodity or product are becoming more crucial.

Inventory management: The goal of inventory management is to maintain the lowest possible carrying cost while maintaining sufficient inventory to satisfy customer demands. This system is effective at managing inventory to meet customer demands in addition to accounting tasks like inventory valuation and management.

Product quality: A commodity is considered good quality if it fulfills the customer's expectations. Customers request certain goods in different quantities when placing an order. Companies typically use facilities they hire out to create their goods in batches. [8]. Storage can ensure that the order is fulfilled quickly and help reduce the amount of time for the product flow.
Customer service: Getting goods to a specific place and time is the logistic responsibility. Supply warehouses are essential because they facilitate the scheduling of a manufacturing schedule (while preserving stock levels), resulting in expedited order fulfillment. It is imperative to bear in mind that supplier costs associated with stock shortages are unpredictable and client demand is erratic.
Protection from unexpected events: A company would not want the production process to come to a stop due to delays in the supply of raw materials. In distribution warehouses, unforeseen circumstances can also arise [9]. Items damaged in transport have an effect on the inventory level and the efficiency of the fulfillment of the order smooths out the flow of the commodities which is especially crucial when there are seasonal variations.

14.1.8 Purpose of Logistical Warehousing

The purpose of logistical warehousing is classified into three groups. The realization of warehousing tasks and activities includes all activities that go into completing warehouse projects, including receiving and storing goods, recording and storing stocks, releasing supplies, preparing paperwork, and reporting losses that occur during the warehousing process. The warehouse management procedure allows supplies to be received, stored, and released as long as material turnover, supply protection, ongoing supply maintenance that takes sensitive goods into account, all material records, and stock control are documented [10]. Warehouse management encompasses organizing, coordinating, controlling, and planning. These activities relate to:

- Organizing a warehouse operation and completing tasks according to different deadlines.
- Operations related to technical and human resource management and procedures.
- Controlling the Warehouse operation (documentation, specification, and placement).
- Putting organizational authorizations into practice.

Location registration and building warehouses: A warehouse's position within a supply chain affects its setup, tools, and operational planning. It also plays a significant role when it comes to the location and operations of warehouses. For example, product conditioning in conventional warehouses shortens the amount of time that items are stored in the warehouse by placing them close to facilities or in logistic centers.

14.1.9 Kinds of Warehouse Logistics

Public warehouses: Public warehouses are owned by government organizations and are open to commercial enterprises. The use of these warehouses for both personal as well as business is allowed.

Private warehouses: Private warehouses are those that are owned by wholesalers, distributors, or manufacturers in private. Big-box retailers as well as internet marketplaces own their personal owned warehouses [11]. For eCommerce SMBs looking to establish a significant, extended strategic presence in a prominent area, private warehouses can still be a good option, despite the fact that they are frequently more expensive than public warehouses.

Bonded warehouses: Bonded warehouses are one kind of warehouse that can hold imported goods until customs duties are paid on them. To ensure that businesses renting space from them won't lose money when their goods are distributed, the government posts bonds. Until the goods are released, businesses using them to store goods are exempt from paying customs duties. Restricted items can also be stored there while their documentation is being completed. Goods can be stored for a long time in a bonded warehouse.

Smart warehouses: Automated fulfillment, management, and storage procedures are exploited by AI. Automation can take the form of management software, robots, or drones that perform tasks like weighing, packing, moving, and storing goods. Large-scale smart warehouses are utilized by Amazon and Alibaba, for instance, to speed up order fulfillment and lower the possibility of human error.

Consolidated warehouse: Consolidated warehouses collect small shipments from multiple suppliers and combine them into one large shipment prior to distributing larger shipments to clients. The only problem is that all shipments go to the same place. Conversely, consolidated warehouses are a very inexpensive way to fulfill orders, especially for startups and small businesses. Additionally, consolidated warehouses require less inventory volume and an initial capital investment, which makes them a great option for newly established eCommerce SMBs.

Distribution Center: It is a kind of storage facility that is specifically generated to fulfill specific needs. Only temporary needs are satisfied by the item and storage is moved suddenly throughout the supply chain. A substantial amount of merchandise is received and delivered to retailers and resellers in a brief amount of time. Distribution centers usually distribute some food and perishable goods in a single day. Distribution centers can range widely in price depending on the products they house, but they are generally cheap to rent.

14.2 Drones

14.2.1 Classification of Drones

The development of the micro air vehicle, a smaller version of an aerial drone, has led to an increase in the requirement for intelligence operations during the past few decades. Subsequently, proper efforts are made these days to create and construct tiny air drones for specific purposes [12]. Subsequent of these efforts, numerous small drone varieties with a variety of shapes and flight modes have been developed. Figure 14.1 depicts a complete classification of all drones that are currently in operation.

Drones can be divided into categories based on how they perform. Weight, wingspan, range, wing loading, range, greater altitude, speed, endurance,

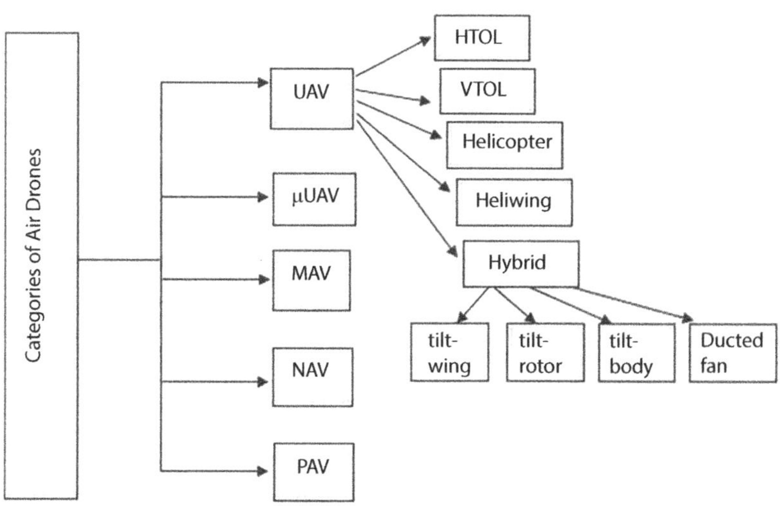

Figure 14.1 Classification of the air drones.

and manufacturing costs are important design parameters that distinguish various drone types and offer useful categorization systems. They can also be divided into groups based on the types of engines they have. For example, Unmanned Aerial Vehicles (UAVs) usually use fuel engines, but MAVs use electrical motors. Depending on the model, different propulsion systems are employed in different drones.

UAV

The intended use of the vehicle, the materials equipped in the construction, and the intricacy and expense of the control system are the major features which differentiate small drones like MAVs and NAVs from UAVs. The sizes and configurations of UAVs vary widely. Their wingspan, for example, could be as big as a Boeing 737 or as small as a drone controlled by radio. Different mission requirements led to the creation of various types of UAVs. So, it is always beneficial to segregate UAVs based on their capabilities. Figure 14.1 illustrates the different classifications of UAVs: helicopter, heliwing, tilt-wing, tiltrotor, tilt-body, and ducted fan, Horizontal Take Off Landing (HTOL), Vertical Take Off Landing (VTOL), and unconventional types.

μUAV

A UAV that is manageable for one person to carry is called a little UAV. It takes off without a runway and is manually launched. UAVs are smaller than hand-launched UAVs but larger than Micro Air Vehicles (MAVs) that a soldier can carry [13]. UAV configurations differ greatly. UAVs can be classified as HTOL, VTOL, helicopter, ornithopter (flapping wing), ornicopter, cyclocopter, hybrid model (tilt-wing, tiltrotor, tilt-body, and ducted fan), or unconventional forms. Although they are often smaller in size and weight, HTOL, VTOL, tiltrotor, tilt-wing, tilt-body, ducted fan, and unusual μUAVs are equivalent to UAV types.

MAV

Micro MAVs, or micro aircraft, are usually smaller than 100 cm and weigh less than 2 kg. These drones are divided into nine categories: VTOL, rotary, tiltrotor, ducted, fixed, flapping, ornicopter, helicopter, and other unique wing types. These drones are equipped with optical, acoustic, chemical, and biological sensors. Various engineering disciplines, such as computer, mechanical, electrical, and aerospace engineering, are becoming interested in different kinds of micro air vehicles. These aerial drones are limited by the Defence Advanced Research Projects Agency program to a length, width, or height of less than 150 mm and a weight of between 50 and 100 g. Nevertheless, the definition of MAV is modified with the introduction of NAVs and PAVs [14]. Subsequently, according to the review, these drones weigh between 50 and 2 kg and range in size from 15 to 100 cm. Due to their smaller size, MAVs may perform in a wider range than UAVs. The RAND Institute conducted the first thorough investigation on MAV in 1993. Because of the rapid advancements in microtechnology over the last ten years, MAVs have gained a lot of attention. Consequently, a number of research studies on microplanes were conducted in the years that followed. These kinds of planes are small in size and have the ability to fly slowly. MAVs primarily fly at low altitudes for a variety of purposes, including mapping, tracking particular targets, and monitoring hazardous areas. For this reason, precision in the design and construction of these aerial vehicles is necessary. Because micro air vehicles have non-traditional flight missions and require less time to design, produce, and evaluate than conventional UAVs, their conceptual designs typically differ from conventional UAV designs.

NAV

Apart from the micro air vehicles, the Defence Advanced Research Projects Agency initiated a Nano Air Vehicle (NAV) program. NAVs are defined as tiny, light drones that weigh less than 50 g. The drones can fly up to 100 meters above the ground and have a range of less than one kilometer. The various configurations of NAVs include rotary wings, flapping, and fixed wings.

PAV

Researchers have attempted to design and build drones the size of insects in the last few years. To achieve it, a brand-new class of drones known as Pico Air Vehicles (PAVs) was established. There are only a few different kinds of PAVs due to their small size and lightweight. PAV class designs include flapping wings and quadrotors [15]. Recent research has focused more on flapping wing PAVs than rotary wings (quadrotor) because of the remarkable ability of flapping insects to fly, including hovering, sudden acceleration, and quick turning.

14.2.2 Application Areas of Drones

The development of new services and a wide range of unmanned operations applications is facilitated by the improvements in the possibility of sensor deployment on UAVs. A few well-known UAV application areas are mentioned below.

Military

UAVs play a crucial part of military monitoring operations. Several countries have included UAVs in their plans for defense strategies. Nations use these aerial robotic devices for marine monitoring of vital shipping routes, border security, enemy detection, and anti-poaching operations. Affordable, dependable, and versatile UAVs are currently changing the game when it comes to aerial surveillance, monitoring, and surveying any given area to prevent any illegal activity. For example, drones can be used for surveillance to spot possible threats and monitor any movement inside a forbidden area. By automatically identifying alerts and requiring little manual labor, UAVs can complete these tasks.

Construction

As-built mapping, construction monitoring, and site inspection have all become easier, faster, and more efficient thanks to UAVs. Ensuring the caliber of the work being done on the site is guaranteed by managing building projects from start to finish. It can provide reports with images, videos, and three-dimensional mapping to potential stakeholders. This technique can significantly help applications for building and infrastructure inspection. UAVs are used in power line construction project monitoring, gas pipeline monitoring, GSM tower inspection, and power line monitoring.

Remote Sensing

Currently, amateur drone technology is used to collect high-resolution image data of remote areas, mountaintops, and coastlines. UAV technology can be integrated with data from ground-based, space-based, and airborne remote sensing systems [16]. High-quality, temporally and spatially resolved observation is made possible by UAVs' affordable and lightweight construction. Disease detection may benefit from the usage of UAVs with distant sensing capabilities, locating oil and gas, estimating yields, hydrological modeling, conserving biodiversity, surveying terrain, mapping forests, and monitoring crops, water quality, and drought.

With the use of this technology, archaeology and cartography are now actively involved. It can also be used to create 3D environmental maps and crowdsource mapping. Drones that are reasonably priced can assist in providing up-to-date mapping information without breaking the bank when it comes to land planning.

Precision Farming

Precision agriculture can make use of UAVs to collect water quality, schedule irrigation, identify weeds, monitor and control crops, apply pesticides, and detect diseases. Using UAVs for precision agriculture is a time and cost-effective method that boosts crop yields and farming systems' profitability and efficiency. Furthermore, UAVs aid in the control of infestations, Farming management, chemical-based spraying, and weed monitoring. This results in increased crop yields to meet specific production requirements. Precision farming could change significantly with the use of UAVs and remote sensing. In addition to providing temporal, spatial, and spectral resolution, it can also provide detailed data on vegetation height and multi-angular observation. The agriculture industry can be greatly impacted by UAVs through intelligent aerial reconnaissance. Sensors and cameras equipped with UAVs can keep an eye on crop health, including temperature, chlorophyll content, foreign contaminants, and leaf thickness.

Disaster Management

UAVs are able to reach disaster truck locations that are dangerous for human safety in the case of a natural or man-made disaster, such as a terrorist attack, flood, or tsunami. These disasters might cause significant damage to transportation, water, electrical, and telecommunications utilities. UAVs can traverse debris, help collect information, and offer emergency repairs when necessary. It can also cooperate with effective countermeasures, timely disaster alerts, and disaster estimation. In the event of a wildfire, drones fitted with fire extinguishers can track, monitor, and investigate any area without endangering human life.

Real-Time Monitoring of Road Traffic

One area of application where the UAV integration created a lot of focus is road traffic monitoring systems. UAVs can be used to fully automate the transportation sector. Cops and support field teams will all be automated as part of this. UAVs that are dependable and intelligent can help automate these components. UAVs are a new and promising tool for collecting data on traffic conditions on highways. Drones can monitor large road segments at a lower cost than traditional monitoring equipment like looping detectors, CCTV cameras, and sensors [17].

Local law enforcement can employ drones to observe traffic accidents clearly or to carry out a comprehensive security operation against crimes related to traffic, such as auto theft. Additional uses include identifying vehicles, conducting raids on allegedly suspicious vehicles, pursuing armed thieves, and pursuing people who disobey traffic laws. Additionally, it is helpful to prevent traffic jams and mass congestion by detecting accidents and vehicle overspeeding.

14.2.3 Areas of Indoor Drone Use Cases

The automation of modern warehouses has begun to heavily rely on drones. Drones are becoming more and more popular because they are autonomously able to fly and hover, prevent challenges in various storage facility layouts, maneuver inside, soil accurately, and possibly work in fleets. Managing the intra-logistics of items, and review and monitoring are among the greatest potential applications for indoor drones in warehouses.

Inventory Management

The following tasks related to inventory management can be carried out by drones: cycle estimation, item search, stock audit, buffer inventory maintenance, and taking stock. Stock taking is the process of physically verifying the quantity of items stored in warehouses. Usually, taking stock is done once a year or at the end of the fiscal year. On the other hand, cycle counting refers to the practice of counting a subset of a warehouse's inventory more regularly. To do this, inventory control employees usually work in small groups two or three days a week.

There are still a few drawbacks, despite the fact that in contrast to once-a-year one-time inventory checks, this method increases inventory accuracy. Counting cycles is labor-intensive (requiring multiple inventory staff members), dangerous (involving high-altitude operations), costly (labor expenses), labor-slow (manual), and prone to error (involving repetitive tasks), among other drawbacks. This process can be streamlined with the help of drones. The three primary objectives of deploying drones for inventory management are to increase inventory accuracy, reduce labor expenses, and lessen worker hazard.

Intra-Logistics of Items

Another application for drones is intra-logistics. The ability of drones to transport objects and navigate prearranged paths presents a great potential of opportunities for indoor applications, including the prompt delivery of tools, spare parts, and lubricants to location. However, there are significant intralogistics limitations with regard to payloads, gripping/placing motions, and navigating.

Surveillance and Inspection

In smart factories, drones are a useful alternative to manual surveillance and assessment work. Drones are already used for inspection in a wide range of industries, including petrochemical, oil and gas extraction, development, and energy production. The usage of drones for inside inspections is growing in popularity. Drones can inspect storage facilities' walls, roofs, racks, pallet positioning, and other surfaces.

The development of storage facilities and client demand has made inspection processes expensive and difficult. Expert examiners are usually required for duties related to indoor assessments. Indoor drones are perfect for operations that need to be seen and inspected at high altitudes or dangerous places. In order to discourage theft and other undesirable behavior, drones can also be utilized for routine surveillance missions.

14.2.4 State of Drone Technology

The most difficult aspect of reaching level 5 automation is indoor navigation. Due to insufficient navigation accuracy, it is unlikely that drones will be able to fly themselves in almost any of the fields of application in the near future. However, recent developments indicate that high-precision indoor navigation by drones is imminent. Algorithms based on vision offer a possible path toward 100% accuracy. One of the most sophisticated visual Simultaneous Localization and Mapping (SLAM) algorithms available today has a 5 cm accuracy. However, visual-based SLAM performs better than other localization methods like UWB.

Radio frequency-based technology is not a good substitute for indoor drone localization because it is frequently used to track floor conveyors, like Pallet trucks or forklifts with a relatively low precision of ten to thirty cm. It's possible that additional cutting-edge technologies will improve the precision of visual-based SLAM. One promising technology for indoor navigation is light detection and ranging, or LiDAR [18]. It is a high-precision method that uses pulsed laser light to illuminate a target to determine its distance. Leica Geosystem and DJI collaborated to develop a drone that combined cameras and LiDAR sensors. The so-called "Aibot" covers a 10-hectare (100,000 square meters) area with an accuracy of 2.5 cm. Nevertheless, the drone's nearly 10-kilogram weight places limitations on indoor spaces.

Vtrus Technology is an example of a cutting-edge indoor navigation system that is currently available. To achieve the highest accuracy possible, this technology combines 360° wide-angle cameras with 3D scanners and depth sensors. Up to 30 times per second, SLAM processes trillions of pixels in a camera concurrently. Furthermore, they can replicate a three-dimensional map to find the drone within it. Similar in technology to the Vtrus, PINC offers navigational indoor remedies, inventory, and vision-based eyesee. For instance, the PINCs solution can localize inventory in addition to checking it. According to InventAIRy, their software can perform vision-based inspections and provide details on the quality of pallets, packaging, and potential product damage. The secret to successfully implementing indoor drones in warehouses and other indoor environments is to master the problem of indoor navigation.

To further increase localization accuracy, Infinium Robotics and Geodis-Delta Drone interact drones with a programmed/automated ground vehicle. The fact that wired drones can connect to a ground vehicle *via* a cable extends the life of the batteries. Subsequently, the flying time of this technology can reach four hours, while that of technologies (like inventAIRy) without an automated ground vehicle can reach up to thirty minutes. However, using wired drones limits their ability to hover and maneuver, and their integration into warehouse environments becomes more challenging.

14.3 Usage of Drones in Inventory Management

The following are the several advantages of drones in inventory management.

- ✓ Because counting cycles and audits of inventories are so quick and efficient, they are utilized more frequently than manual labor. This makes it possible to update inventory accuracy more frequently and make inventory errors easier to track.
- ✓ Drones operating on their own cannot be manually controlled, unless they are in an emergency. The automation technology that comes with the drone can be used to complete the flight mission by someone with little to no prior drone flying experience. Therefore, additional material handling equipment with related facilities are not required for anyone to perform tasks like cycle counting.
- ✓ Pilots will be working from a distance or when they can see the drone. As a result, no one is actively working in the area. Hence, the workplace security is raised.
- ✓ Even though drones are effective tools for accurately controlling labor, their accuracy can disqualify them from consideration. Drone accuracy can reach 97% to 99% because of physical changes made during storage, human error, or other factors. The ability of technology to examine errors, such as products that are not registered in the system or labels that

are not readable is also helpful. When the technology is first put into use, these incidents might occur frequently, but with time, the system becomes more accurate.

14.3.1 Challenges in Drone Technology Implementation in Warehouses

- *Incompatible WMS*: Some of the current WMS can't be fully customized to take advantage of this new technology, or it may not be able to handle it. Therefore, before choosing the best drone technology partner, suitability needs to be verified.
- *Labeling accuracy*: Labels are not evenly distributed throughout a given racking or shelving bay due to different stock keeping unit box sizes. Pasting those labels to each pallet, box, or bin is inconsistent; that is, they might not always be pasted in the same direction. These labels may have a build-up of dust or printing problems. Scanning may be hampered by such problems. These difficulties are substantially overcome by the most recent developments in scanning, cameras, and machine learning-based applications. Drones must be trained to interpret these events and transform them into the intended result.
- *Inventory view*: The capabilities of current technology only allow us to select the pallet, box, or bin's face. It is therefore only effective for single deep storage.
- *Disorganized storage*: Drone use in messy storage can be very challenging. If the correct procedure for label maintenance to face drones is not followed, errors will occur, and the drone will not be able to capture stock keeping units. Furthermore, it has been observed that incorrect box stacking can lead to crumpling and partially or completely invisible labels, which reduces scan accuracy. If any boxes protrude too far into the aisle, there may be a safety risk. Experience flying could reduce these kinds of situations.
- *Drone flying time*: Drone flying time is restricted by current battery-technology. This means that after brief cycles, the drone would need to return for charging. Multiple drones deployed sequentially could solve this problem; for example, when a drone's battery runs low, it returns for charging

and sends out another drone to take over the task. Another option is to use a cable to connect the drone to the mother autonomous vehicle, which moves on land. Drone deployment takes longer with this arrangement because larger vehicles with batteries are available to support the drone's power. Another option is to have the drone fly back to a wirelessly charged landing pad. This allows for even more autonomy by removing the need to go outside and replace dead batteries with new ones.

14.3.2 How Drones can Simplify Inventory Management

Companies that use drones in their warehouses stand to gain a lot. Drones can first expedite warehouse operations by automating repetitive tasks like picking, scanning, counting, and replenishing inventory. Drones can quickly inventory all of the stock in a large warehouse in a few hours. Drones are capable of recording product temperature, humidity, and other relevant information in addition to taking images of the inventory that is kept on the shelves. With the use of all this data, one can evaluate the risk of perishable product deterioration and implement the appropriate countermeasures to minimize waste. Additionally, by automating the gathering of visual data, drones can streamline warehouse inspections. This implies that in order to examine the state of the products, inspectors won't need to scale a ladder or enter small areas of the warehouse. Alternatively, they can just go over the information that the drone has gathered.

Drone usage also improves worker safety for humans. Inventory kept in odd locations throughout the warehouse that are out of the direct line of sight or inaccessible to humans can be readily scanned by them. Additionally, they ease the flow of handling equipment by reducing traffic in working aisles. Drones are also used extensively in intralogistics, which is the goods transportation inside warehouses or between warehouses and nearby workshops. Drones can transport cargo between locations in warehouse operations and follow pre-planned flight routes.

14.3.3 Drones Impact in the Product Delivery Ecosystem

Drones are predicted to have an important effect on the environment surrounding product delivery in addition to inventory management within warehouses. Prominent retailers have already begun utilizing drones to complete same-day deliveries in an effort to shorten delivery times. Up to 36 Wal-Mart locations across seven states currently provide 30-minute drone delivery services. At the moment, the retailer offers drone delivery for up to 20,000 items. Drones can reduce delivery costs in addition to delivery times. This can be a big help for businesses trying to make their way through a very uncertain business climate. Lastly, considering the persistent labor shortage and growing labor expenses in many areas, using drones makes financial sense. The swift expansion of warehouses has posed a challenge for companies seeking to staff empty positions.

14.3.4 Technical Challenges

Routing

Routing is a crucial problem to solve because it seeks to identify the best delivery areas which enable last mile drone delivery from an operational and financial standpoint. Vehicles are required for transporting parcels, supplies, and drones. Thus, routes for both trucks and drones should be considered when planning a last-mile drone delivery. Drone routing could include a single drone along with a truck, several drones and a single truck, or several drones and numerous vehicles. The best set of connections used by a fleet of automobiles to transport goods to a particular customer base long before drones were used in last-mile delivery. Their concern was the best path for a fleet of delivery trucks carrying gasoline from the petroleum product storage to the stations.

Cargo Distribution Optimization

Since it allows for the appropriate package distribution for last-mile deliveries using transport vehicles, drones, bikes, and machines, cargo distribution optimization is an important problem to solve. Delivery cars vary in terms of energy consumption, safety ratings, dependability, and cargo capacity limitations. Conversely, packages vary in terms of their weights, sizes, locations, needs, safe handling specifications, and costs. Combinatorial optimization techniques such as the knapsack problem, bin packing problem and cargo distribution scheduling, may be used in cargo distribution optimization. Two of the main supply-chain management optimization issues are allocation of cargo among delivery vehicles and vehicle routing.

Battery Management

Drone battery management is another important problem that must be fixed to achieve dependability and energy economy. Drone delivery in the last mile for battery management can take two forms: (i) battery replacement/charging strategies, or (ii) drone conditions and settings optimized to save energy for sustainable usage of battery.

Numerous studies focus on the subject of drone battery consumption. It is contingent upon various factors, including package weight, drone speed, flight paths, altitudes, weather, energy save mode, and additional drone conditions. In general, optimization of route, speed of drone, altitude and cargo optimizing, and weather-based scheduling are the main areas of focus for battery consumption optimization.

Researchers recommend a variety of battery replacement and charging techniques. Battery stations, for instance, could be incorporated into the delivering areas such that drones could halt by and swap out or perform battery recharging. As seen in Figure 14.2, battery stations can typically handle three different methods of recharging drone batteries: Battery swapping, battery recharging *via* contact, and wireless recharging.

An operator must handle the battery swapping process by hand. In this instance, a major obstacle is the battery limitation issue. They offered three options for fixing the issue. Drone swapping is the first solution. This implies that a new drone assumes its tasks when a functioning drone needs to be recharged. This enables the drone to reach the charging area and use a charger based on contact to finish up. When it comes to drone swapping, the second solution is comparable to the first. Nevertheless, battery replacement stations have taken the place of battery charging stations. These stations call for human operators to use mechanical tools to interchange the battery of the drone. The last method is comparable to the prior one in that drone batteries are refilled using non-contact wireless chargers.

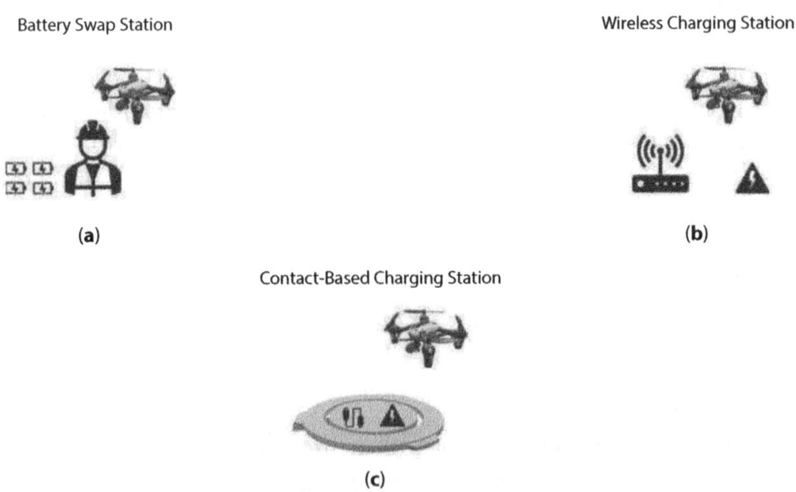

Figure 14.2 Three different kinds of drone battery recharge stations.

Data Communication

Data communication is a critical challenge that must be addressed to enable data collection, sharing, communication, routing, and management. It is possible to communicate data between devices, humans, devices, and humans, among other combinations. The Global Positioning System, cellular networks, flying *ad hoc* networks (FANETs), and networking are the key communication challenges.

The Internet of Things (IoT) attracts interest from a diversity of industries and has enormous future potential [19–21]. Unified computing, which includes cloud, fog, and edge computing, and ubiquitous sensing, also known as ubiquitous geo-sensing, which makes use of ubiquitous sensor networks and their real-time geographic phenomenon probing capabilities, were made possible by the IoT [22]. Control systems for lighting and the surroundings in a home ubiquitous computing system could be linked to wearable personal biometric monitors, enabling continuous and imperceptible modulation of the lighting and ambient temperature. Industrial cloud computing has the processing ability to analyze data and make decisions in addition to storing data [23]. It functions as an Internet-based data center equipped with a range of computing instruments for data analysis. This platform gathers data from devices, processes it, and presents it in a way that makes sense to people [24].

Unpredictable circumstances might arise while a drone is operating, so it must be able to sense its surroundings, process information, and make decisions on its own. By pushing the limits of technology, drones equipped with Smart services are made possible by sensors, actuators, digital control systems, drive mechanisms, propellers, and electronics. Figure 14.3 shows the utilization of various devices in data communication models.

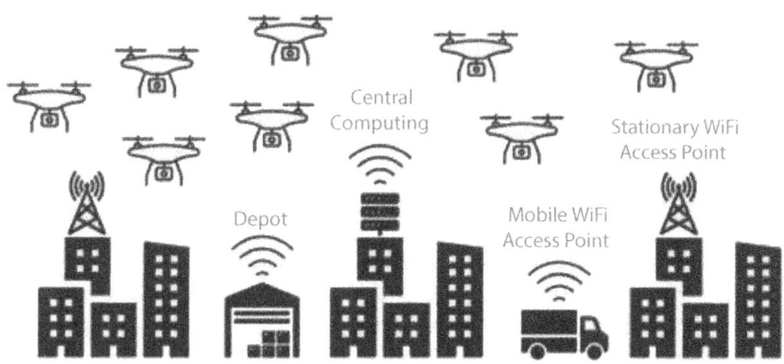

Figure 14.3 Secure data communication model.

Environmental Protection

Since last mile drone delivery encourages environmentally friendly methods for protecting the environment and ecosystems, security of the environment is an important issue to tackle. There are many ways to protect the environment, including cutting down on pollution and carbon footprints, increasing the delivery system's energy efficiency, using renewable green power sources etc. For last mile drone delivery, vehicles are required for transporting parcels, supplies, and drones. However, the majority of fuels used by trucks are greenhouse gases produced by fossil fuels. The emissions of greenhouse gases can be decreased by using alternative energy sources like hydrogen fuels, electricity, renewable natural gas, and biofuels.

14.4 Application of Drone in Warehouse Logistics

Data was collected in a warehouse setting by taking random paths through the simulation. The drone was designed to be able to move in both yaw and pitch in addition to three axes. The drone was navigated through the simulation by moving it in increments of 5 cm in any direction, and each movement was captured by the drone's camera at least once. The ArUco tag was identified as being present in the image, and its area, pixel-by-pixel coordinates in the camera, and presence or absence were all calculated using the tag ID [25].

In machine learning prediction, the aircraft's actual location data is used as the output value. For every location data set, a regression model was constructed. Machine learning and location prediction were carried out using the Virtual Fiducial Markers (VFM) collected information, where the real point of location is pre-defined because of the system simulation. Data was gathered for this procedure from 11,703 points in total. Test and training data were separated out of these datasets. A predetermined route was used to test the system's functionality using location prediction, with 8778 data for training and 2325 data for testing used.

14.4.1 Simulation Environment and Data Collection

A simulation setting measuring 20m * 25m * 15m was created to replicate the system's functioning within the CopelliaSim simulation program; 200 rounds of flights were conducted in the warehouse model at arbitrary heights, rotations, directions, and start-stop locations in order to prepare the dataset for supervised learning. Automated VFM data collection and feature extraction were carried out during the flights. The following were used: VFM, marker size, marker separators, and the camera's x and y pixel marker coordinates.

The warehouse was set up with a total of 27 racks arranged in six compartments around the center. In two of the warehouse's corners, the racks were positioned at 90° angles, and in the other two, at 135° angles. Boxes of various sizes, colors, and counts were added to every shelf. Every one of the 265 boxes in warehouses has a barcode tag appended to it. Every box had a barcode tag affixed to the upper right corner. Additionally, a 6*6 VFM ArUco type was used to label each of the 27 racks that were used in the simulation. The ArUco markers were numbered counterclockwise, with their location positioned in each rack's top-right edge.

The drone, which was supposed to move in three dimensions, moved in five-centimeter increments on one axis only in the simulation scenario. At every step, the drone yaw angle was also adjusted to 5°. Put differently, it completes a full 360-degree rotation in 67 steps. The drone was permitted to move as it would in actual dynamics while in flight. Drone camera data, pose, and position were all received while the drone was in motion. A picture was captured by the camera at every turn and axis that the drone made. Its coordinates in pixels within the camera, and its area was deduced based on tag ID, which was read from the VFM markers in the captured image [26].

The camera image is utilized to read the barcodes on the shelves' boxes. Stock control and warehouse space occupancy detection were two applications of the barcode data. The aircraft's compass information was also acquired in addition to this data. In machine learning prediction, the aircraft's actual location data is used as the output value. For every location data set, a regression model was constructed. Regression models were used to predict the aircraft's immediate location and pose in every area.

14.4.2 Identification and Positioning of Fiducial Markers in the Simulation Environment

Characteristics and contrasts between the devices equipped with cameras also make use of images labeled tracking methods. It is not necessary to know any of the features that were extracted ahead of time in marked tracking. Nevertheless, a great deal of computation is needed to complete the feature extraction and matching procedures. It takes a long time to complete this calculation. On the other hand, pose details using markers can be reliably and rapidly realized.

The visual indicators are intended to be applied on a level surface. Computer vision algorithms can identify each marker based on its unique identity and corresponding graphical representation. They are employed in physics, meteorology, aviation, robotics, augmented reality, and medicine. In the studies, a wide variety of VFMs may be found. We are able to generate VFM tags in every kind and form if we choose.

14.4.3 Machine Learning Techniques

ArUco markers are unique reference markers with a multitude of applications, such as robotics, machine vision, and augmented reality. ArUco markers have several unique features that make them ideal for this use, and they are made to be quickly identified by machine learning algorithms. The following is a list of ArUco marker attributes that are used as machine learning inputs.

- *ArUco ID*: A machine learning algorithm can read the distinctive identifier code that each ArUco beacon has, enabling it to discern between different beacons [27]. Applications where numerous markers must be defined and tracked at the same time will find this especially helpful.
- *ArUco area*: There are numerous uses for ArUco markers due to their diverse size range. This is especially beneficial for applications that need the placement of markers in various environments. The study's 20 × 20 cm dimensions were expanded to include markers. Nevertheless, the pixel areas of these markers differ based on how far away from the camera.
- *ArUco camera coordinates*: It is the center of the marker's vertical and horizontal axis coordinates, expressed in pixels, when the camera detects ArUco markers. Regression outputs included the vehicle's actual positions in the simulator (x, y, and z) and the rotation on the z-axis, or pose angle (yaw) γ.

14.4.4 Regression Algorithms

K-Nearest Neighbor

An algorithm for learning according to the k-nearest neighbors' values is called the K-Nearest Neighbors algorithm (KNN). A non-parametric method for either regression or classification is the KNN algorithm. The categorization of news articles was its initial use. Prior to using the KNN algorithm for learning, the separation of every data point in an examined dataset is computed.

The Euclidean, Manhattan, or Hamming distance functions are used to calculate this distance [28]. Next, for every point of data, the average of the k-nearest neighbors is determined. If the k value is set low, overfitting will happen, and the boundaries will jitter. If k value is set excessively high, the lines dividing the two will be less sharp, which will lead to underfitting. The distance calculation process, which increases the computational load owing to the increased amount of data, is the disadvantage of the KNN algorithm.

For classification and regression, the algorithm known as KNN is a non-parametric technique. By employing a distance function like Manhattan, Hamming, or Euclidean distance, the KNN algorithm determines each data point's distance from every other data point in the dataset. Next, for every data, the average of the K nearest neighbors is computed. The hyper parameter in the KNN algorithm is the value of K, and choosing a suitable value is essential to prevent either underfitting or overfitting. When dealing with big datasets, the process of calculating distance can be computationally expensive. The simplicity of the KNN algorithm is one of its main benefits. It requires little data preprocessing and is simple to implement and understand.

Adaptive Boosting

A particular kind of adaptive Boosting (AdaBoost), an ensemble strategy, leverages weak learners to create a stronger learning structure. Due to its low computational cost, a single level Decision Tree Algorithm is typically used as a slow learner. The four stages of AdaBoost are as follows: running n slow learners, computing the error values, assigning higher weights to learners with high error, and adding up all of the learners with their weights. Learning performance and computational cost both rise with the number of learners.

Random Forest (RF) Algorithm

By choosing certain features and samples at every decision node, the RF creates a forest of decision trees. A randomized subset of the information and a randomized subset of the information's features are chosen in order to apply the RF algorithm [29]. An example of a model that uses data to generate a set of rules that enable it to make predictions, and it is trained using these features and data. A forest of decision trees is the end result of repeating this process numerous times with various portions of characteristics and data. A new data point is predicted by running it through every forest's decision tree, and the average of every tree's prediction yields the RF algorithm's final prediction.

Extreme Gradient Boosting (XGBoost)

One kind of Gradient Boosting Algorithm which is intended to be quick as well as effective is called XGBoost. One decision tree at a time is fitted using a gradient descent algorithm, and each tree is made to make up for the errors made by the earlier trees in the ensemble. Initializing the hyperparameters of the model, like the learning rate and the maximum tree depth, is the first rule in implementing the XGBoost algorithm. The first decision tree must then be trained using the Gradient Descent Algorithm on the data. Next, using the prediction this tree produced, the remaining error among the true and the prediction value is calculated.

The second decision tree is then trained using this error as the target variable. The error is considered as the objective changeable every time, and this procedure is replicated for more trees. The last prediction is produced by adding the predictions from every tree after they have all been trained. XGBoost is a quick and effective algorithm that works well with complicated and big data sets. It has a great degree of customization, and tweaking the hyperparameters will boost performance. If the hyperparameters are not selected carefully, it is challenging to interpret the results and susceptible to over fitting.

Artificial Neural Networks

Computational models modeled after the nervous system are called neural networks. They are information-based systems made up of artificial synapses, which are connections between processing units known as artificial neurons [30]. Numerous disciplines, including, and optimization, pattern recognition, use Neural Networks. Multilayer neural networks are used to solve a variety of issues in domains like pattern recognition and system identification. They are composed of one or more cerebral layers concealed.

A neural network type with at least a private layer positioned among the input/output layer is called a multilayer perceptron. The input data and corresponding desired output are fed into the network during the supervised training process. To reduce the error between the desired and the predicted output, the neutron connection weightage can be changed. Conventional multilayer perceptron networks lack feedback from the hidden or output layers, and through the hidden layers, information is transferred from the input layer to the output layer. Because of its adaptability, this kind of network finds many uses in fields like system optimization and pattern recognition. The hidden output layer weightage is documented independently during the multilayer perceptron algorithm backward phase and yields successful outcomes in non-linear systems. The algorithm is made more general by using the momentum coefficient.

14.5 Conclusion

The anticipated case study provided the use of drones in warehouse logistics. Drone-based applications like inventory management, indoor intra-logistics, and surveillance were examined in this chapter. It summarized the characteristics of several drones such as UAV, μUAV, MAV, NAV & PAV. Further, the numerous challenges and constraints of warehouse logistics were discussed. The benefits of drone technology in inventory management have been scrutinized in detail. Finally, the machine learning and regression algorithms including KNN, Adaptive Boosting, RF, XGBoost, etc., were elaborated in this chapter. Machine learning-based drones can utilize their sensors to identify obstacles and evade them. This proper identification aids in mitigating accidents and increasing the efficiency of flights.

References

1. Thibbotuwawa, G.A., Bocewicz, P., Nielsen, P., Unmanned aerial vehicle routing problems: A literature review. *Appl. Sci.*, 10, 4504, 1–18, 2020.
2. Rejeb, A., Rejeb, K., Simske, S., Humanitarian drones: A review and research agenda. *Internet Things*, 16, 100434, 1–14, 2021.
3. Lin, I.C., Lin, T.H., Chang, S.H., A decision system for routing problems and rescheduling issues using Unmanned Aerial Vehicles. *Appl. Sci.*, 12, 1–20, 2022.
4. Almuhaideb, S., Alhussan, T., Alamri, S., Optimization of truck-drone parcel delivery using metaheuristics. *Appl. Sci.*, 11, 1–12, 2021.
5. Mohsan, S.A.H., Zahra, QA., Khan, M.A., Role of drone technology helping in alleviating the COVID-19 pandemic. *Micromachines*, 13, 1593, 1–17, 2022.
6. Szász, L., Bálint, C., Csíki, O., The impact of COVID-19 on the evolution of online retail: The pandemic as a window of opportunity. *J. Retail. Consum. Serv.*, 69, 103089, 2022.
7. Bilinska-Reformat and Dewalska-Opitek, K., E-commerce as the predominant business model of fast fashion retailers in the era of global COVID 19 pandemics. *Procedia Comput. Sci.*, 192, 2479–2490, 2021.

8. Raj, A., Mukherjee, A.A., de Sousa Jabbour, A.B.L., Supply chain management during and post-COVID-19 pandemic: Mitigation strategies and practical lessons learned. *J. Bus. Res.*, 142, 1125–1139, 2022.
9. Gupta, A., Afrin, T., Scully, E., Advances of UAVs toward future transportation: The state-of-the-art, challenges, and opportunities. *Future Transp.*, 1, 326–350, 2021.
10. Suguna, M., Shah, B., Raj, S.K., A study on the influential factors of the last mile delivery projects during COVID-19 era. *Oper. Manage. Res.*, 15, 1–14, 2021.
11. Sorooshian, S., Sharifabad, S.K., Parsaee, M., Towards a modern last-mile delivery: Consequences and obstacles of intelligent technology. *Appl. Syst. Innov.*, 5, 82, 2022.
12. Merkert, R., Bliemer, M.C., Fayyaz, M., Consumer preferences for innovative and traditional last- mile parcel delivery. *Int. J. Phys. Distrib. Logistic Manage.*, 52, 261–284, 2022.
13. Mohsan, S.A.H., Khan, M.A., Noor, F., Towards the unmanned aerial vehicles (UAVs): A comprehensive review. *Drones*, 6, 147, 2022.
14. Ayamga, M., Akaba, S., Nyaaba, A.A., Multifaceted applicability of drones: A review. *Technol. Forecast. Soc. Change*, 167, 120677, 2021.
15. Hassanalian, M. and Abdelkefi, A., Classifications, applications, and design challenges of drones: A review. *Prog. Aerosp. Sci.*, 91, 99–131, 2017.
16. Beloev, I.H., A review on current and emerging application possibilities for unmanned aerial vehicles. *Acta Technologica Agriculturae*, 19, 70–76, 2016.
17. Merkert, R. and Bushell, J., Managing the drone revolution: A systematic literature review into the current use of airborne drones and future strategic directions for their effective control. *J. Air Transport Manage.*, 89, 101929, 2020.
18. Benarbia, T. and Kyamakya, K., A literature review of drone-based package delivery logistics systems and their implementation feasibility. *Sustainability*, 14, 360, 2021.
19. Aruchamy, P., Sabeena, G., Sowndarya, Pushpalatha, N., An artificial intelligence approach for energy-aware intrusion detection and secure routing in internet of things-enabled wireless sensor networks. *Concurrency Comput. Pract. Exper.*, 35, 1–21, 2023.
20. Sankar, S., Ramasubbareddy, S., Dhanaraj, R.K., Cluster Head Selection for the Internet of Things Using a Sandpiper Optimization Algorithm (SOA). *J. Sens.*, 1, 1–11, 2023.
21. Prasanth, A., Lakshmi, D., Dhanaraj, R.K., Balusamy, B., Technological Advancement in Internet of Medical Things and Blockchain for Personalized Healthcare: Applications and Use Cases. *CRC Press*, 1–290, 2024.
22. Aruchamy, P., Lavina, B., Sowndarya, K., An energy-aware link fault detection and recovery scheme for QoS enhancement in Internet of Things-enabled wireless sensor network. *Comput. Electr. Eng.*, 123, 1–20, 2025.

23. Sui, Q., Wang, Y., Liu, C., Interpretable Switching Deep Markov Model for Industrial Process Monitoring *via* Cloud-Edge Collaborative Framework. *IEEE Trans. Instrum. Meas.*, 74, 1–14, 2025.
24. Balraj, L. and Prasanth, A., An energy-aware software fault detection system based on hierarchical rule approach for enhancing quality of service in internet of things-enabled wireless sensor network. *Trans. Emerg. Telecommun. Technol.*, 35, 1–29, 2024.
25. Moshref-Javadi, M. and Winkenbach, M., Applications and research avenues for drone-based models in logistics: A classification and review. *Expert Syst. Appl.*, 177, 114854, 2021.
26. Rejeb, A., Rejeb, K., Simske, S.J., Treiblmaier, H., Drones for supply chain management and logistics: A review and research agenda. *Int. J. Logist. Res. Appl.*, 26, 1–24, 2021.
27. Macrina, G., Di Puglia Pugliese, L., Guerriero, F., Drone-aided routing: A literature review. *Transp. Res. Part C: Emerg. Technol.*, 120, 102762, 2020.
28. Devakirubai, N., Velu, A., Sumathi, D., Optimizing Vehicle-to-Vehicle (V2V) Communication Efficiency with KNN-based Dynamic Time slot Allocation, in: *Proceedings in 2024 International Conference on Social and Sustainable Innovations in Technology and Engineering*, pp. 1–6, 2024.
29. Tranfield, D., Denyer, D., Smart, P., Towards a methodology for developing evidence-informed management knowledge by means of systematic review. *Br. J. Manage.*, 14, 207–222, 2003.
30. Watts, A.C., Ambrosia, V.G., Hinkley, E.A., Unmanned aircraft systems in remote sensing and scientific research: classification and considerations of use. *Remote Sens.*, 4, 1671–1692, 2012.

15

Data Driven Emergency Response Management for UAV-Based Future Transportation – A Case Study

D. Vinodha[1]*, J. Jenefa[1], E.A. Mary Anita[1] and Sameer Al-Dahidi[2]

[1] *School of Engineering and Technology, CHRIST (Deemed to be University), Bengaluru, India*

[2] *School of Applied Technical Sciences, German Jordanian University, Amman, Jordan*

Abstract

The transportation sector plays a vital role and is heavily involved in emergency response and disaster relief because of the necessity for quick deployment, evacuation and rescue operations, supply chain support, infrastructure restoration, aerial support and surveillance, traffic management, and route optimization, among other things. All of these are areas that can greatly benefit from the use of drones. The capabilities of drones provide an efficient and speedy option for surveying impacted regions, spotting possible dangers, finding survivors, accessing dangerous or inaccessible locations, delivering supplies, and generally facilitating relief efforts with speed and minimal human risk. UAV (Unmanned aerial vehicle)-based transportation can become a precise substitute for time-sensitive items, emergency relief, medical supplies, and more by overcoming obstacles like traffic and geographical accessibility. The success of emergency response management for UAV-based transportation is dependent upon the quality of the data collected and the effectiveness of its analysis and interpretation. Hence, in this chapter, we propose to analyze and explore the challenges and issues involved in the design and implementation of a data-driven emergency response management system for UAV-based future transportation and its applications.

***Keywords*:** Unmanned aerial vehicle, transportation, emergency response management, remote monitoring, UAV mapping, UAV operational challenges, UAV interoperability, secure communication

**Corresponding author*: vinodha.d@christuniversity.in; vinodha.rajkumar@gmail.com; ORCID: https://orcid.org/0000-0003-1852-6257

A. Prasanth, Rajesh Kumar Dhanaraj, Munish Sabharwal, Vandana Sharma and Seifedine Kadry (eds.) Drones for Transportation Logistics and Disaster Management, (329–370) © 2026 Scrivener Publishing LLC

15.1 Scope

As inevitable as they are, disasters represent the pinnacle of natural events symbolically. Disasters plant the seeds of loss by virtue of their inherent merits. The amount of the failure, which results in the destruction of infrastructure, vegetation, life, and much more, depends on the type of the disaster. Disasters are unavoidable, which leads to the implementation of controversial safety protocols. Technology improvements have made it clear that a revolution in catastrophe management and control is greatly needed.

In its most basic form, one of the biggest responsibilities and difficulties facing the public sector has been disaster management. The fundamentals of disaster management mostly hinge on the quick actions and relief efforts undertaken; the kind and intensity of the calamity necessitate different safety precautionary measures, which ultimately decide the impacted party's fate. Using a variety of instruments and methods, new opportunities and advancements enable us to understand the magnitude of the calamity. In this way, technology enters the heart of the tragedy to offer ingenious ways to bring assistance and relief. Given that the majority of disasters are unexpected and unplanned, the best way to deal with the many kinds of catastrophic events is to act quickly.

Drone technology has advanced significantly in the field of disaster management. UAVs offer exciting benefits such as precise data and findings in a timely manner. Drones with sophisticated sensors and imaging capabilities in the years to come will be able to deliver increasingly precise and comprehensive data, facilitating quicker and more efficient response times during emergencies. Compared to regular cellular networks, UAVs can provide interconnection in areas where technology is impacted or where communications infrastructure is lacking [1], supplying materials that are urgent. Delivering emergency supplies is another way that UAVs' capacity to embody intricacy validates their significance in disaster management. Emergency supplies include first aid, medications with a quick onset of action, and tiny aid items like ropes and lanterns.

Despite the fact that crisis and catastrophe management procedures have delays and are resistant to interference, there may be a barrier to maintaining the correct operation of IoT devices in the restrictions supplied by driven structures, encompassing a tragic natural catastrophe. While drones' dynamic features are intended to deliver a superior level of services with a huge mistake rate, UAVs offer interconnection. UAVs can quickly travel to certain locations without encountering any obstacles. One of the main advantages of UAV mesh nodes is their extreme adaptability, which allows the network architecture to change depending on the situation to enhance the intended. The major contributions of this case are given below.

- → To highlight and discuss the various applications domain which require emergency response.
- → To provide insight on the various challenges in implementing emergency response systems based on UAV transportation and their existing solutions.

The case study will help the researchers to gain specific, relevant, and in-depth understanding of the applications and importance of UAV based transportation to deliver emergency response. The reader also gains insight on the various challenges in implementing UAV based transportation and the pros and cons of existing solutions. This helps the reader to carry the research further in the direction of finding better solutions to meet the challenges in the UAV transportation based emergency response system.

The chapter is organized as follows: Various applications of UAV during emergency response management are discussed in Section 15.2. Section 15.3 presents the challenges and issues in designing a data-driven emergency response management system for UAV-based future transportation. Section 15.4 presents the learnings and knowledge outcomes.

15.2 Applications of UAV During Emergency Response Management

Over the last ten years, drone use has expanded quickly, impacting numerous distinct industries including agriculture, commerce, and disaster relief and management.

Any kind of disaster has an adverse effect on the wellbeing and health of the impacted community, but massive catastrophes claim an inevitably high toll in lives. Disasters, whether natural or man-made, are categorized based on their cause: environmental, commercial, healthcare, or terrorist [2]. Therefore, a quick and efficient reaction is crucial to aiding the public, minimizing the level of casualties, and lessening the financial burden of a disaster. The method or practice of disaster management includes preparedness, mitigation measures, reaction, and restoration, independent of the kind of incidents or disasters that took place [3]. The role UAV in provisioning emergency response in various use cases are discussed as follows.

15.2.1 Remote Monitoring

Drones' excellent agility has made it easier to monitor and survey disaster sites quickly. It makes sense that disasters are immense to size, spanning vast areas of land or water. Nonetheless, UAVs and quadcopters' versatility has facilitated the examination of the affected areas and remote places of the disaster. Nevertheless, the sharply focused, high-resolution cameras enable drones to record photographs of the accident in real time and gather pertinent data. As a result of the drones' smooth hovering movement, the data is obtained more quickly than with helicopters or any other way. After the data has been gathered, it can be analyzed and plotted on 3D scales to enable prompt and efficient action.

15.2.2 UAV Mapping

UAV mapping is the process of providing precise geospatial intelligence about a region by using drones equipped with cameras and photogrammetry software. To create 3D models and 2D orthomosaics that resemble maps, images are first gathered and then stitched together. Users can access almost real-time information about the area because the maps are photo-based.

15.2.3 Delivering Emergency Products

Drones are now being used by hospitals to quickly and safely move lab samples [4], and organizations that aid humans are using them to transport medical supplies, blood samples and other related products, vaccines, medications, and organs to underdeveloped or faraway places at a low cost [5].

15.2.4 Forest Fire Rescue Operation

Drones are primarily employed to obtain situational awareness during a fire, which can be utilized to guide the firefighters' efforts in identifying and managing hot spots. Forest fires are more dangerous to persons and property because they are more difficult to contain—and perhaps impossible to do so. In this case, drones can be extremely helpful in spotting, containing, and putting out forest fires.

In a matter of minutes, drones may take off, fly to a place, map the fire's impacted area, and communicate the information to all pertinent agencies. Numerous sensors, including a thermal camera with different color palette capability, can be carried by a drone. A more accurate image of the fire's speed and spread is provided by the combination of all these sensors, which can assist civil authorities in developing a relief strategy.

15.3 Challenges and Solutions

The following are some challenges and issues in designing a data-driven emergency response management system for UAV-based future transportation. Figure 15.1 exhibits the several challenges and issues of UAV.

- **Data Security and Privacy Concerns:** Ensuring confidentiality and privacy is essential because emergency response activities need the transmission of critical data. It helps protect the data and prevent it from unauthorized access and misuse. It aids in safeguarding the information and keeps it safe from abuse and illegal access. Compliance with Data Protection Regulations, Regular Security Audits, Assessments, and Employee Training and Awareness.
- **Real-Time Data Processing:** Developing systems capable of processing real-time data to enable timely decision-making and quick response to dynamic emergency scenarios.
- **Regulatory Compliance and Legal Frameworks:** Adhering to regulatory standards and legal frameworks to ensure compliance with privacy laws, aviation regulations, and data protection guidelines for UAV operations in emergency response situations.
- **Data Accuracy and Reliability:** Ensuring the precision and dependability of the information collected and processed to avoid erroneous choices and improve the overall effectiveness of the emergency response management system.
- **Integration of Diverse Data Sources:** In an emergency, managing and combining data from a variety of sources guarantees a thorough understanding of the scenario and makes choice-making easier.
- **Interoperability and Communication:** Establishing seamless communication and interoperability between different UAVs, ground control stations, and emergency response teams to facilitate efficient coordination and information exchange during critical situations.

Figure 15.1 Numerous challenges of UAV.

- **Ethical and Social Implications:** Addressing ethical considerations and potential social implications associated with using UAVs for emergency response, including public concerns, community acceptance, and ethical guidelines for UAV operations.
- **UAV Operational Challenges:** Overcoming obstacles to the effective deployment of UAVs during emergency response operations, such as limited battery life, flight range, payload capacity, and navigation in inclement weather.

15.3.1 Data Security and Privacy Concerns

The Federal Aviation Administration reportedly predicted that, compared to present levels, the number of commercial UAVs might increase dramatically to 835,000 by 2025. However, as UAV technology develops and their use in commercial and military applications grows, there is a greater chance that nefarious actors may try to exploit them.

UAVs are susceptible to a range of assaults due to the rise of novel technology, like the ability of autonomous flight. This section examines the elements that put UAVs at danger of cyberattacks, the possible consequences of those assaults, and ways to reduce the risk.

The main causes of the increasing threat posed by drones include:

- **The increasing complexity of UAV technology:** With the continued advancement of UAV technology, Malicious actors have more options for acquiring and exploiting targets.
- **Inadequate security measures:** A lot of UAV systems don't have enough security features like authentication or encryption.
- **High mobility and low cost:** UAVs are appealing because of their great mobility and inexpensive cost.
- **Lack of awareness:** Many people do not understand or are unaware of the possible risks to their security.
- Governments are increasingly concerned about the security issues posed by UAVs. Privacy implications of UAVs being deployed into national airspace as the threat landscape surrounding UAVs grows. Furthermore, the majority of commercial UAVs on the market today lack security features like Intrusion Detection Systems (IDS). As such, they make ideal targets for enemies. It is important to note that the widespread and ongoing usage of civilian UAVs for various purposes may present new privacy and security risks.

In [6], the authors proposed a model by incorporating modern methods such as secure multi-party computation, differential privacy, and federated learning. The level of accuracy of intrusion detection and data confidentiality is strengthened by the integrating Deep Neural Networks (DNNs). This allows accurate threat identification and real-time anomaly detection. This enables UAVs to decide quickly in changing situations. In order to prevent security breaches, a real-time decision mechanism is proposed that sets off notifications and starts the automatic blacklisting process. Additionally, the IDS database's access security is strengthened with multi-factor authentication.

In order to identify data examples for improved integrity and confidentiality in drones, the authors of [7] presented a hybrid machine learning strategy that combines logistic regression and random forest. The method showed an astounding 98.58% accuracy. Nevertheless, the authors failed to adequately demonstrate how their methodology can safeguard integrity and confidentiality in UAVs, which is necessary to enable proper performance evaluation.

For safe internet of UAV environments, the authors of [8] suggested the robust deep learning based blockchain IDS by representing the deep learning part of the technique by using Reinforcement Learning and Neural Network with Radial Basis function as agents. It outperformed state-of-the-art methods in identifying attacks with high accuracy, even with a greater number of drones.

In [9], Anomaly based intrusion detection is proposed by monitoring the features that can be observed and trained machine learning model to detect unexpected modification in the data. In [10], mobile edge computing technology-enabled IDS for anticipating and identifying a variety of assaults within the UAV network is proposed. The Framework has shown effective assault detection across a range of UAV network zones. But because of the usage of one DNN algorithm, it faces the issue of less accuracy. The authors of [11] proposed additional improved hierarchical anomaly-based IDS intended to identify and alert users to potentially fatal attacks during military activities using Internet of Drones networks. The foundational mechanism for the IDS was developed as an effective random forest classifier by utilizing randomized search cross-validation and an optimized hyperparameters approach.

A new framework with layered broad learning systems and federated continuous learning was proposed by the authors of [12] for detecting intrusion. They use DNN to decentralize the learning process and improve attack detection. Usage of a discriminative learning system enables the framework to provide high accuracy and training efficiency while addressing data silos and guaranteeing privacy and security. However, performance might be improved much further by combining this strategy with additional safe and private methods.

Securing the transfer of data and blocking unauthorized access have become crucial concerns as UAVs are incorporated more and more into urban contexts. The system's actual efficiency and its effect on hardware components in real-world circumstances may not be fully assessed in the existing methods. Furthermore, testing the IDS with a large group of UAVs and investigating blockchain technology further for improved privacy might be taken into consideration. There is a need for ensemble technologies and systematic performance evaluation metrics are required to improve the accuracy level.

15.3.2 Integration of Diverse Data Sources

Integrating various data sources in UAV-based transportation is a complex and pivotal factor that significantly influences the system's safety, efficiency, and dependability. The diverse data sources are illustrated in Figure 15.2. Factors such as Navigation and flight Data, Weather Data, Emergency Response Data, Infrastructure and terrain Data, Sensor Data, Regulatory Compliance Data, Supply Chain and Logistics Data, Energy and Battery Data, Public Safety and Privacy Data, User and Customer Data need to be considered when developing an efficient UAV-based transportation system. Successfully integrating these diverse data sources in the context of UAV-based future transportation necessitates a robust data infrastructure, advanced analytics capabilities, and strict adherence to privacy and regulatory standards. Cooperation between the government, business partners, and technological companies, among other stakeholders, is indispensable for creating a comprehensive and effective UAV transportation system. Detailed explanations of these data sources and integration approaches are provided below.

- **Navigation and Flight Data**
  Establishing efficient communication between UAVs requires gathering data from UAV sensors, GPS systems, and onboard navigation systems. It helps prevent collisions as well. For UAVs to navigate precisely and in real time, avoid obstacles, and optimize routes, data from many sources must be integrated.

Figure 15.2 Diverse data sources.

- **Weather Data**
  Meteorological sources' weather updates are crucial since they offer critical alerts regarding the state of the weather. Weather data integration makes it easier to modify UAV flight plans and schedules in response to both past and future weather. This guarantees operational efficiency and safety.
- **Emergency Response Data**
  Information from emergency services, disaster response organizations, and other relevant sources is included in the data about emergency response. UAVs can respond to emergencies like natural disasters and accidents more skillfully when this emergency response data is integrated.
- **User and Customer Data**
  This involves information about users, customers, and their preferences. Customer data can be leveraged for personalized services, optimizing UAV operations in response to user demand, and improving the overall user experience.
- **Infrastructure and Terrain Data**
  Terrain and infrastructure data, including maps, geographical data, and specifics on the infrastructure that is currently in place, are essential. Optimizing flight paths, avoiding obstructions, and enhancing overall navigation all depend on integrating this landscape and infrastructure data.
- **Regulatory Compliance Data**
  This includes details on airspace limitations, legal requirements, and aviation laws. UAV operations are guaranteed to comply with regulations when relevant compliance data is incorporated into the decision-making process.
- **Supply Chain and Logistics Data**
  Logistics operations, cargo, and package tracking information are important concerns. Using UAVs to move items efficiently, tracking cargo status, and optimizing delivery routes are all made possible by the integration of supply chain data.
- **Public Safety and Privacy Data**
  This includes details on privacy laws and issues related to public safety. It helps in ensuring compliance with privacy laws, addressing public safety concerns, and integrating pertinent data to mitigate potential risks associated with UAV operations.

- **Sensor Data**
  Throughout UAV missions, information from a variety of onboard sensors—such as gyroscopes, photographic equipment like cameras, magnetometers Light Detection and Ranging (LiDAR), and thermal imaging—is used for situational awareness, obstacle detection, and environmental monitoring.
- **Energy and Battery Data**
  When managing UAV energy consumption and route optimization, data about battery levels, energy usage, and charging infrastructure is taken into account. Furthermore, data from charging stations can be combined to enable longer-lasting and more effective operations.

Large-scale data availability has the potential to revolutionize the development of Intelligent Transportation Systems (ITS), transforming them from traditional technology-driven systems into more potent, multifunctional, data-driven ITS [13]. Large-scale data availability has the power to revolutionize the design and creation of ITS, transforming them from traditional technology-driven systems into more potent, versatile, data-driven ITS. In order to make an UAV to serve as an automated catastrophe dispatcher for a user inside an automobile experiencing a hazardous situation, an algorithm is proposed in [14]. When a predefined dangerous situation is recognized and the UAV is situated inside an area of coverage, it takes off and sends distress signals to a communication network. Information related to location of vehicle, level of injury, video, images to capture the status of passengers' health are vital. Since some fields are inside coverage gaps, a group of cooperative UAVs gathers data from these fields and works together to send the data in order to extend the UAVs' lifespans [15]. The various data to be collected while deploying UAV for logistic related applications are surveyed in [16]. The data gathering techniques used when a vast sensor network is installed in places where terrestrial mobility is challenging, as well as the steps involved in the aerial data collection procedure and the main issues that need to be resolved are discussed in [17].

15.3.3 Real-Time Data Processing

Real-time data processing is a pivotal element in UAV based transportation systems, significantly enhancing their efficiency, safety, and overall effectiveness. The process is imperative for enabling UAVs to navigate dynamic environments, utilizing data from onboard sensors like GPS, accelerometers, gyroscopes, and cameras to ascertain the UAV's precise position and orientation.

Path planning algorithms leverage real-time data to dynamically adjust the UAV's route based on obstacles, weather conditions, and other variables subject to change. The system must also actively identify and evade obstacles in real-time to guarantee secure and reliable transportation. Sensors like LiDAR, radar, and cameras are crucial in providing data that can be processed to recognize obstacles and strategize alternative routes.

Efficient management of multiple UAVs within the same airspace is essential, involving coordinating flight paths, maintaining safe distances between UAVs, and optimizing traffic flow to avert collisions [18]. The impact of rapidly changing weather conditions on UAV flights underscores the need for real-time weather data processing. This capability enables UAVs to adapt their routes or operations based on prevailing weather conditions, ensuring safety, and optimal performance.

Maintaining communication links between UAVs and ground control stations is a critical aspect. This encompasses processing data from communication systems to guarantee reliable and secure data exchange. In transportation, UAVs may carry diverse payloads, such as goods, medical supplies, or sensors. Real-time monitoring of payload data, including weight and status, ensures that the UAV operates within predetermined limits.

Real-time data processing is pivotal in responding promptly to emergencies by providing situational awareness. This involves processing data from various sensors to detect emergencies, assess the situation, and take appropriate actions. Efficient data transmission and storage are crucial because UAVs generate substantial data. Timely and relevant information is transmitted to ground stations or stored for subsequent analysis.

Ensuring compliance with aviation regulations is paramount, and real-time data processing proves invaluable in monitoring and ensuring adherence to regulatory requirements, such as no-fly zones or altitude restrictions. Equally important is ensuring the integrity and privacy of information collected by UAVs. This encompasses implementing measures such as encrypting communication channels, safeguarding sensitive information, and adopting secure data storage practices. In summary, real-time data processing is a foundational component within UAV-based transportation systems, contributing significantly to their safety, efficiency, and overall success across diverse applications.

15.3.4 Regulatory Compliance and Legal Frameworks

Regulatory compliance and legal frameworks are crucial in shaping the UAV based transportation landscape. As the utilization of UAVs continues to increase, governments and aviation authorities worldwide are implementing and refining regulations to ensure these aerial systems' safe and responsible operation. Some key aspects of regulatory compliance and legal frameworks in UAV-based transportation are exposed in Figure 15.3.

- **Registration and Licensing**
  Regulation: Many countries which are using UAVs for commercial business purposes require the registration of UAVs. In order to guarantee that operators possess the required abilities and expertise, licensing may also be required.
  Compliance: The registration and licensing procedures established by the aviation authorities in their various regions must be followed by UAV operators.
- **Operational Limitations**
  Regulation: Operational constraints include the maximum altitude, the minimum distance from airports, and the prohibition on flying in specific airspace zones are defined by the authorities.
  Compliance: To maintain safety and regulatory compliance, UAV operators need to be aware of these operational constraints and follow them to the letter.
- **No-Fly Zones and Restricted Areas**
  Regulation: Certain sites are marked as no-fly zones or restricted regions for UAVs, including government buildings, airports, and important infrastructure.
  Compliance: To reduce security concerns and adhere to rules, UAV operators must respect and stay away from certain restricted zones.
- **Remote Pilot Certification**
  Regulation: In certain areas, UAV operators must get a remote pilot certificate as proof of their familiarity with aviation laws and safety protocols.
  Compliance: In order to legally fly UAVs, operators must complete the required training and certification.

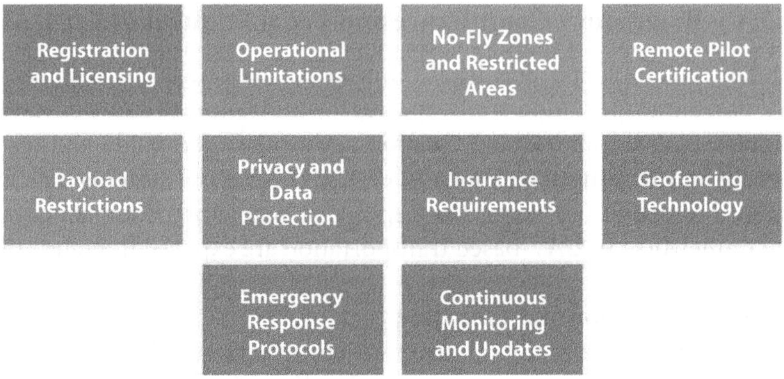

Figure 15.3 Key aspects of regulatory compliance and legal frameworks.

- **Continuous Monitoring and Updates**
  Regulation: Regulatory frameworks for UAVs are subject to evolution and updates to address technological advancements and emerging challenges.
  Compliance: UAV operators must stay informed about regulation changes and ensure that their operations align with the latest legal requirements.
- **Insurance Requirements**
  Regulation: To reduce the danger of mishaps or property damage, certain areas may mandate that UAV operators carry insurance.
  Compliance: In order to comply with legal requirements and reduce financial risk, operators must seek insurance coverage.
- **Geofencing Technology**
  Regulation: In order to enforce no-fly zones and operational restrictions, geofencing technology may be required in UAVs.
  Compliance: Geofencing technology must be included into UAV systems by manufacturers, and operators are responsible for making sure it operates properly while in flight.
- **Emergency Response Protocols**
  Regulation: Authorities may establish protocols for reporting accidents or incidents involving UAVs, and operators may be required to have emergency response plans.
  Compliance: UAV operators must know and follow established emergency response protocols to report incidents promptly and appropriately.

- **Privacy and Data Protection**
  Regulation: The usage of UAVs may be impacted by privacy laws and data protection rules, particularly when it comes to the gathering and processing of personal data.
  Compliance: In order to ensure that they handle data responsibly and ethically, UAV operators must abide by privacy and data protection legislation.
- **Payload Restrictions**
  Regulation: When UAVs are used for commercial purposes, there may be restrictions on the kind of payloads they are allowed to carry. For example, transporting dangerous goods may not be permitted.
  Compliance: Payload requirements must be followed by UAV operators in order to safeguard the environment and the general public.

In summary, regulatory compliance and legal frameworks are integral to the safe and responsible integration of UAVs into transportation systems. Operators must proactively understand and adhere to these regulations to foster a secure and efficient UAV-based transportation environment.

15.3.5 Data Accuracy and Reliability

Data accuracy and reliability are critical factors in the successful operation of UAV based transportation systems. The collection, processing, and utilization of accurate and reliable data are essential for ensuring the safety, efficiency, and effectiveness of UAV operations. Figure 15.4 enumerates

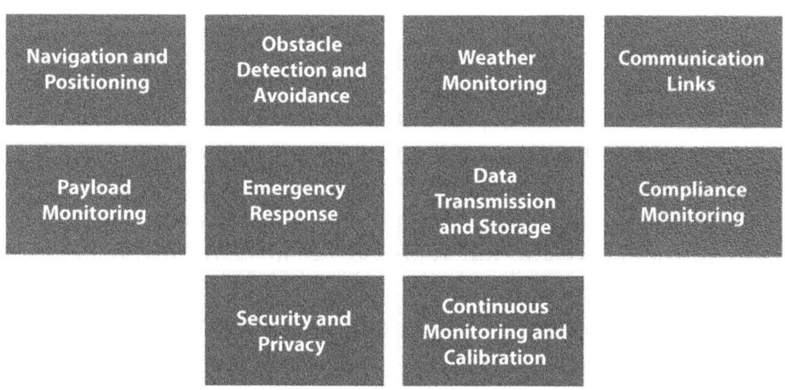

Figure 15.4 Key considerations related to data accuracy and reliability.

some of the key considerations related to data accuracy and reliability in UAV-based transportation.

- **Navigation and Positioning:** Accurate GPS data is necessary for accurate flight routes in volatile circumstances, which are essential for UAV navigation and placement. The accuracy of positioning data may be impacted by issues such as GPS errors, signal interference, and environmental conditions.
- **Obstacle Detection and Avoidance:** Reliability of information from sensors like LiDAR, radar, and cameras is essential for recognizing obstacles, which are critical for preventing collisions. Obstacles include ambient circumstances, sensor constraints, and occlusions.
- **Weather Monitoring:** A key component of modifying UAV paths based on current conditions and improving operational safety and efficiency is real-time meteorological data. The maintenance of dependable meteorological data is hampered by difficulties with quickly shifting weather conditions and the requirement for fast updates.
- **Data Transmission and Storage:** Efficient data transmission and storage are essential for managing the large volumes of UAV-generated data. Challenges, such as bandwidth limitations, transmission delays, and data integrity issues, can impact reliability.
- **Security and Privacy:** Paramount in UAV data collection, ensuring the security and privacy of data requires reliable encryption and protection measures. Risks include cybersecurity threats and vulnerabilities in communication channels compromising data security and privacy.
- **Emergency Response:** Swift and effective emergency response relies on reliable data and situational awareness from accurate data processing. Challenges, such as delayed or inaccurate data, can impede emergency response efforts and decision-making.
- **Payload Monitoring:** Accurate monitoring of payload data, including weight and status, is essential for guaranteeing UAVs function within predetermined limitations. This is especially important when transporting sensitive equipment or cargo. Errors in the instrumentation and calibration are challenges.

- **Compliance Monitoring:** Accurate data is vital for monitoring and ensuring compliance with aviation regulations, including adherence to no-fly zones and altitude restrictions. Regulatory violations may occur due to inaccurate data reporting or non-compliance from faulty sensors.
- **Communication Links:** Reliable communication lines between UAVs and ground control stations are essential for real-time data sharing and control, but they can be compromised by issues including interference, signal degradation, and distance restrictions, which can cause operational disruptions.
- **Continuous Monitoring and Calibration:** It is essential to continuously check sensor calibration and system performance in order to maintain data accuracy over time. Sensor accuracy can be affected by ageing equipment, environmental factors, and wear and tear if frequent calibration and monitoring aren't performed.

Because of costly sensors, complex data processing, and susceptibility to weather and environmental conditions, existing survey tools for inspecting overhead transmission lines cannot guarantee the security of UAVs and electrical systems. Hence, a mathematical model of this transmission line is developed in [19] to look at the information's field power distribution surrounding transmission lines. The positioning system's data is used to feed the model. The information detected by the model helps the UAV to avoid obstacles. UAV with onboard GPS is suggested in [20] for establishing ground point contact. By establishing various mathematical equations, a relative positioning method is suggested in [21] with improved accuracy. The method applies fusion over the information received from three different types of sources.

Rather than using a spatial device, a LiDAR sensor was utilized to estimate the distance between the quadcopter and landing pad [22]. The quadcopter demonstrated autonomous takeoffs and landings based on the results. But this system's drawback is that it can't accurately determine the quadcopter's attitude because of errors in the Inertial Measurement Unit and GPS measurements. Because of the limitation of GPS based positioning systems, positional measurements based on computer vision have become vital in recent days [23]. Approaches based on reinforcement learning have gained popularity recently for UAVs with little processing power. In both indoor and outdoor situations, several optimization techniques are used for local and global path-planning tactics. These reinforcement learning techniques let the UAV navigation system adopt a policy that generally generalizes over changing power restrictions.

The authors in [24] decided to investigate a partially viewable environment that simulated real-world challenges while quickly increasing the ability of UAV agents to learn and understand. They used a two-state input technique, which combines location information from a map with the knowledge gleaned from the raw image. While crowded areas should be avoided according to the feature map of the current scene, these positional data improve our understanding of the UAV's location and the separation between itself and the target location. Neural network based solutions are proposed for unconstructed environments.

The authors of [25] suggested artificial intelligence technologies for spacecraft systems, like deep learning and neural networks, to address issues with their dynamics and navigation. The many artificial network topologies, techniques to train them, the benefits and drawbacks of applying them to different kinds of situations are covered in the article. It also looks at how machine intelligence can be incorporated into space systems for optical navigation, control synthesis, system identification, and other purposes.

Localization performance is enhanced through multi-sensor fusion, which integrates data from many sensors, including GPS, LIDAR, and cameras. In the future, floating-base manipulators for solo or collaborative job completion will be investigated. It is not an easy task to operate an aerial vehicle with a manipulator; there are many obstacles to overcome due to the manipulator's shifting center of gravity and external disturbances created by the interaction. This capacity will likely transform the use of UAVs and involve difficult vision-based tasks. Additional investigation in this field is required to surmount these obstacles and diminish the constraints of the existing methodologies.

15.3.6 UAV Operational Challenges

UAVs have the ability to transform transportation in various ways, including the delivery of goods, passenger transport, and more. However, several safety concerns must be addressed before UAV-based transportation can become widespread. Here are some of the key safety concerns:

15.3.6.1 Collision Avoidance

One of the most significant safety concerns is the risk of mid-air collisions between UAVs or between UAVs and traditional aircraft. Drones run the risk of crashing into another aircraft or damaging people or property when they land in a populated area. These mishaps may result from a drone losing contact with its ground-based operator or from hackers infiltrating the software and causing it to deviate from its intended flight route. Effective collision avoidance systems and regulations are essential to prevent accidents. The major procedures adopted for anti-collision mechanisms are sensing obstacles, collision prediction, and avoidance.

The UAV flying in 3 dimensional spaces faces various challenges. UAVs often follow complex flight paths, including maneuvers like hovering, circling, and rapid changes in direction. These complex trajectories can make it challenging to predict their future positions accurately.

Here are some common collision avoidance techniques and technologies used for UAVs:

- **GPS and GNSS:** UAVs often use GPS or Global Navigation Satellite System (GNSS) to determine their position and navigate through predefined routes or waypoints. This information can be used for basic collision avoidance by avoiding predefined no-fly zones or restricted areas and planning efficient flight paths. GPS-based collision avoidance systems can significantly reduce the risk of mid-air collisions by providing real-time data on the positions of other aircraft in the vicinity. This is particularly crucial for UAVs operating in shared airspace with manned aircraft. Also, It enables UAVs to operate with greater autonomy by reducing the need for constant human intervention. This is especially valuable for long-duration missions or when multiple UAVs need to operate simultaneously. It can be integrated with other technologies like ADS-B (Automatic Dependent Surveillance-Broadcast) to provide more comprehensive situational awareness and collision avoidance capabilities. This makes them suitable for use in complex and crowded airspace. Though GPS based collision avoidance systems support enhanced safety with improved navigation and increased autonomy, it also comes with limitations such as altitude constraints, signal dependency, cost, and complexity. The effectiveness of such systems depends on the specific use case, the environment in which the UAV operates, and the quality of the equipment and software used. UAV operators should carefully evaluate their needs and consider these pros and cons when implementing collision avoidance technology.

- **Obstacle Detection Sensors:** Ultrasonic sensors, LiDAR, and radar are just a few of the sensors that may be installed aboard UAVs to help them identify obstructions in their route of flight. The UAV can alter its flight route to prevent collisions with objects thanks to the real-time data these sensors offer.
- **AIS and ADS-B:** Systems like ADS-B and the Automatic Identification System (AIS) are used to track and identify other aircraft. These systems enable UAVs to detect other aircraft in their area and, if necessary, conduct evasive maneuvers.
- **Artificial Intelligence and Machine Learning:** Using the taught learning models, the sensor data may be processed to make collision avoidance judgments in real time. With time, these systems' capacity for making decisions can be enhanced by learning from historical data.
- **Path Planning Algorithms:** Algorithms for path planning can be used to calculate safe flying routes that steer clear of obstructions and possible collision hazards. These algorithms consider the environment, destination, and current position of the UAV.
- **Geofencing:** Geofencing technology sets virtual boundaries that UAVs are not allowed to cross. This can be used to prevent UAVs from entering restricted areas or flying too close to sensitive locations.
- **Fail-Safe Systems:** UAVs can be equipped with fail-safe systems that trigger automatic return-to-home functions or emergency maneuvers when they detect potential collisions or technical failures.
- **Computer Vision:** It is possible to use computer vision technology to identify and detect obstacles or other aircraft. This technology typically uses cameras and image processing. This can assist the UAV in making well-informed decisions regarding path correction.
- **Communication and Sense-and-Avoid Systems:** UAVs can communicate with other aircraft or ground stations to coordinate their movements and avoid collisions. Sense-and-avoid systems rely on cooperative communication between aircraft to enhance safety.

- **Regulations and Standard Operating Procedures:** Adherence to regulations and standard operating procedures, such as maintaining line of sight and flying at appropriate altitudes, is considered as a fundamental part of collision avoidance for UAVs.

Existing Solutions

Collision avoidance methods for UAVs can be categorized into Centralized and Decentralized methods. In centralized methods, UAVs are aided by central control units to oversee the routes and the current state of the environment. On the other hand, centralized approaches cannot prevent collisions if unmanaged UAVs appear in the vicinity since the centralized control network cannot identify them. Furthermore, the employment of active technologies is restricted in remote places and during natural catastrophes due to their requirement for a continuous cellular network connection. Decentralized approaches, on the other hand, give UAVs flight control capabilities so they may avoid collisions on their own. It can be used even outside of cellular service zones because they don't require a centralized infrastructure on the ground.

Alanezi et al. in [26] developed an obstacle avoidance-based autonomous navigation of quadrotors which are rotary-wing UAVs capable of vertical take-off and landing. To guarantee the quadrotor's autonomous navigation, two genetic algorithm-based PID controllers for attitude and altitude control were created. An obstacle avoidance algorithm was also used. *Shan et al.* in [27] proposed a novel method that uses D2D information exchange to prevent collisions between UAVs, is a distributed-type collision avoidance method based on the elastic collision concept. They also proposed a configurable weighting technique for every UAV in order to further enhance efficiency. The quantity of UAVs used in low-altitude space during an emergency situation increases the risk of collision and poses a significant risk to public safety. To address this, the authors in [28] proposed a deep learning based two layer framework. The model avoids collision by training the UAVs to make a collaborative decision. However, just the velocity vector and distance were taken into account while making the decision, which could affect the avoidance of subsequent collisions.

15.3.6.2 Air Traffic Management

Air Traffic Management (ATM) in UAV transportation is a critical aspect that involves the coordination, monitoring, and control of UAVs operating in shared airspace. Unlike traditional manned aircraft, UAVs introduce unique challenges and considerations to air traffic management due to their diverse applications, varying sizes, and the need for safe integration into existing airspace. Integrating UAVs into existing airspace and air traffic management systems is complex. UAVs must be tracked and managed to avoid conflicts with other aircraft, which requires robust infrastructure and technology.

Existing Solutions

Path planning plays an important role in traffic management. *Sathyaraj et al.* [29] compared various algorithms that can be used for planning UAV paths. The comparison is done between includes Dijkstra's shortest path algorithm, Bellman-Ford algorithm, algorithm by Robert Floyd and Stephen Warshall and A* algorithm. Their findings indicate that the performance of A* algorithm is better than the other path algorithms. Further advancements in path planning algorithms, such as A* with Jump Point for Search operation and Rectangular Symmetry Reduction proposed by *Botea et al.* [30], *Harabor et al.* [31], and Harabor and Grastien [32], build upon the A* algorithm. It is determined that A* using jump point outperforms A* with rectangular symmetry reduction.

Additionally, Fast-Marching methods, known for providing solutions that are reliable, precise, and incredibly effective in obstacle-ridden path-planning scenarios [33], have been identified as suitable for UAV path-planning applications [34]. Random Trees using rapid exploration have been effectively employed in searching for paths for multirotors, as demonstrated by *Janjoš et al.* [35]. *Ng et al.* [36] leverage the Optimal control which is recognized as a popularly utilized trajectory-optimization approach in aviation, to create flight trajectories aiming to reduce both fuel utilization and travel time of flight. In an earlier study, *Li and Hansen* [37] compared various path planning algorithms, ultimately determining that methods based on Fast-Marching outperformed several others in terms of efficacy.

Ang Li [38] introduces a framework for managing traffic in Unmanned Aircraft Vehicle systems, specifically focusing on delivery of packages within urban low-altitude airspace. The framework encompasses path planning for UAVs based on cluster, Unmanned Aircraft System Traffic Management featuring conflict detection and resolution (CD&R), and the development of mechanisms for allocating airspace resources.

15.3.6.3 *Weather Conditions*

Weather can significantly impact the safety of UAV operations. Strong winds, thunderstorms, and other adverse weather conditions can pose challenges for safe flight. UAVs need to be equipped with technology to handle these situations and have the ability to land safely.

Drones are not reliable for time-sensitive operations since they cannot fly in all types of weather. There is a lack of knowledge regarding the extent and variability of weather influence worldwide. Weather-limited drone flyability (the amount of time drones can fly without risk) is investigated in [39] by using a computer simulation to compare historic wind velocity, humidity, and precipitation information with manufacturer-reported thresholds of well-known commercial weather-resistant drones. It is also demonstrated that high latitudes and areas above oceans have the least flyability, while it is strongest in warm, dry continental locations.

15.3.6.4 *Cybersecurity*

UAVs are vulnerable to hacking and cyberattacks, which can result in loss of control, data theft, or other malicious activities. Ensuring the security of UAV systems and communication networks is crucial.

15.3.6.5 *Human Error*

Even with advanced automation, human error can still play a role in accidents. Operators and remote pilots must be well-trained and adhere to safety protocols.

15.3.6.6 *Battery Life and Range*

Limited battery life and range pose safety concerns for long-distance flights and passenger transport. A UAV running out of power mid-flight could lead to a crash or emergency landing.

15.3.6.7 *Mechanical Failures*

Like any machine, UAVs can experience mechanical failures that may lead to accidents. Routine maintenance and stringent quality control measures are necessary to prevent such failures.

Privacy Problems: Since UAVs can be used for data collection and monitoring, their widespread use poses privacy problems. People's right to privacy must be safeguarded by laws and limits.
Noise Pollution: Communities may be disturbed by the noise pollution that certain UAVs can produce. To allay these worries, regulations should include aircraft lanes and noise levels.
Emergency Response: Having efficient procedures for emergency response and recovery is crucial to reducing risks and damage in the case of an emergency, such as a malfunction or collision.
Infrastructure Difficulties: One of the biggest challenges is creating the landing pads, charging stations, and maintenance facilities required for UAV-based transportation. Accidents and delays might result from inadequate infrastructure.
Regulatory Structure: establishing a thorough and adaptable regulatory framework that takes into account.

Addressing these safety concerns will be crucial for the successful integration of UAV-based transportation into our future transportation systems. It will require collaboration between governments, industry stakeholders, and technology innovators to establish clear standards and protocols to ensure safe and efficient operations.

15.3.7 Interoperability and Communication

Interoperability and communication are pivotal aspects of UAV-based transportation systems, ensuring seamless integration and coordination among various components. These elements enhance UAV operations' efficiency, safety, and overall effectiveness. Some key considerations related to interoperability and communication in UAV-based transportation are shown in Figure 15.5.

15.3.7.1 System Integration

Interoperability, the seamless interaction between different subsystems, stands as a cornerstone in the integration of diverse UAV systems. These subsystems encompass vital components such as navigation, communication, and payload systems, each playing a crucial role in the UAV's

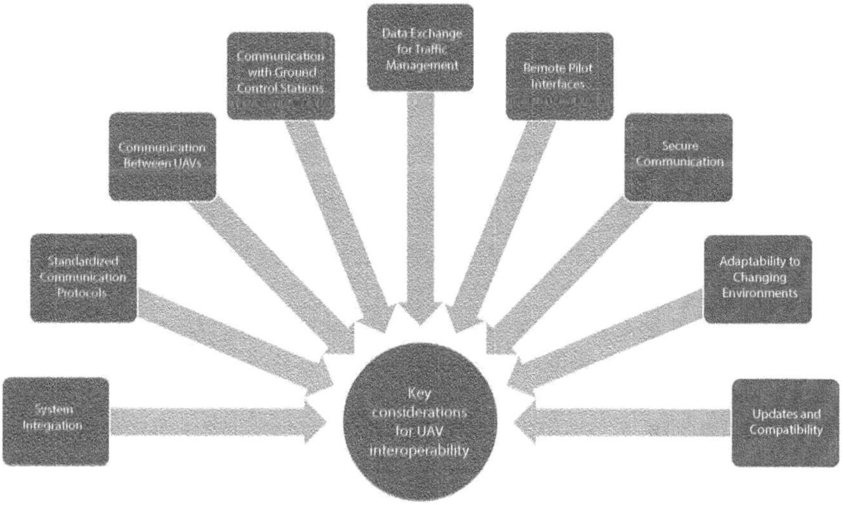

Figure 15.5 Key consideration of UAV interoperability.

overall functionality. However, achieving such seamless integration is not without its challenges. One of the primary hurdles in ensuring interoperability among UAV subsystems lies in the variations present in communication protocols and hardware interfaces. These differences can lead to compatibility issues, making it difficult for the subsystems to communicate effectively with one another.

For instance, one subsystem may utilize a specific communication protocol or hardware interface that is incompatible with another subsystem, hindering its ability to exchange data and coordinate actions. In addition, these challenges are compounded by the diverse nature of UAV applications and the multitude of manufacturers producing UAV systems. Each manufacturer may employ different standards and technologies, further complicating efforts to achieve interoperability. As a result, integrating UAV subsystems from different manufacturers into a cohesive and harmonious system becomes a complex task.

Addressing these challenges requires concerted efforts in standardization and compatibility testing. Establishing common communication protocols and hardware interfaces can facilitate interoperability among UAV subsystems, enabling them to communicate and collaborate seamlessly. Additionally, rigorous testing procedures can help identify and rectify compatibility issues early in the integration process, ensuring smooth system operation. In conclusion, while interoperability is vital for the effective integration of diverse UAV subsystems, challenges such as differences in communication protocols and hardware interfaces must be addressed to achieve seamless system integration. Through standardization efforts and thorough compatibility testing, the UAV industry can overcome these challenges and realize the full potential of integrated UAV systems in various applications.

15.3.7.2 Standardized Communication Protocols

Standardized communication protocols serve as the linchpin for establishing a seamless exchange of data between UAVs and ground control stations, fostering a cohesive ecosystem that enhances operational efficiency and effectiveness. By adhering to common standards, UAVs can communicate with ground control stations in a uniform manner, ensuring that data transmission is consistent and reliable across various platforms. This uniformity not only simplifies the integration process but also promotes interoperability, enabling UAVs from different manufacturers to collaborate seamlessly on missions ranging from surveillance and reconnaissance to disaster response and infrastructure inspection. Moreover, standardized protocols facilitate the development of robust communication networks, providing UAV operators with greater flexibility and scalability to adapt to evolving operational needs while maintaining compatibility with existing systems. Nevertheless, the absence of universally accepted standards in UAV communication presents a formidable obstacle, impeding the seamless exchange of information and hindering collaboration between different UAV platforms.

Without standardized protocols, UAVs may utilize proprietary communication methods or incompatible formats, leading to fragmentation within the UAV ecosystem. This fragmentation not only complicates the coordination of multi-UAV missions but also limits the interoperability between UAVs and ground control stations, potentially compromising the efficiency and safety of operations. Addressing these interoperability challenges requires concerted efforts from industry stakeholders to establish common standards and protocols that can serve as the foundation for a cohesive UAV communication framework. By fostering collaboration and standardization, the UAV industry can unlock new opportunities for innovation and advancement, enabling UAVs to realize their full potential in diverse applications across various sectors.

15.3.7.3 Communication Between UAVs

In dynamic and crowded airspace scenarios, establishing reliable communication links between UAVs becomes imperative for ensuring safe and coordinated operations. However, this task presents significant challenges, primarily due to interference and coordination issues inherent in such environments. Interference from other UAVs, manned aircraft, or external sources can disrupt communication signals, leading to data loss or delays in transmission. Moreover, coordinating communication between multiple UAVs in real-time adds another layer of complexity, as each UAV must share pertinent information while avoiding conflicts and collisions.

One of the primary challenges in establishing reliable communication links between UAVs in dynamic airspace is managing frequency interference. With an increasing number of UAVs operating in the same airspace, frequency bands become congested, making it difficult for UAVs to find available channels for communication. This congestion can result in signal degradation or complete loss of communication, jeopardizing the safety and efficiency of UAV operations. Additionally, coordinating communication protocols and addressing compatibility issues among different UAV platforms further complicates the task, as each UAV may operate on different frequencies or utilize proprietary communication methods. It ensures effective coordination between UAVs to maintain safe distances and optimize traffic flow requires robust communication protocols and advanced coordination algorithms.

UAVs must exchange real-time data regarding their positions, velocities, and intended flight paths to avoid collisions and navigate through congested airspace safely. However, achieving seamless coordination among multiple UAVs in dynamic environments requires sophisticated communication systems capable of handling high volumes of data with minimal latency. Additionally, implementing collaborative decision-making algorithms and adaptive control strategies can enhance coordination and improve the overall efficiency and safety of UAV operations in crowded airspace. Addressing these challenges necessitates interdisciplinary collaboration and ongoing research efforts to develop innovative solutions that can support the growing demand for UAV applications in dynamic and congested environments.

15.3.7.4 Communication with Ground Control Stations

Real-time communication between UAVs and ground control stations serves as a critical component for effective mission monitoring, control, and updates. It enables operators to maintain situational awareness, issue commands, and receive real-time data from the UAV, facilitating prompt decision-making and ensuring mission success. However, several factors can significantly impact the reliability of communication links between UAVs and ground control stations, posing challenges that need to be addressed to maintain seamless connectivity.

Distance plays a crucial role in determining the reliability of communication links between UAVs and ground control stations. As the distance between the UAV and the ground control station increases, the strength of the communication signal diminishes, potentially leading to signal degradation or loss. This degradation can result in latency issues, data packet loss, or even complete loss of communication, impeding the ability of operators to effectively monitor and control the UAV in real-time. To mitigate the effects of distance on communication reliability, UAVs often employ communication systems with sufficient transmission power and antenna configurations optimized for long-range communication.

Signal degradation caused by environmental factors such as atmospheric interference, electromagnetic interference, and terrain obstacles further compounds the challenges associated with maintaining reliable communication links between UAVs and ground control stations. Atmospheric conditions like rain, fog, or strong winds can attenuate radio signals, reducing signal strength and impairing communication reliability. Additionally, electromagnetic interference from other electronic devices or radio frequency sources in the vicinity can disrupt communication signals, leading to interference and signal distortion. Moreover, terrain obstacles such as buildings, mountains, or vegetation can obstruct line-of-sight communication between the UAV and the ground control station, exacerbating communication challenges.

To address these challenges and enhance the reliability of communication links between UAVs and ground control stations, various strategies can be employed. This includes utilizing advanced communication technologies such as frequency hopping spread spectrum, adaptive modulation and coding, and error correction techniques to improve signal robustness and mitigate the effects of interference and signal degradation. Moreover, employing redundant communication links, such as satellite communication or mesh networking, can provide backup options in case of primary link failure, ensuring continuity of communication even in challenging environments. Additionally, conducting thorough site surveys and implementing proper antenna positioning and orientation strategies can optimize signal propagation and enhance communication reliability, ultimately enabling seamless real-time communication between UAVs and ground control stations across diverse operational scenarios.

15.3.7.5 *Data Exchange for Traffic Management*

Interoperable communication stands as a cornerstone for enabling the seamless sharing of real-time data concerning UAV positions, flight routes, and operational intentions among various stakeholders, thereby fostering collaborative traffic management in airspace systems. By adhering to common communication standards and protocols, UAVs, air traffic control agencies, and other relevant entities can exchange crucial information swiftly and accurately, facilitating proactive decision-making and enhancing overall airspace safety and efficiency. This interoperability allows for the integration of UAVs into existing airspace infrastructure seamlessly, enabling effective coordination and collaboration between manned and unmanned aircraft. With interoperable communication, stakeholders can gain comprehensive situational awareness, identify potential conflicts or hazards, and implement timely interventions to mitigate risks and ensure smooth and orderly UAV operations.

Coordinating communication among multiple UAVs to prevent collisions and optimize airspace usage demands efficient data exchange mechanisms and robust coordination strategies. UAVs must communicate their positions, velocities, flight intentions, and sensor data in real time to effectively coordinate with other airspace users and ground control systems. Efficient data exchange enables UAVs to share critical information, such as planned routes, altitude changes, and emergency situations, allowing for dynamic adjustments to flight paths and airspace assignments to avoid conflicts and maintain safe separation distances. Additionally, coordinated communication facilitates collaborative decision-making processes, enabling UAVs to work together to optimize airspace usage, minimize congestion, and maximize operational efficiency. By leveraging interoperable communication frameworks and coordination mechanisms, the UAV industry can realize the full potential of collaborative airspace management, paving the way for safe, scalable, and sustainable integration of UAVs into airspace systems worldwide.

15.3.7.6 *Interagency Collaboration*

Effective collaboration among different agencies, including aviation authorities, law enforcement, and emergency services, during UAV operations hinges on interoperability to facilitate seamless communication. Interoperability ensures that diverse agencies can exchange vital information in real time, enabling coordinated responses and informed decision-making during various UAV missions, including surveillance, search and rescue, and disaster management. However, the presence of varying communication protocols and operational procedures among agencies poses a significant challenge to achieving seamless collaboration and interoperability.

The existence of disparate communication protocols and operational procedures among different agencies can create communication barriers and hinder the exchange of critical information during UAV operations. Incompatible communication systems may result in data silos, where agencies are unable to share information effectively, leading to delays in response times and potentially compromising mission outcomes. Moreover, differences in operational procedures and protocols can lead to misunderstandings or misinterpretations of information, further exacerbating coordination challenges and impeding collaborative efforts. As a result, achieving interoperability among diverse agencies becomes paramount to ensure effective communication and coordination during UAV operations, especially in complex and time-sensitive scenarios.

To address these challenges, efforts are needed to standardize communication protocols and harmonize operational procedures among different agencies involved in UAV operations. Establishing common communication standards and protocols allows for seamless integration of disparate systems, enabling agencies to communicate and share information effortlessly. Additionally, promoting interoperability through training, exercises, and joint operational planning fosters a culture of collaboration and enhances coordination among agencies. By aligning communication systems and operational practices, agencies can overcome interoperability barriers and improve their ability to collaborate effectively during UAV operations, ultimately enhancing overall mission success and ensuring the safety and security of airspace operations.

15.3.7.7 Remote Pilot Interfaces

Standardized interfaces for remote pilots play a crucial role in enhancing interoperability and fostering seamless communication across different UAV platforms. By adhering to common interface standards, remote pilots can access consistent communication tools and displays regardless of the specific UAV platform they are operating. This standardization not only streamlines the pilot's workflow but also facilitates efficient information exchange and collaboration among operators, ground control stations, and other relevant stakeholders. Moreover, standardized interfaces promote ease of integration with existing systems and software, enabling interoperability between UAVs from different manufacturers and ensuring compatibility with diverse operational environments.

The presence of differing user interfaces and control systems among UAV manufacturers can pose significant challenges for pilot training and operation. Without standardized interfaces, remote pilots may encounter difficulties in transitioning between different UAV platforms, as each may have its unique set of controls, displays, and operating procedures. This variability not only complicates pilot training and certification processes but also increases the risk of human error during UAV operations. Standardized interfaces mitigate these challenges by providing a common framework for pilot interaction, simplifying the learning curve and promoting consistency in operation across different UAV platforms. This, in turn, enhances operational efficiency, reduces training costs, and improves overall safety by ensuring that pilots can focus on mission objectives rather than struggling with unfamiliar control systems or interfaces. Ultimately, standardized interfaces empower remote pilots to operate UAVs more effectively and confidently, thereby maximizing the potential of unmanned aerial systems in various applications, from aerial photography and mapping to surveillance and emergency response.

15.3.7.8 Secure Communication

Ensuring secure communication channels is paramount to safeguarding data integrity and preventing unauthorized access, particularly when transmitting sensitive information over networks. Cybersecurity threats such as data breaches, interception, and manipulation underscore the importance of employing robust encryption and authentication mechanisms to protect data in transit. Encryption standards play a pivotal role in securing communication channels by encoding data in a manner that can only be decrypted by authorized recipients, thereby thwarting eavesdropping and data tampering attempts. Implementing strong encryption algorithms, such as Advanced Encryption Standard or Rivest-Shamir-Adleman, ensures that transmitted data remains confidential and tamper-proof, even if intercepted by malicious actors.

Effective vulnerability management practices are essential for identifying and mitigating potential weaknesses in communication systems that could be exploited by cyber threats. Regular security assessments, vulnerability scans, and penetration testing help organizations proactively identify security vulnerabilities and address them before they can be exploited. Additionally, implementing secure communication protocols such as Transport Layer Security or Secure Shell adds an extra layer of protection by encrypting data and authenticating communication endpoints, thereby reducing the risk of unauthorized access and data compromise. By continuously monitoring and updating security measures, organizations can stay ahead of emerging threats and maintain the integrity and confidentiality of their communication channels.

In conclusion, ensuring secure communication channels involves a multi-faceted approach that encompasses robust encryption standards, stringent vulnerability management practices, and adherence to secure communication protocols. By implementing comprehensive cybersecurity measures, organizations can mitigate the risk of data breaches, protect sensitive information, and uphold the trust and confidentiality of their communication networks. This proactive approach to cybersecurity not only safeguards critical data assets but also strengthens the resilience of communication infrastructure against evolving cyber threats in an increasingly interconnected digital landscape.

15.3.7.9 Adaptability to Changing Environments

Interoperable communication systems play a pivotal role in facilitating seamless data exchange across diverse operational scenarios, necessitating adaptability to changing environmental conditions to ensure reliable communication. However, adapting communication protocols to dynamic and unpredictable environments, such as urban settings or adverse weather conditions, presents significant challenges. In urban environments, for instance, densely populated areas, tall buildings, and electromagnetic interference pose obstacles to signal propagation, potentially causing signal degradation or loss. Adapting communication protocols to mitigate these challenges requires robust mechanisms for signal optimization, frequency management, and interference mitigation to maintain reliable data exchange amidst environmental variability.

Adverse weather conditions, including rain, fog, snow, and strong winds, can further exacerbate communication challenges by attenuating radio signals and impairing signal strength and reliability. Adapting communication protocols to address weather-induced signal degradation necessitates advanced modulation techniques, error correction mechanisms, and dynamic frequency allocation strategies to optimize signal transmission and minimize the impact of environmental disturbances. Additionally, leveraging alternative communication channels, such as satellite links or mesh networking, can enhance resilience and redundancy, ensuring continuous data exchange even in the face of adverse weather conditions.

The dynamic nature of operational scenarios, such as rapidly changing terrain, varying altitudes, and unpredictable movement patterns, demands flexibility and agility in communication protocols to accommodate evolving communication requirements. Adaptable communication systems should be capable of dynamically adjusting transmission parameters, such as bandwidth, modulation schemes, and transmission power, to optimize communication performance and adapt to changing environmental conditions in real time. By integrating adaptive communication protocols with intelligent networking algorithms and cognitive radio technologies, interoperable communication systems can effectively navigate the complexities of diverse operational environments, ensuring reliable data exchange and seamless connectivity across a wide range of scenarios.

15.3.7.10 Updates and Compatibility

Regular updates to communication systems are crucial for accommodating technological advancements and maintaining interoperability across UAV fleets and ground infrastructure. However, upgrading communication protocols while ensuring backward compatibility with existing systems poses significant challenges. One of the primary challenges is the need to balance innovation and compatibility, as newer communication protocols may offer enhanced features and performance but may not be compatible with legacy systems.

To address this challenge, careful planning and phased implementation strategies are essential. Organizations must conduct thorough compatibility assessments to evaluate the impact of protocol upgrades on existing UAV fleets and ground infrastructure. This includes assessing hardware and software compatibility, identifying potential interoperability issues, and developing migration plans to minimize disruptions to operations. Additionally, organizations may need to invest in retrofitting or upgrading legacy systems to ensure compatibility with updated communication protocols.

Another challenge is the risk of system downtime or operational disruptions during the transition period. Organizations must carefully manage the upgrade process to minimize downtime and ensure continuity of operations. This may involve conducting pilot tests or simulations to identify potential issues and refine implementation strategies before full-scale deployment. Additionally, organizations should provide training and support to personnel to familiarize them with the new communication protocols and minimize the impact of operational changes. Furthermore, ensuring backward compatibility while upgrading communication systems requires collaboration and coordination among stakeholders, including UAV manufacturers, communication technology vendors, regulatory authorities, and end-users.

Open communication channels and stakeholder engagement are critical for addressing concerns, resolving compatibility issues, and ensuring a smooth transition to updated communication protocols. Overall, while upgrading communication systems poses challenges, careful planning, stakeholder engagement, and phased implementation strategies can help mitigate risks and ensure a successful transition while maintaining interoperability and compatibility with existing infrastructure. By embracing technological advancements and investing in communication system upgrades, organizations can enhance operational efficiency, resilience, and capabilities in the rapidly evolving UAV ecosystem.

Existing Solutions

A few solutions are proposed to deal with these challenges, which are discussed below. *Scarlato et al.* [40] proposed designing a blockchain that cannot be accessed publicly without permission to avoid and recover from collision in an enhanced manner for UAVs. Their vision involved creating a collaborative ecosystem where participating UAVs constantly exchange collision data and obstacle information. *Rahman et al.* [41] also introduced a UAV network to guarantee an environment without collision. Routes are carefully designed to avoid prohibited places, like privately owned land. and flight altitudes are mentioned to minimize collision risks by reducing the number of drones operating at the same altitude. A smart contract is employed by the authors to log the movement and location information of drones throughout the operation to ensure adherence to the specified route. Any deviation from the intended flying path tarnishes the drone's reputation.

In a different approach, *Kuzmin et al.* [42] suggested a route-sharing system where cooperative drones use route data stored in blockchains to automatically resolve conflicts on their paths. The utility of this approach is highlighted, particularly when the contact is lost between a manually operated drone and its base station. Similarly, *Allouch et al.* [43] introduced, implemented, and evaluated a permissioned blockchain for secure path planning and data sharing among participating drones. To address UAVs' limited computation power and storage resources, they offloaded computations to a cloud server while employing a decentralized off-chain storage system, OrbitDB. Furthermore, they excluded participating UAVs from the peer-to-peer network, designating solely ground control stations as peers responsible for storing ledger copies. The solution was deployed on the Hyperledger Fabric platform to assess their architecture. Lastly, they assessed the delay and resource consumption of a single transaction. The latency of a network with 50 users was 455 milliseconds. Although the latency observed in this study may be considered relatively high compared to existing network delays, the findings demonstrate promise for applying blockchain-based UAV networks in real-time scenarios.

In summary, achieving effective interoperability and communication in UAV-based transportation requires a concerted effort to establish standardized protocols, address technical challenges, and promote stakeholder collaboration. These efforts are essential for creating a cohesive and efficient ecosystem that can unlock the full potential of UAV operations.

15.3.8 Ethical and Social Implications

Addressing ethical considerations and potential social implications associated with using UAVs for emergency response, including public concerns, community acceptance, and ethical guidelines for UAV operations. The widespread adoption of UAV-based transportation introduces ethical and social implications that need careful consideration. As this technology continues to evolve and integrate into daily life, it brings forth various ethical challenges and impacts on society. Here are key considerations:

Privacy Concerns
Utilizing UAVs outfitted with advanced cameras and sensors triggers profound ethical deliberations regarding the infringement of privacy, as these technologies have the capacity to indiscriminately capture imagery and data spanning both public and private spheres. Such capabilities raise significant societal concerns, potentially undermining individuals' fundamental rights to privacy and emphasizing the urgent need for regulatory frameworks to curtail unauthorized surveillance and data collection practices. The pervasive deployment of these surveillance technologies underscores the critical importance of safeguarding personal privacy in an increasingly digitally interconnected world, compelling policymakers and stakeholders to enact stringent measures to uphold individual freedoms and protect against unwarranted intrusions into personal lives.

Security Risks
Employing UAVs for various purposes, ranging from surveillance to payload delivery introduces ethical dilemmas concerning the potential for misuse, criminal activities, or security breaches. Public concerns regarding security and misuse have the potential to influence societal attitudes,

highlighting the essential need for regulatory frameworks to mitigate risks and foster responsible UAV deployment. These regulations are vital to ensuring that UAV technologies are utilized in a manner that prioritizes safety, security, and the protection of individual rights and freedoms.

Noise Pollution

UAVs, particularly those equipped with rotor-based propulsion systems, can exacerbate noise pollution in urban and residential areas, posing significant ethical challenges. The amplified noise levels from these UAVs can potentially disrupt communities, adversely impacting their overall quality of life. Furthermore, these disturbances raise valid concerns about the environmental consequences of increased noise pollution. As such, addressing the ethical implications of noise pollution stemming from UAV operations becomes imperative in preserving societal well-being and environmental integrity.

Job Displacement and Economic Impact

The increasing automation of transportation *via* UAVs may lead to job displacement in traditional delivery services or other transportation-related sectors, prompting ethical reflections. Displaced workers may face challenges transitioning to new roles, highlighting societal responsibilities to address potential economic disruptions. This underscores the need for proactive measures to support affected individuals and mitigate adverse impacts on livelihoods.

Equitable Access

Guaranteeing fair access to UAV-based transportation services is crucial to prevent the widening of societal inequalities. Restricted access for specific demographics or geographical regions can exacerbate existing disparities in transportation accessibility, thereby impeding opportunities for marginalized communities. It is imperative to address these disparities through proactive measures that promote inclusivity and ensure equal access to emerging transportation technologies.

Environmental Impact

The environmental repercussions of UAVs, encompassing factors such as energy consumption and the possibility of ecological disturbances, prompt ethical reflections concerning sustainability. Public apprehensions regarding the environmental footprint of UAV operations may significantly influence societal acceptance and spur the advancement of eco-friendly UAV technologies. This underscores the imperative of prioritizing environmental sustainability in developing and deploying UAVs to mitigate adverse ecological impacts and foster public trust in these innovative technologies.

Data Security and Ownership
UAVs generate and transmit substantial volumes of data, raising apprehensions regarding data security, ownership, and the potential for misuse. Public awareness of these data security concerns can significantly impact trust in UAV-based services, highlighting the need for transparent data governance and user rights. Ensuring clear guidelines for data management and safeguarding user privacy is essential to maintain public confidence and uphold ethical standards in UAV operations.
Regulatory Compliance and Accountability
Responsible utilization of UAVs entails adherence to regulations and accountability for any unintended consequences of accidents that may arise. Ensuring compliance with ethical standards and legal regulations is essential to uphold public trust and safety. By maintaining rigorous adherence to established guidelines and ethical principles, stakeholders can mitigate risks and promote the safe and responsible deployment of UAV technologies.
Cultural Sensitivity
UAV operations have the potential to intersect with cultural and religious sensitivities, particularly in regions where traditional practices or beliefs are prevalent. To ensure responsible UAV deployment in diverse communities, it is imperative to understand and respect cultural nuances. By acknowledging and accommodating these sensitivities, stakeholders can avoid unintended conflicts and foster positive relationships within the communities they serve. This approach promotes inclusivity and ensures that UAV operations are conducted in a culturally sensitive and respectful manner.
Public Perception and Acceptance
Transparency, addressing concerns, and actively involving the public in decision-making processes are crucial ethical considerations for fostering societal acceptance of UAV technology. Public perception plays a significant role in determining the success of UAV-based transportation initiatives, underscoring the importance of open communication and engagement. By prioritizing transparency and soliciting feedback from stakeholders, UAV operators can build trust, address concerns, and ensure that their initiatives align with the needs and expectations of the communities they serve. This collaborative approach promotes accountability and inclusivity, ultimately contributing to the responsible adoption and integration of UAV technology into society.

In summary, the ethical and social implications of UAV-based transportation underscore the importance of responsible development, transparent deployment, and continuous engagement with the public and regulatory bodies. Balancing technological advancements with ethical considerations is crucial for creating a positive societal impact and ensuring the long-term sustainability of UAV technology.

15.4 Learning and Knowledge Outcomes

As UAVs are being used in more and more urban settings, protecting data transmission and avoiding unauthorized access are critical considerations. The existing approaches may not properly measure the system's true efficiency and its impact on hardware components in real-world scenarios. Additionally, a swarm of UAVs might be used to test the IDS, and further research into blockchain technology for enhanced privacy could be considered. To increase the accuracy level, ensemble technologies are needed, as well as systematic performance evaluation criteria.

Two crucial sources of data utilized in remote sensing of structures and man-made objects are RGB and LiDAR. Smaller object's position, velocity, and activity can be observed with greater ease as image sensor resolution increases. In the future, more varied and smaller fake targets can be detected using data from UAVs. Another crucial area of advancement is data fusion from several sources. Utilizing diverse UAV remote sensing data from multiple sources and types, including LiDAR, multi-spectral technology, etc.—is a less common subject of current research. Therefore, another area of research focus is the methods for data fusion from various sources.

Real-time data processing is a foundational component within UAV-based transportation systems, contributing significantly to their safety, efficiency, and overall success across diverse applications. Hence, it can be considered as one of the vital research focus areas. Regulatory compliance and legal frameworks are integral to the safe and responsible integration of UAVs into transportation systems. Operators must proactively understand and adhere to these regulations to foster a secure and efficient UAV-based transportation environment. More frameworks need to be developed to foster a secure and efficient UAV-based emergency response system. Achieving effective interoperability and communication in UAV-based transportation requires a concerted effort to establish standardized protocols, address technical challenges, and promote stakeholder collaboration. These efforts are essential for creating a cohesive and efficient ecosystem that can unlock the full potential of UAV operations.

15.5 Conclusion

The case study assisted the researchers in gaining a detailed, pertinent, and precise grasp of the uses and significance of UAV-based emergency response transportation. The various application domains which require emergency responses were highlighted and discussed. Also, insight on the various challenges in implementing emergency response systems based on UAV transportation and their existing solutions were provided. The reader also learned about the advantages and disadvantages of current options and the many difficulties in implementing UAV-based transportation. This aids the reader in continuing the investigation to identify more effective ways to address the problems with the emergency response system based on UAV transportation.

References

1. Nguyen, N.T., Liu, B.H., Wang, S.Y., Network under limited mobile sensors: New techniques for weighted target coverage and sensor connectivity, in: *Proceedings of the 2017 IEEE 42nd Conference on Local Computer Networks (LCN)*, pp. 471–479, Singapore, 9–12 October 2017.
2. Kannan, S.S.D., Kumar, V.S., Rathinavelu, P.K., Indiran, M.A., Awareness and attitude towards mass disaster and its management among house surgeons in a dental college and hospital in Chennai, India. *WIT Trans. Built Environ.*, 175, 121–129, 2017.
3. Park, J.H., Park, S.H., Kim, K.A., Disaster management and land administration in South Korea: Earthquakes and the real estate market. *Land Use Policy*, 85, 52–62, 2019.
4. Cohen, J.K., WakeMed health & hospitals joins forces with UPS, FAA for drone pilot, in: *Modern Healthcare*, Available: https://www.modernhealthcare.com/care-delivery/wakemed-health-hospitals-joins-forces-ups-faa-drone-pilot. [Accessed: September 30, 2021].
5. Bright, J., Zipline begins US medical delivery with drone program honed in Africa, in: *TechCrunch*, Available: https://techcrunch.com/2020/05/26/zipline-begins-us-medical-delivery-with-uav-program-honed-in-africa/. [Accessed: September 30, 2021].
6. Ntizikira, E., Lei, W., Alblehai, F., Saleem, K., Lodhi, M.A., Secure and privacy-preserving intrusion detection and prevention in the Internet of Unmanned Aerial Vehicles. *Sensors*, 23, 2023.
7. Aldaej, A., Ahanger, T.A., Atiquzzaman, M., Ullah, I., Yousufudin, M., Smart cybersecurity framework for IoT-empowered drones: Machine learning perspective. *Sensors*, 22, 2022.

8. Heidari, A., Navimipour, N.J., Unal, M., A secure intrusion detection platform using blockchain and radial basis function neural networks for Internet of Drones. *IEEE Internet Things J.*, 1, 2023.
9. Da Silva, L., Ferrão, I.G., Dezan, C., Anomaly-based intrusion detection system for in-flight and network security in UAV swarm, in: *Proceedings of the 2023 International Conference on Unmanned Aircraft Systems (ICUAS)*, pp. 812–819, Warsaw, Poland, 6–9 June 2023.
10. Ihekoronye, V.U., Ajakwe, S.O., Kim, D., Cyber edge intelligent intrusion detection framework for UAV network based on random forest algorithm, in: *Proceedings of the 2022 13th International Conference on Information and Communication Technology Convergence (ICTC)*, pp. 1242–1247, Jeju Island, Republic of Korea, 19–21 October 2022.
11. Ihekoronye, V.U., Ajakwe, S.O., Kim, D., Lee, J.M., Hierarchical intrusion detection system for secured military drone network: A perspicacious approach, in: *Proceedings of the MILCOM 2022–2022 IEEE Military Communications Conference (MILCOM)*, pp. 336–341, Rockville, MD, USA, 28 November–2 December 2022.
12. He, X., Chen, Q., Tang, L., Wang, W., Liu, T., Li, L., Liu, Q., Federated continuous learning based on stacked broad learning system assisted by digital twin networks: An incremental learning approach for intrusion detection in UAV networks. *IEEE Internet Things J.*, 1, 2023.
13. Zhang, J., Wang, F.Y., Wang, K., Lin, W.H., Xu, X., Chen, C., Data-driven intelligent transportation systems: A survey. *IEEE Trans. Intell. Transp. Syst.*, 12, 1624–1639, 2011.
14. Almasoud, A.M., UAV-based intelligent transportation system for emergency reporting in coverage holes of wireless networks. *Sensors*, 21, 2021.
15. Abdullah, A.M., Mohamed, Y.S., Alqasir, A., Energy-efficient data forwarding in disconnected networks using cooperative UAVs, in: *Proceedings of the 2018 IEEE Global Communications Conference (GLOBECOM)*, Abu Dhabi, United Arab Emirates, 9–13 December 2018.
16. Li, Y., Liu, M., Jiang, D., Application of unmanned aerial vehicles in logistics: A literature review. *Sustainability*, 14, 2022.
17. Wang, C., Ma, F., Yan, J., De, D., Das, S.K., Efficient aerial data collection with UAV in large-scale wireless sensor networks. *Int. J. Distrib. Sens. Netw.*, 11, 2015.
18. Huang, Y., Tang, J., Lao, S., Cooperative multi-UAV collision avoidance based on a complex network. *Appl. Sci.*, 9, 2019.
19. Li, Y., Zhang, W., Li, P., Ning, Y., Suo, C., A method for autonomous navigation and positioning of UAV based on electric field array detection. *Sensors*, 21, 2021.
20. Tahar, K.N. and Kamarudin, S.S., UAV onboard GPS in positioning determination. *Int. Arch. Photogrammetry Remote Sens. Spatial Inf. Sci.*, XLI-B1, 1037–1042, 2016.

21. Song, H., Hu, S., Guo, Q., Jiang, W., Relative positioning method for UAVs based on multi-source information fusion. *Math. Probl. Eng.*, 2022, 2022.
22. Diels, L., Vlaminck, M., De Wit, B., Philips, W., Luong, H., On the optimal mounting angle for a spinning lidar on a UAV. *IEEE Sens. J.*, 22, 21240–21247, 2022.
23. Pavel, M.I., Tan, S.Y., Abdullah, A., Vision-based autonomous vehicle systems based on deep learning: A systematic literature review. *Appl. Sci.*, 12, 2022.
24. Maciel-Pearson, B.G., Marchegiani, L., Akcay, S., Abarghouei, A., Garforth, J., Breckon, T.P., Online deep reinforcement learning for autonomous UAV navigation and exploration of outdoor environments, 2019.
25. Silvestrini, S. and Lavagna, M., Deep Learning and Artificial Neural Networks for Spacecraft Dynamics, Navigation and Control. *Drones*, 6, 2022.
26. Alanezi, M.A., Haruna, Z., Sha'aban, Y.A., Bouchekara, H.R.E.H., Nahas, M., Shahriar, M.S., Obstacle Avoidance-Based Autonomous Navigation of a Quadrotor System. *Drones*, 6, 2022.
27. Shan, L., Li, H.B., Miura, R., Matsuda, T., Matsumura, T., A Novel Collision Avoidance Strategy with D2D Communications for UAV Systems. *Drones*, 7, 2023.
28. Zhang, J., Zhang, H., Zhou, J., Hua, M., Zhong, G., Liu, H., Adaptive Collision Avoidance for Multiple UAVs in Urban Environments. *Drones*, 7, 2023.
29. Sathyaraj, B.M., Jain, L.C., Finn, A., Drake, S., Multiple UAVs Path Planning Algorithms: A Comparative Study. *Fuzzy Optim. Dec. Making*, 7, 257–267, 2008.
30. Botea, A., Mueller, M., Schaeffer, J., Near Optimal Hierarchical Path-Finding. *J. Game Dev.*, 1, 1–30, 2004.
31. Harabor, D., Botea, A., Kilby, P., Path Symmetries in Undirected Uniform-Cost Grids, in: *SARA 2011 - Proceedings of the 9th Symposium on Abstraction, Reformulation, and Approximation*, 2011.
32. Harabor, D. and Grastien, A., Online Graph Pruning for Pathfinding on Grid Maps, in: *National Conference on Artificial Intelligence (AAAI)*, 2011.
33. Sethian, J.A., Fast-Marching Methods. *SIAM Rev.*, 41, 199–235, 1999.
34. Valero-Gomez, A., Gómez, J., Garrido, S., Moreno, L., Fast Marching Methods in Path Planning. *IEEE Rob. Autom. Mag.*, 20, 111–120, 2013.
35. Janjoš, F., Reichart, R., Niermeyer, P., Smooth Path-Generation Around Obstacles Using Quartic Splines and RRTs. *IFAC-PapersOnLine*, 50, 9108–9113, 2017.
36. Ng, H.K., Sridhar, B., Grabbe, S., Optimizing Aircraft Trajectories with Multiple Cruise Altitudes in the Presence of Winds. *J. Aerosp. Inf. Syst.*, 11, 35–47, 2014.
37. Li, A. and Hansen, M., Obstacle Clustering and Path Optimization for Drone Routing, in: *9th International Conference for Research in Air Transportation*, 2020.

38. Li, A., Hansen, M., Zou, B., Traffic Management and Resource Allocation for UAV-Based Parcel Delivery in Low-Altitude Urban Space. *Transp. Res. Part C Emerging Technol.*, 143, 2022.
39. Gao, M., Hugenholtz, C.H., Fox, T.A., Weather Constraints on Global Drone Flyability. *Sci. Rep.*, 11, 2021.
40. Scarlato, M., A Blockchain for the Collision Avoidance and the Recovery of Crashed UAVs, in: *Proceedings of the 2019 IEEE International Conference on Robotics and Automation (ICRA)*, pp. 463–467, 2019.
41. Rahman, M.S., Khalil, I., Atiquzzaman, M., Blockchain-Powered Policy Enforcement for Ensuring Flight Compliance in Drone-Based Service Systems. *IEEE Netw.*, 35, 2021.
42. Kuzmin, A. and Znak, E., Blockchain-Based Structures for a Secure and Operate Network of Semi-Autonomous Unmanned Aerial Vehicles, in: *2018 IEEE International Conference on Service Operations and Logistics, and Informatics (SOLI)*, IEEE, pp. 32–37, 2018.
43. Allouch, A., Cheikhrouhou, O., Koubâa, A., Toumi, K., Khalgui, M., Nguyen Gia, T., UTM-Chain: Blockchain-Based Secure Unmanned Traffic Management for the Internet of Drones. *Sensors*, 21, 2021.

16

Remotc Assessments and Aerial Imaging Using UAV for Disaster Management and Precision Agriculture with Immediate Response – A Case Study

Kalaivanan Karunanithy[1]*, Bhanumathi Velusamy[2] and A. Prasanth[3]

[1]School of Electronics Engineering, Vellore Institute of Technology, Chennai, India
[2]Department of ECE, Anna University, College of Engineering, Guindy, Chennai, India
[3]Department of Computer Science and Engineering, Vel Tech Rangarajan Dr. Sagunthala R&D Institute of Science and Technology, Chennai, India

Abstract
Advancement in Remote Sensing (RS) and Geographic Information System (GIS) allows non-destructive and timely observation of particular events. Precise and timely assessment of natural disasters such as earthquakes, landslides, forest fires, and volcanic eruptions is essential in giving an operational challenge for the research community. Unmanned Aerial Vehicles (UAVs) play a major role in remote sensing, which provides a feasible, accurate, and safe tool for analyzing the impacts of natural disasters and agriculture in real time. The sensor nodes and various imaging techniques are utilized to access the geographic information using UAVs and monitor the environmental consequences across different management practices. This chapter analyzes the use of various Machine Learning (ML) algorithms with the integration of UAVs to fetch valuable insights from the target field and provide accurate information. This helps the scientific community to extrapolate disaster management and reduce the risk of physical onsite inspection of humans. Also, it can reduce the human errors in physical inspection and predict the diseases at earlier stages. The insights detection of the captured images and the prediction model results are analyzed by categorizing them into worst, average, and best cases to understand the challenging real-world situation.

**Corresponding author*: kalaivaanankk@yahoo.com

A. Prasanth, Rajesh Kumar Dhanaraj, Munish Sabharwal, Vandana Sharma and Seifedine Kadry (eds.) Drones for Transportation Logistics and Disaster Management, (371–398) © 2026 Scrivener Publishing LLC

The outcome of this study is to provide a framework for developing an autonomous decision-making system to fully automate the process that is used to get the hazardous and agricultural field assessment effectively and reduce the human errors and risks in physical inspection.

Keywords: Disaster management, machine learning, remote assessment, satellite, unmanned aerial vehicle

16.1 Introduction

Natural disasters such as landslides, earthquakes, forest fires, storms, floods, and volcano eruptions have occurred due to climate changes and global warming. The disasters often threaten human lives and economic stability. More than 4.8 million people were evacuated, and 18,000 deaths were caused from 1998 to 2017 due to the landslides [1]. Currently, 1350 volcanoes are active worldwide, and 45 volcanoes are continuously erupting, like Sakurajima, Japan, and Tavurvur, Papua New Guinea. According to recent University of Maryland studies, the annual loss of tree cover (measured in hectares) has increased by more than three million since 2001 [2]. The wildfire and volcano eruption killed 4,000 people and affected about 4.8 million others [1]. Additionally, 389 climate-related disasters with significant financial losses were reported in 2020 (26% and 23% more hurricanes and floods, respectively) [3].

Disaster management is responsible for five major activities: prediction, assuagement, readiness, rehabilitation, and responsibility. Remote sensing plays a vital role in predicting the occurrence of the event, assessing disaster-affected areas, and estimating economic losses. Remote sensing is essential for estimating the likelihood of an event, evaluating areas impacted by disasters, and calculating economic losses. Also, it is used in the agriculture field to make automation in disease detection, prediction of nutrient requirements, yield and harvesting stage, and crop land survey. Our daily lives are being rearranged in a variety of ways with 6542 satellites in operation worldwide. Research in computer vision with RS is made possible by the effective utilization of satellite data. Typically, remote sensing data, particularly images and videos, is used for many purposes, including object tracking, land surveying, disaster management, and so forth. Due to the foggy and low-light conditions, satellite imagery's time interval may be affected. This will create opportunities for remote sensing influenced by drone technology. Drones can capture higher-resolution images than satellites since they can fly at low altitudes. This will facilitate a methodical examination of the terrain, including the topography, rivers, buildings, and disaster sites. Additionally, it offers an exact measurement for impact evaluations, catastrophe planning both before and after, and repossession operations [4].

High-quality imagery is captured by drones, covering a vast region of the affected area quickly and efficiently. By doing this, the authorities may be able to update the community on damage estimates. Additionally, the information from the drones is gathered and made centrally available to all teams, preventing misunderstandings and confusion amongst them. The centralized data is available to the disaster management team for use in forecasting affected areas and deploying operations. Additionally, it can be used to identify areas where infrastructure is susceptible to destruction and conduct a land survey for possible victims. Aerial imagery guarantees the safety of people and their property while providing numerous insights into pre- and post-disaster management [5]. These images have been utilized to identify the exact location and extent of the flood, forest fire, landslide, and volcanic eruption. This ensures the preparedness activities to handle the pre- and post-disaster. Because of their large coverage area, easy deployment, high spatial resolution, high-quality images, and inexpensive

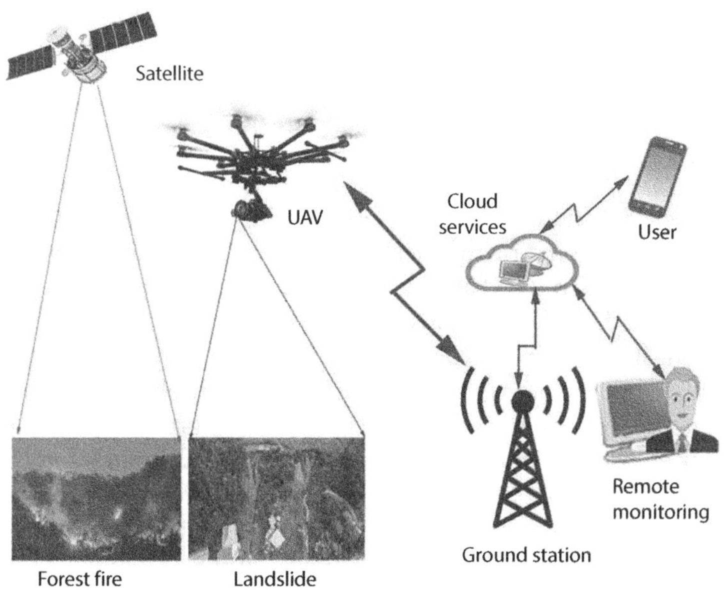

Figure 16.1 General diagram for disaster management with IoT.

operating costs, UAVs are extensively used in disaster management, as shown in Figure 16.1. Drone/UAV-captured images have a great possibility of detecting ground-level objects like plants, rivers, vehicles, buildings, etc.

The traditional methods of detecting and monitoring forest fires involve the use of watchtowers and human observers, both of which create a significant risk to their safety. Because the ground monitoring systems are typically set in one location, their restricted coverage may cause problems. Manned aircraft are also used to monitor disaster sites, which can quickly cover a broad region and extend the surveillance range. However, dangerous environments and weather can endanger the lives of pilots [6]. Another approach to current and conventional disaster management is UAV-based remote sensing, which pays close attention to computer vision to realize the hazardous environment. Also, it is capable of providing high spectral resolution and meeting the spatial-temporal requirements of remote sensing.

Computer vision-based forest fire detection considers three important features, such as color, geometry, and gesture. Most of the researchers focused on both color and gesture features for giving an accurate detection of forest fires.

A drone equipped with a camera or remote sensing sensor flies at low altitudes to cover a large area in any circumstance. Various techniques and components like a ground control station, electro-optical system, Global Positioning System (GPS), digital magnetic compass, gyro-meter, inertial navigation system, and attitude heading reference system are equipped in the autonomous UAV to find the correct location of the target [7].

16.1.1 Various Remote Sensing Techniques

Satellite RS, optical RS, Unmanned Aerial Vehicle (UAV), Global Navigation Satellite System (GNSS), Light Detection and Ranging (LiDAR), Interferometric Synthetic Aperture Radar (InSAR), and Global Information System (GIS) are used to extract the information from disaster area and damaged infrastructure as shown in Figure 16.2. It also provides fast and non-destructive information sharing for disaster management. Remote sensing has been used for earth observation in many different contexts, most notably precision agriculture, geographic conditions and disaster relief [8]. The satellite's spatial and temporal images are used in numerous applications, including land mapping, flood monitoring, landslide prediction, crop growth monitoring, crop area calculation, and volcano eruption detection. Currently, airborne remote sensing is used instead of satellite technology, which offers temporal flexibility and high-resolution photos. UAVs are the most cost-effective real-time alternative to satellites for obtaining high-spectral images.

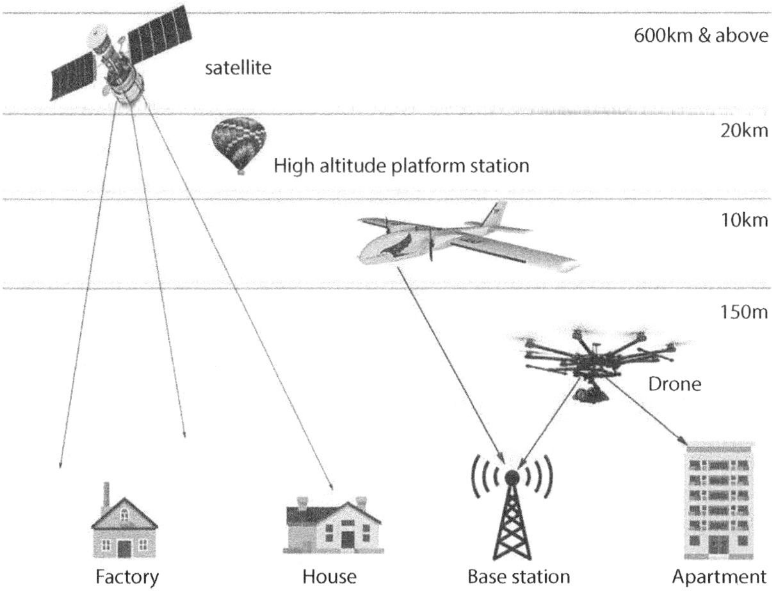

Figure 16.2 Remote sensing techniques.

Unmanned Aerial Vehicles (UAVs) rely on sensors and cameras to carry out tasks independently through machine learning algorithms, eliminating the need for human intervention. UAVs equipped with cameras are divided into two categories: (i) UAV type and (ii) camera type. The purpose of the application and terrestrial region determine the choice of UAV platform deployment. Typically, the RGB, thermal, and spectral cameras equipped on UAVs/Drones are utilized to capture the disaster field. Multi-rotor UAVs can fly at very low altitudes and provide high spectral images. However, it can't carry high-weight cameras. The fixed-wing UAV is more suitable for carrying hyper-spectral cameras than multi-rotor UAVs. However, the resolution of the image is severely distorted and tainted when it is captured from a drone at a very high speed [9].

16.2 Various Types of Cameras Used in Remote Sensing

16.2.1 RGB Camera in Remote Sensing

A commonly used camera in RS with a drone (around 87%) is Red-Green-Blue (RGB). This is because RGB cameras have excellent spatial resolution, are easy to deploy, inexpensive, lightweight, and user-friendly. Additionally, the RGB camera-equipped drone is a better option for remote sensing, making it the right choice for many applications. The visible information is used to detect the symptoms of crop diseases visually. The camera operation is limited in the daytime due to smoke, fog, and rain. However, RGB cameras provide a lower performance in detecting plant diseases at early stages than thermal and spectral cameras due to being highly susceptible to weather and environmental conditions. RGB cameras operate in visible regions, which are used to detect the features of the plant growth and diseases.

16.2.2 Thermal Camera in Remote Sensing

The thermal cameras perform admirably in poor visibility conditions due to precipitation and smog. It also has advantages in providing observations both during the day and at night time. Most studies utilized thermal cameras for monitoring volcano eruptions, forest fires, human detection, and rescue in disaster places. Additionally, plant stress and illnesses can be identified by using thermal cameras, which function in the infrared spectrum. Also, it provides a heat-map assessing the radiation released by the plants. Thermal cameras are very sensitive to the infrared spectra that give highly valuable features of the crop. These features are not obtained using RGB spectral cameras. But the thermal camera is costlier than the RGB camera. A few points to keep in mind while mounting the thermal cameras with UAV: (i) flying height; (ii) viewing angle; (iii) spectral and temporal resolution; (iv) weather condition; and (v) plant status.

16.2.3 Hyper, Multi-Spectral Camera in Remote Sensing

Hyper/multispectral remote sensing expands beyond aerial platforms to include UAVs and drones, mostly useful for precision agriculture, disaster relief, and search and rescue operations. Based on spectral resolution, hyper-spectral and multispectral images are categorized. The electromagnetic spectrum, which contains the visible and near-infrared regions, allows us to compute various vegetation indices. This characteristic is well suited to agriculture with the ability to find plant growth and illnesses. The spectral cameras offer a range of illumination conditions, guaranteeing an accurate determination of the stress level. In comparison to multi-spectral images, hyper-spectral images offer more information on the vegetation indices, making it easier to precisely determine the stress state of the plant.

Table 16.1 Performance measurement metrics for various types of camera inputs.

References	Performance metrics	Formula	Purpose
[10]	Reflectance value of red band	$\frac{R_R}{R_R + R_G + R_B}$	Measure a reflectance value
[10]	Reflectance value of green band	$\frac{R_G}{R_R + R_G + R_B}$	Measure a reflectance value
[10]	Reflectance value of blue band	$\frac{R_B}{R_R + R_G + R_B}$	Measure a reflectance value
[11]	normalized difference vegetation index (NDVI)	$\frac{R_{NIR} - R_{RED}}{R_{NIR} + R_{RED}}$	R_{NIR} is the reflectance of near infrared spectral band, R_{RED} is the reflectance of red spectral band, R_{RE} is the reflectance of red edge spectral band
[12]	Green normalized difference vegetation index	$\frac{R_{NIR} - R_{GREEN}}{R_{NIR} + R_{GREEN}}$	Health monitoring *via* changes in vegetation
[13]	ratio vegetation index	$\frac{R_{NIR}}{R_{RED}}$	Indicate the healthy vegetation or not
[13]	difference vegetation index	$R_{NIR} - R_{RED}$	Find the density of vegetation
[11]	normalized difference red edge	$\frac{R_{NIR} - R_{RE}}{R_{NIR} + R_{RE}}$	vegetation indices based on the red edge spectral

(Continued)

Table 16.1 Performance measurement metrics for various types of camera inputs. (*Continued*)

References	Performance metrics	Formula	Purpose
[11]	renormalized difference vegetation index	$\frac{R_{NIR}-R_{RE}}{\sqrt{R_{NIR}+R_{RE}}}$	It indicates a healthy vegetation of the plant
[10]	Vegetation fraction	$\frac{N_S}{N_{tp}}$	It can measure the canopy coverage. N_s is the number of pixel segmented, N_{tp} is total number of pixel
[11]	Normalized relative canopy temperature	$\frac{T_{canopy}-T_{min}}{T_{max}+T_{min}}$	T_{canopy} is the mean temperature of canopy, T_{min} and T_{max} is the minimum and maximum temperature of the thermal map.
[12]	Crop water stress index	$\frac{T_{canopy}-T_{max}}{T_{min}-T_{max}}$	Measure a water status
[12]	Stomatal conductance index	$\frac{T_{canopy}-T_{max}}{T_{min}-T_{canopy}}$ Or $\frac{T_{min}-T_{canopy}}{T_{canopy}-T_{max}}$	Measure a water status
[10]	canopy variation	2G-R-B	Measure canopy variation in green biomass
[14]	Length ratio	$\frac{\sum_{j=1}^{m} L_r(j)}{\sum_{j=1}^{n} L_a(j)}$	Effective detection of object through length ratio. $L_r(j)$ and $L_a(j)$ is the received length and actual length of the power line, n, m are the received and actual number of power line.

16.2.4 LiDAR in Remote Sensing

Optical remote sensing is capable of obtaining reflectance spectral parameters, including biomass, vegetation index, and textural aspects. One significant property of LiDAR is its great penetration, which allows it to provide information about vertical structure, including topographic parameters, biomass, and height. LiDAR is also primarily used in obstacle identification due to its high degree of precision in direction and distance readings. However, the low-resolution of space-borne data and the high expense of processing data in the air restrict the usefulness of LiDAR deployment.

In comparison to other sensor technology, which offers a high spatial and temporal solution in many aspects, aerial photographs are a viable option for getting the essential parameters. A crucial element of precision agriculture farm management in agriculture is the monitoring of biomass. Additionally, it offers expanded understanding of weed condition, soil fertility, pest, water stress, and yield projection [15]. Performance measurement metrics for various types of camera inputs are given in Table 16.1.

16.3 Machine Learning-Based Object Identification

A set of methods, resources, and techniques known as machine learning enables a machine to learn without explicit human direction, as shown in Figure 16.3. In order to improve predictions, it can also be used to analyze and find the data's hidden patterns. The ML technique is classified into (i) supervised learning, (ii) unsupervised learning, and (iii) reinforcement learning.

Drones/UAVs are deployed to collect the data and image from the target area by installing a suitable sensor and camera. The drone is capable of flying at low altitude, so that it can take high quality images from the field. Also, it can cover a huge area easily and update the images at regular intervals. Data pre-processing is a procedure that are used to take out the undesirable signal from the pictures. The images are also cleaned and organized before they can be utilized for image categorization. The augmentation and normalization of the data are integrated to complete the data pre-processing. Image segmentation divides the images into smaller parts according to the similarities and differences between the image sections. Additionally, it uses edge detection and thresholding to carry out image segmentation. Feature extraction is the key element of the objective analysis that is extracted from the photos using texture analysis and shapes. In image classification, a trained model is used to classify the photos.

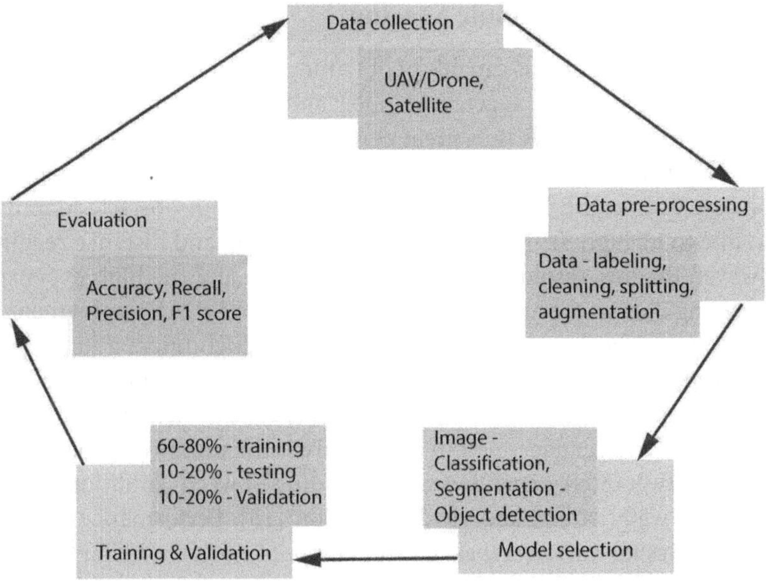

Figure 16.3 Machine learning with image-based performance analysis.

For instance, a real-time model ought to determine whether a forest fire is there or not based on the given image. The flow diagram of the ML-based classification and prediction is shown in Figure 16.4.

In data processing, supervised learning is an important tool for finding the connection between the input and output layers. The supervised learning utilizes a set of desired input and output information to create a training data set and produce the outcome towards the target. The supervised learning uses iterative optimization techniques to acquire the desired objective function. Here, data are mentioned with labels in order to analyses and classify the information and use it in a custom model effectively. The supervised learning produces a model to represent the dependency between the input data and the prediction of the objective function. Unsupervised learning considers only available input data and learns the training data samples. Here, there is no desired output label. The output is associated with distinctive sets that are obtained from its own input data structure. And it is highly preferable when data sets are produced without a label in the target value. Usually, unsupervised learning provides data sets in easier representation in the form of lower-dimensional, sparse, and independent data representation. Reinforcement learning uses an agent which performs smartly to learn from the environment. It generates a

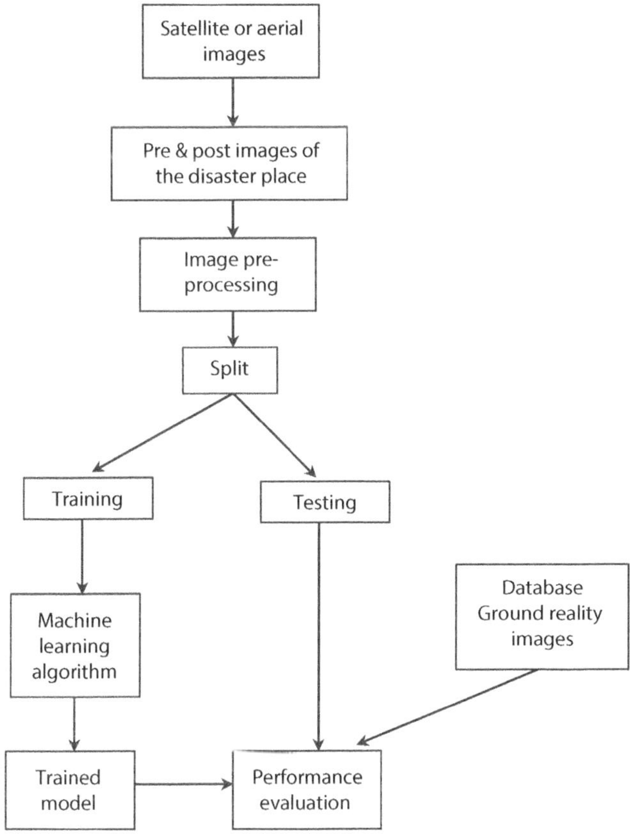

Figure 16.4 Flow diagram of ML-based classification and prediction.

perfect behavior model by interacting directly with the environment. This learning is achieved through "actions" and "rewards" policies. A summary of remote sensing techniques in precision agriculture [10, 14, 23–31] and disaster management [32–39] is depicted in Tables 16.2 and 16.3. Performance evaluation metrics in machine learning models are enumerated in Table 16.4.

16.3.1 Support Vector Machine

Support Vector Machine (SVM) is introduced to achieve binary classification. Later, it is extended to the non-parametric model, which means that it does not utilize the fixed model while designing the model. And also, it effectively performs the regression and multi-task classification.

Table 16.2 Summary of remote sensing techniques in precision agriculture.

References	Image	Parameters	Methodology	Plant	Application	Vehicle
[23]	RGB	Green canopy cover, crop height, canopy volume	RGB color ratio based canopy volume prediction.	Onion	Biomass monitoring	UAV
[24]	RGB	Tree crown, trees discolored by PWD, and forest gaps	Multi-scale segmentation and object-oriented classification Method	Pine	Pine Wilt disease	UAV
[25]	RGB	Dice coefficient	Genetic algorithm	Sugarcane	Sugarcane plant line detection	UAV
[26]	RGB	Vegetation index, canopy height, and canopy reflectance	k-fold cross validation procedure	Rice	Measuring of Leaf Area Index	UAV
[27]	Color infrared image	NDVI, ground normalized difference vegetation index	Empirical line calibrations	Potato	Monitoring nitrogen status	UAV
[10]	RGB	Canopy coverage and plant height	Multiple linear regression models	onion and garlic	Growth monitoring	UAV

(Continued)

Table 16.2 Summary of remote sensing techniques in precision agriculture. (*Continued*)

References	Image	Parameters	Methodology	Plant	Application	Vehicle
[28]	RGB	Vegetation index, textual information, plant height, canopy coverage	Ensemble learning models	Bean	above-ground biomass and bean yield monitoring	UAV
[29]	RGB	Leaf area index, plant height	Linear model	Cotton	Monitoring of cotton canopy parameters	UAV
[30]	RGB	Normalized difference index , green index , and green leaf index	Linear model	Wheat	Monitoring of wheat foliage disease severity	UAV
[31]	RGB and multispectral images	Plant height, NDVI, Normalized difference red-edge index , plant density, and plant volume	Deep learning	Corn	Monitoring of corn growth	UAV, GIS
[14]	LiDAR with camera	Length ratio, over segmentation rate	Position and angle measurement using LiDAR and fusion filters	Power line detection	UAV flying height control	UAV

Table 16.3 Summary of remote sensing techniques in disaster management.

References	Image/sensor	Parameters	Methodology	Purpose	Application	Vehicle
[32]	Lidar, image sensor	Intrinsic camera parameters, computed ROI of the detected target	Deep reinforcement learning	Search and rescue	Indoor application	Autonomous Aerial robot
[33]	Thermal infrared remote sensing	Ground temperature	Linear regression, radiation conduction and the Plank function	Coal fire	Coal mine	UAV
[34]	electrical resistivity tomography imaging	Morphometry and areal changes, surface displacement, movement rate and direction, spatial variability	getting displacement and internal structure of landslides from Digital Elevation Model (DEM), orthophotos, ERT images	landslide	Natural disaster	UAV
[35]	hyperspectral remote sensing image	Geographical structure, geomorphic features, surface environments	Machine learning model	landslide	Natural disaster	UAV

(Continued)

Table 16.3 Summary of remote sensing techniques in disaster management. *(Continued)*

References	Image/sensor	Parameters	Methodology	Purpose	Application	Vehicle
[36]	Postseismic aerial images	Appearance and motion features extracted from successive images	online clustering algorithm	Collapsed building detection	Natural disaster	UAV
[37]	UAV-based aerial photos	High-resolution (DEMs), orthophotos, and density point clouds	Multiscale model-to-model cloud comparison algorithm	landslide	Natural disaster	UAV
[38]	optical RGB, thermal and spectral sensor	Orthomosaics elaboration, DEMs, 3D and probabilistic models, chemical composition of surface seawater.	DEMs with probabilistic simulation	Volcanic eruption	Natural disaster	UAV
[39]	Satellite image (Google earth)	In-house aerial video dataset	Convolutional Neural Networks	Detecting the ground objects	Natural disaster	Satellite

Table 16.4 Performance evaluation metrics in machine learning model.

References	Parameter	Formula	Definition
[40]	Mean absolute error	$\frac{1}{N}\sum_{i=1}^{N} x_i - \tilde{x}_i$	The mean of the differences between the desired (x_i) and expected values ($\tilde{x}_i$).
[40]	Mean square error	$\frac{1}{N}\sum_{i=1}^{N} (x_i - \tilde{x}_i)^2$	The average of the squared difference between the actual and expected value
[11]	Root mean square error	$\sqrt{\frac{1}{N}\sum_{i=1}^{N} (x_i - \tilde{x}_i)^2}$	The square root of the average of the squared difference between the actual and expected value.
[13]	Mean absolute error (%)	$\frac{\sum_{i=1}^{N} \lvert\frac{x_i - \tilde{x}_i}{\tilde{x}_i}\rvert}{N} \times 100$	Measure the prediction accuracy
[11]	Relative root mean square error (%)	$\frac{\sqrt{\frac{1}{N}\sum_{i=1}^{N} (x_i - \tilde{x}_i)^2}}{x} \times 100$	x is the mean value of n samples
[11]	R^2	$1 - \sum_{i}^{n} (x_i - \tilde{x}_i)^2 / \sum_{i}^{n} (x_i - x)^2$	Measure the variation in the dependent variable

(Continued)

Table 16.4 Performance evaluation metrics in machine learning model. (*Continued*)

References	Parameter	Formula	Definition
[39]	Sensitivity or Recall	$\frac{TP}{FN+TP}$	It calculates how frequently a machine learning algorithm finds true positives from an actual positive sample
[42]	Specificity	$\frac{TN}{FP+TN}$	It calculates the percentage of true negatives that the model identifies accurately
[22]	ROC	Two dimensions of event and non-event phenomena	It is used for validation and assessment
[41]	Accuracy	$\frac{TN+TP}{FP+FN+TN+TP}$	It calculates how frequently the ML algorithm predicts an outcome correctly
[39]	Precision or Positive Predictive value	$\frac{TP}{FP+TP}$	It calculates how frequently the ML algorithm predicts the positive class correctly
[39]	F1 score	$2\times\frac{Precision\times Recall}{Precision+Recall}$	It computes how many times the ML performed a correct prediction across the complete dataset

Linear regression is one of the most utilized linear predictive models in the machine learning and statistical community. The well-known logistic regression algorithm adapts to the SVM to perform the classification tasks. But SVM takes a long time to execute large datasets. An Apache Spark-based parallel SVM was developed to forecast forest fires. This method yields 100% accuracy, compared to 94% for linear SVM [15].

16.3.2 Decision Tree

A Decision Tree (DT) is a supervised learning method that is applied for classification depending on if-then rules. The decision tree uses three types of nodes (i) leaf nodes that lie at the bottom of the tree which have one incoming edge and no outgoing edges, (ii) intermediate nodes that have one or more outgoing edges and one incoming edge, (iii) root node that lies at the top of the tree and has no incoming edge. The root and intermediate nodes are called decision nodes that consist of various attribute conditions with different characteristics (i.e., decision rules) to classify the data. ID3 utilizes information gain value to determine whether it goes to the decision node or not. From the root node, the appropriate test condition is applied to the test data. The outcome of the appropriate branch enters into a new intermediate node in the tree until it reaches the leaf node, for which a new test condition is applied to the test data. The label connected with the leaf nodes is allocated to the record and produces the final outcome. The decision tree makes predictions about the target by creating a training model using the rules extracted from the training data. The main advantages of the DT are (i) making transparency, (ii) enabling comprehensive analysis, and (iii) reducing ambiguity in decision making.

16.3.3 Random Forest

The Random Forest (RF)-based classifier is comparable to that of DT, in addition to ensemble learning that consists of a collection of trees in the forest, and each tree provides a classification. This classifier generates forests that contain a group of decision trees, in which each tree predicts the data class based on the data feature set. Once the decision trees predict the data class, find the majority for the correct data classification based on the voting method. Random forest performs two major operations (i) creation of a classifier (ii) the prediction of outcomes. The main advantages of random forest are (i) efficiently working in huge size datasets (ii) efficiently handling heterogeneous data, (ii) accurately predicting the missing values. RF trees perform better when dealing with highly linked data and the curse of dimensionality. Since a randomly generated training sample's subsets and isolating features will produce a huge number of decision trees, ensuring the required training samples are available on the decision tree. In [16], a greedy approach was utilized to obtain feature ranking in recursive feature elimination with random forest technology-based landslide detection, yielding an accuracy of 87.2%. The random forest technique utilized features including height, slope angle, vegetation index difference, red over green difference, and brightness difference to predict landslides in [17]. And it can achieve precision above 90%.

16.3.4 K-Nearest Neighbor

K-Nearest Neighbor (KNN) algorithm predicts the numerical targets through similarity measures. The distance functions including Euclidean distance, Taxicab distance, and Minkowski distance, are used to classify the data from k nearest neighbors. KNN is an instance-based learning system that uses the distance between training and testing samples to accomplish data regression and classification. This algorithm reduces the dimensionality and finds the missing value in the feature space. In [18], the landslide prediction was carried out based on the geological units, slope angle and shape, distance from the river, and tectonic plates. Each parameter was examined using KNN, and an accuracy of 80% was obtained.

16.3.5 Naive-Bayes Classifier

This supervised machine learning technique is based on a statistical learning approach and uses statistical analysis, such as the t-test, binomial test, Kruskal Wallis test, chi-square test, and others, to learn conditional independence and determine the links between the datasets. The sensor module (TCS3200) measures the soil's nitrogen content with an accuracy of 87.5% and transmits the data to the cloud server. The geographic information system is used to map the region and train the dataset using the Naïve-Bayes theorem [19].

16.3.6 Convolutional Neural Networks (CNN)

CNN is a fully interconnected network structure that consists of various layers, including the convolutional, pooling, spatial feature, flatten, and concatenate layer. These layers carry out the convolution processes to extract the features from the data set. The convolution layer detects the feature maps with various kernel sizes and then it is shared with the pooling layer.

Various pooling techniques, including max pooling, stochastic, average, spatial pooling, and so on have been utilized in the CNN. The activation unit values are calculated at each layer to identify the required data pattern from the input data set. A CNN-based forest fire detection algorithm is presented in [20], which provides an accuracy of 80.4% for original images and 90.7% for segmented images. With the use of real-world images, a transfer learning-based convolutional NN model was developed to locate forest fires with an accuracy of 95.4% [21].

16.3.7 K-Means Clustering

The K-means algorithm uses a predefined size of clusters and their function depends on Euclidean distance. It initially considers k number of random location points and the remaining location points are connected with nearest center points and form the clusters. The centroid of the clusters is changed until it gets the same centroid value of all clusters. In this method, each object goes to the clusters with the closest mean and reduces the variance of the intra-cluster. In [22], each cluster was statistically examined using logistic regression after forming a cluster using the K-means clustering technique. Factors such as slope gradient, angle, susceptibility map, and distance to rivers and cities were identified as causative variables to predict the occurrence of landslides. Additionally, it provides 91.76% prediction accuracy.

16.4 Challenges and Limitations of Technology Deployment in Disaster Management

Urban monitoring systems are not appropriate for single color-based object identification and classification because the color appearance of undesirable and normal incidents varies and is more complex than in rural areas. Instead of concentrating on single-color results, the researchers should focus on multi-color similarity findings in order to obtain high accuracy [36].

The principal uses of UAVs include radiation detection, building collapses, landslides, and cracks in bridges. However, the image resolution, image blurriness, and ground vegetation all have an impact on the classification process. In addition, the visual interpretation of the early landslide is quite intricate since it is determined by the deformation of the earth.

For satellite images, it might be difficult to identify and detect microscopic objects like cracks while complicated environmental conditions and dense vegetation [43].

The model needs sufficient photos for prediction and classification because the UAV cannot cover the entire target place [31]. So, multiple UAVs are deployed to avoid the geometric noises that occur due to the low altitude and overlapped images. Also, the UAV movement causes low quality photos that are blurry and have a lot of noise. Multiple recognition techniques are integrated into UAV functions to attain real-world efficiency [44].

The fixed-wing UAV is made up of more sophisticated and specialized equipment that is expensive and requires technical expertise. Furthermore, its functioning is restricted to very low altitudes due to their large size. The multi-rotor drones are not suitable for large-scale application deployment owing to their constrained payload, range, and flight hours [45]. The image quality is severely compromised by the poor temporal and spatial resolutions. Additionally, gathering data in a variety of environmental conditions is highly challenging [46].

The model cannot be trained directly on the high-quality remote sensing photos due to their limited resources. So, the images should be pre-processed before being given to the classification model. In addition, poor illumination conditions make it difficult to categorize the objects since multiple surfaces can appear to be the same, which lowers the model's accuracy rate [47]. The visual appearance and spatial placement of the objects in the low-altitude aerial photos vary significantly from those in the standard collection [48]. Due to their high cost and intricate data processing methods, hyper- and multi-spectral cameras are not widely used [29].

The challenges associated with using low-altitude aerial photos for object recognition and classification in disaster areas are as follows: (i) the object's background appearance varies rapidly and is not static; (ii) the performance is unpredictable due to factors such as camera view angle, height, movement, and brightness, etc. [36]. Automatic object classification and identification in disaster management has addressed various issues and constraints associated with conventional machine learning algorithms, especially in state vector machines, decision trees, Naive-Bayes classifiers, random forests, and K-nearest neighbors. Also, these algorithms are limited in their ability to locate small items in crowded areas [46]. CNNs are used in many of the image processing techniques for classification; however, their large dataset requirements and high power consumption result in low accuracy [49].

16.5 Case Studies

16.5.1 Cyclone Triggered Landslide – Ischia (2022)

A mudslide on Ischia Island, southern Italy, 2022 occurred due to the cyclone and illegal building construction. More than seven people were killed and many of them lost their belongings when mud submerged the village. Several vehicles on the shoreline were reportedly submerged and thrown into the sea by the cyclone. Several vehicles were submerged on the shoreline and apparently pushed into the sea by the cyclone. A LiDAR technique was deployed in the UAV/Drone to obtain the three dimensional surface characteristics data. This information can be used to predict the damage and economic loss.

16.5.2 Forest Fire - Spain (2022), Argentina (2021) and Portuguese (2023)

More than 250,000 hectares were burnt and many wild animals lost their lives due to wildfire in Spain in 2022. The Spanish government has formed the Copernicus Emergency Management Service Rapid Mapping team to monitor the forest fire using satellite images (Sentinel - 2) [50].

The moderate resolution imaging spectro-radiometer on the NASA Terra satellite is used to take visible pictures in order to identify smoke and fire in the forest [51]. The locations of current wildfires are indicated by red dots, which are seen with the Visible Infrared Imaging Radiometer Suite. More than 535,000 hectares of wild-area trees were destroyed in forest fires between 2001 and 2021. Wild regions have been devastated by the forest fire that happened in Portugal, and wild animals no longer have anywhere to live. The wildfire, which has burned 8700 hectares, has resulted in about 35 injuries and 1500 evacuations from over 200 settlements, according to the European Forest Fire Information System. The aerial images are used to estimate the biomass damage and loss of the ecosystem services.

16.5.3 Volcano Eruption – Andaman Territory (2022)

In 2022, an eruption took place on Barren Island in the Andaman Islands, India. Ash, lava, and thermal anomalies are regularly released by the eruption. Satellite imagery, particularly infrared photos from Sentinel-2, is used to track thermal activity [52]. The satellite images were utilized to assess the volcano's eruption's intensity and send a warning to the people in the vicinity. The summary of the case studies in remote sensing is given in Table 16.5.

Table 16.5 Summary of case studies in remote sensing techniques.

S. no.	Outcome	Image	Disaster	Year	Place
1	Obtain three dimensional surface characteristics information using LiDAR	Aerial Image	cyclone-triggered landslide	2022	Ischia
2	Estimated the biomass damage and loss of the ecosystem services	Satellite and aerial images	Forest fire	2022	Spain
3	Estimate the impact on vegetation and infrastructure	Aerial image	Forest fire	2023	Portuguese
4.	Identify the pathways and extent of volcanic lava and ash cloud	Satellite image	Volcano eruption	2022	Andaman Territory
5.	Finding of fire spots and estimation of burnt area	Satellite image	Forest fire	2021	Argentina

16.6 Conclusion

In real time, many applications, including the detection of forest fires, volcano eruptions, landslides, floods, building collapses, and smart agriculture, use machine learning-based computer vision algorithms. The disaster management service performs three key functions, including mapping, monitoring, and early warning. Remote sensing techniques provide the data needed to produce accurate maps for all disaster management levels. Before or after a disaster, it can be used to assess hazards. Firstly, the fixed and multi-rotor UAVs and various types of remote sensing cameras and technologies were discussed. Secondly, the different types of machine learning algorithms that are utilized in the existing works to detect and monitor the natural disasters and humanitarian crises are summarized. Finally, the challenges and limitations of UAV deployment in disaster management are presented to fulfill the need for technological requirements.

References

1. https://www.who.int/health-topics/.
2. https://www.wri.org/insights/global-trends-forest-fires.
3. Linardos, V., Drakaki, M., Tzionas, P., Karnavas, Y.L., Machine Learning in Disaster Management: Recent Developments in Methods and Applications. *Mach. Learn. Knowl. Extr.*, 4, 446–473, 2022.
4. Srivastava, A. and Prakash, J., Techniques, Answers, and RealWorld UAV Implementations for Precision Farming. *Wirel. Pers. Commun.*, 131, 2715–2746, 2023.
5. Giordan, D., Adams, M.S., Aicardi, I., Alicandro, M., Allasia, P., Baldo, M., Berardinis, P.D., Dominici, D., Godone, D., Hobbs, P., Lechner, V., Niedzielski, T., Piras, M., Rotilio, M., Salvini, R., Segor, V., Sotier, B., Troilo, F., The use of unmanned aerial vehicles (UAVs) for engineering geology applications. *Bull. Eng. Geol. Environ.*, 79, 3437–3481, 2020.
6. Sinnemann, J., Boshoff, M., Dyrska, R., Leonow, S., Mönnigmann, M., Kuhlenkötter, B., Systematic literature review of applications and usage potentials for the combination of unmanned aerial vehicles and mobile robot manipulators in production systems. *Prod. Eng.*, 16, 579–596, 2022.
7. Han, D., Kim, J., Min, C., Jo, S., Kim, J., Lee, D., Development of Unmanned Aerial Vehicle (UAV) System with Waypoint Tracking and Vision-based Reconnaissance, International Journal of Control. *Autom. Syst.*, 8, 1091–1099, 2010.
8. Han, W., Zhang, X., Wang, Y., Wang, L., Huang, X., Li, J., Wang, S., Chen, W., Li, X., Feng, R., Fen, R., Zhang, X., Wang, Y., A survey of machine learning and deep learning in remote sensing of geological environment: Challenges, advances, and opportunities. *ISPRS J. Photogramm. Remote Sens.*, 202, 87–113, 2023.
9. Helgesen, H.H., Leira, F.S., Fossen, T.I., Johansen, T.A., Tracking of Ocean Surface Objects from Unmanned Aerial Vehicles with a Pan/Tilt Unit using a Thermal Camera. *J. Intell. Rob. Syst.*, 91, 775–793, 2018.
10. Kim, D., Jeong, S.J., Lee, W.S., Yun, H., Chung, Y.S., Kwon, Y., Kim, H., Growth monitoring of fieldgrown onion and garlic by CIE L*a*b* color space and regionbased crop segmentation of UAV RGB images. *Precis. Agric.*, 24, 1982–2001, 2023.

11. Li, J., Schachtman, D.P., Creech, C.F., Wang, L., Ge, Y., Shi, Y., Evaluation of UAV-derived multimodal remote sensing data for biomass prediction and drought tolerance assessment in bioenergy sorghum. *Crop J.*, 10, 1363–1375, 2022.
12. Baluja, J., Diago, M.P., Balda, P., Zorer, R., Meggio, F., Morales, F., Tardaguila, J., Assessment of vineyard water status variability by thermal and multispectral imagery using an unmanned aerial vehicle (UAV). *Irrig. Sci.*, 30, 511–522, 2012.
13. Yao, X., Yi, Q., Wang, F., Xu, T., Zheng, J., Shi, Z., Estimating rice flower intensity using flower spectral information from unmanned aerial vehicle (UAV) hyperspectral images. *Int. J. Appl. Earth Obs. Geoinf.*, 122, 103415, 2023.
14. Ma, Z., Yao, W., Niu, Y., Lin, B., Liu, T., UAV low-altitude obstacle detection based on the fusion of LiDAR and camera. *Auton. Intell. Syst.*, 1, 1–12, 2021.
15. Singh, K.R., Neethu, K.P., Madhurekaa, K., Harita, A., Mohan, P., Parallel SVM model for forest fire prediction. *Soft Comput. Lett.*, 3, 100014, 2021.
16. Zhou, X., Wen, H., Zhang, Y., Xu, J., Zhang, W., Landslide susceptibility mapping using hybrid random forest with GeoDetector and RFE for factor optimization. *Geosci. Front.*, 12, 101211, 2021.
17. Tehrania, F.S., Santinellia, G., Herrera, M.H., Multi-Regional landslide detection using combined unsupervised and supervised machine learning. *Geomatics Nat. Hazards Risk*, 12, 1015–1038, 2021.
18. Paraskevas, T., Constantinos, L., Dimitrios, R., Ioanna, I., Landslide susceptibility assessments using the k-Nearest Neighbor algorithm and expert knowledge. Case study of the basin of Selinounda river, Achaia County, Greece, in: *SafeChania 2015: The knowledge triangle in the Civil Protection Service Center of Mediterranean Architecture*, Chania, Crete, Greece, pp. 10–14, 2015.
19. Yudhana, A., Sulistyo, D., Mufandi, I., GIS-based and Naïve Bayes for nitrogen soil mapping in Lendah, Indonesia. *Sens. Bio-Sens. Res.*, 33, 100435, 2021.
20. Yuanbin, W., Langfei, D., Jieying, R., Forest fire image recognition based on convolutional neural network. *J. Algorithms Comput. Technol.*, 13, 1–11, 2011.
21. Majid, S., Alenezi, F., Masood, S., Ahmad, M., Gunduz, E.S., Polat, K., Attention based CNN model for fire detection and localization in real-world images. *Expert Syst. Appl.*, 189, 116114, 2022.
22. Wang, Q., Wang, Y., Niu, R., Peng, L., Integration of Information Theory, K-Means Cluster Analysis and the Logistic Regression Model for Landslide Susceptibility Mapping in the Three Gorges Area, China. *Remote Sens.*, 9, 938, 2017.
23. Ballesteros, R., Ortega, J.F., Hernandez, Moreno, M.A., Onion biomass monitoring using UAVbased RGB imaging. *Precis. Agric.*, 19, 840–857, 2018.

24. Sun, Z., Wang, Y., Pan, L., Xie, Y., Zhang, B., Liang, R., Sun, Y., Pine wilt disease detection in highresolution UAV images using objectoriented classification. *J. For. Res.*, 33, 1377–1389, 2022.
25. Silva, R.R., Escarpinati, M.C., Backes, A.R., Sugarcane crop line detection from UAV images using genetic algorithm and Radon transform. *Signal Image Video Process.*, 15, 1723–1730, 2021.
26. Gong, Y., Yang, K., Lin, Z., Fang, S., Wu, X., Zhu, R., Peng, Y., Remote estimation of leaf area index (LAI) with unmanned aerial vehicle (UAV) imaging for different rice cultivars throughout the entire growing season. *Plant Methods*, 17, 88, 2021.
27. Jr., E.R.H., Horneck, D.A., Spinelli, C.B., Turner, R.W., Bruce, A.E., Gadler, D.J., Brungard, J.J., Hamm, P.B., Monitoring nitrogen status of potatoes using small unmanned aerial vehicles. *Precis. Agric.*, 19, 314–333, 2018.
28. Ji, Y., Liu, R., Xiao, Y., Cui, Y., Chen, Z., Zong, X., Yang, T., Faba bean aboveground biomass and bean yield estimation based on consumergrade unmanned aerial vehicle RGB images and ensemble learning. *Precis. Agric.*, 24, 1439–1460, 2023.
29. Wu, J., Wen, S., Lan, Y., Yin, X., Zhang, J., Ge, Y., Estimation of cotton canopy parameters based on unmanned aerial vehicle (UAV) oblique photography. *Plant Methods*, 18, 1–19, 2022.
30. Bhandari, M., Ibrahim, A.M.H., Xue, Q., Jung, J., Chang, A., Rudd, J.C., Maeda, M., Rajan, N., Neely, H., Landivar, J., Assessing winter wheat foliage disease severity using aerial imagery acquired from small Unmanned Aerial Vehicle (UAV). *Comput. Electron. Agric.*, 176, 105665, 2020.
31. Xiao, J., Suab, S.A., Chen, X., Singh, C.K., Singh, D., Aggarwal, A.K., Korom, A., Widyatmanti, W., Mollah, T.H., Minh, H.V.T., Khedher, K.M., Avtar, R., Enhancing assessment of corn growth performance using unmanned aerial vehicles (UAVs) and deep learning. *Measurement*, 214, 112764, 2023.
32. Sampedro, C., Ramos, A.R., Bavle, H., Carrio, A., Puente, P., Campoy, P., A Fully-Autonomous Aerial Robot for Search and Rescue Applications in Indoor Environments using Learning-Based Techniques. *J. Intell. Rob. Syst.*, 95, 601–627, 2019.
33. He, X., Yang, X., Luo, Z., Guan, T., Application of unmanned aerial vehicle (UAV) thermal infrared remote sensing to identify coal fires in the Huojitu coal mine in Shenmu city, China. *Sci. Rep.*, 10, 13895, 2020.
34. Samodra, G., Ramadhan, M.F., Sartohadi, J., Setiawan, M.A., Christanto, N., Sukmawijaya, A., Characterization of displacement and internal structure of landslides from multitemporal UAV and ERT imaging. *Landslides*, 17, 2455–2468, 2020.
35. Qi, J., Chen, H., Chen, F., Extraction of landslide features in UAV remote sensing images based on machine vision and image enhancement technology. *Neural Comput. Appl.*, 34, 12283–12297, 2022.

36. Hua, C., Qi, J., Shang, H., Hu, W., Han, J., Detection of collapsed buildings with the aerial images captured from UAV. *Sci. China Inf. Sci.*, 59, 032102, 2016.
37. Eker, R., Aydın, A., Hübl, J., Unmanned aerial vehicle (UAV)-based monitor ing of a landslide: Gallenzerkogel landslide (Ybbs-Lower Austria) case study. *Environ. Monit. Assess.*, 190, 28, 2018.
38. Román, A., Tovar-Sánchez, A., Roque-Atienza, D., Huertas, I.E., Caballero, I., Fraile-Nuez, E., Navarro, G., Unmanned aerial vehicles (UAVs) as a tool for hazard assessment: The 2021eruption of Cumbre Vieja volcano, La Palma Island (Spain). *Sci. Total Environ.*, 843, 157092, 2022.
39. Pia, Y., Nathb, N.D., Behzadan, A.H., Convolutional neural networks for object detection in aerial imagery for disaster response and recovery. *Adv. Eng. Inf.*, 43, 101009, 2020.
40. Wu, J., Wen, S., Lan, Y., Yin, X., Zhang, J., Ge, Y., Estimation of cotton canopy parameters based on unmanned aerial vehicle (UAV) oblique photography. *Plant Methods*, 18, 129, 2022.
41. Park, J., Cho, Y.K., Kim, S., Deep learning-based UAV image segmentation and inpainting for generating vehicle-free orthomosaic. *Int. J. Appl. Earth Obs. Geoinf.*, 115, 103111, 2022.
42. Gupta, N., Rawal, A., Narasimhan, V.L., Shiwani, S., Accuracy, Sensitivity and Specificity Measurement of Various Classification Techniques on Healthcare Data, IOSR. *J. Comput. Eng.*, 11, 70–73, 2013.
43. Xu-gang, L., Zou-jun, L., Hong-yan, Y., Ji-bo, L., Yan-jun, Z., Xiao-yu, L., Yan-ru, W.,Rapid identification of landslide, collapse and crack based on low-altitude remote sensing image of UAV. *J. Mountain Sci.*, 17, 2915–2928, 2020.
44. Gomez-Candon, D., Castro, A., II, Lopez-Granados, F., Assessing the accuracy of mosaics from unmanned aerial vehicle (UAV) imagery for precision agriculture purposes in wheat. *Precis. Agric.*, 15, 44–56, 2014.
45. Templin, T., Popielarczyk, D., Kosecki, R., Application of Low-Cost Fixed-Wing UAV for Inland Lakes Shoreline. *Pure Appl. Geophys.*, 175, 3263–3283, 2018.
46. Bouguettaya, A., Zarzour, H., Kechida, A., Taberkit, A.M., Deep learning techniques to classify agricultural crops through UAV imagery: a review. *Neural Comput. Appl.*, 34, 9511–9536, 2022.
47. Behera, T.K., Bakshi, S., Sa, P.K., A Lightweight Deep Learning Architecture for Vegetation Segmentation using UAV-captured Aerial Images. *Sustain. Comput. Inf. Syst.*, 37, 100841, 2023.
48. Mittal, P., Sharma, A., Singh, R., Deep learning based high performance classification architecture for lowaltitude aerial images. *Multimed. Tools Appl.*, 83, 16849–16868, 2024.

49. Wang, H., Yao, Z., Li, T., Ying, Z., Wu, X., Hao, S., Liu, M., Wang, Z., Gu, T., Enhanced open biomass burning detection: The BranTNet approach using UAV aerial imagery and deep learning for environmental protection and health preservation. *Ecol. Indic.*, 154, 110788, 2023.
50. https://www.copernicus.eu/en/media/image-day-gallery/spain-plagued-wildfires-summer-2022.
51. https://earthobservatory.nasa.gov/images/150685/fires-in-northwest-argentina.
52. https://volcano.si.edu/showreport.cfm?doi=10.5479/si.GVP.BGVN202206-260010#:~:text=The%20remote%20island%20of%20Barren,(BGVN%2046%3A12).

Index

Also of Interest

Check out these other related titles from Scrivener Publishing

Drone Technology: Future Trends and Practical Applications, Edited by Sachi Nandan Mohanty, J.V.R. Ravindra, G. Surya Narayana, Chinmaya Ranjan Pattnaik and Y. Mohamed Sirajudeen, ISBN: 9781394166653. This book provides a holistic and valuable insight into the revolutionary world of unmanned aerial vehicles (UAV).

Unmanned Aircraft Systems, Edited by Sachin Kumar Gupta, Manoj Kumar, Anand Nayyar, and Shubham Mahajan, ISBN: 9781394230617. This book is an essential resource for anyone looking to understand the cutting-edge applications and evolving technologies of Unmanned Aerial Systems, showcasing how they enhance safety and efficiency in monitoring, emergency response, and smart city development.

Advances in Aerial Sensing and Imaging, Edited by Sandeep Kumar, Nageswara Rao Moparthi, Abhishek Bhola, Ravinder Kaur, A. Senthil, K.M.V.V. Prasad, ISBN: 9781394174690. This groundbreaking book is a comprehensive guide to the technology found in the complex field of aerial sensing and imaging, and the real-world challenges that stem from its growing significance and demand.

Unmanned Aerial Vehicles for Internet of Things: Concepts, Techniques, and Applications, Edited by Vandana Mohindru, Yashwant Singh, Ravindara Bhatt, and Anuj Kumar Gupta, ISBN: 9781119768821. This comprehensive book deeply discusses the theoretical and technical issues of unmanned aerial vehicles for deployment by industries and civil authorities in Internet of Things (IoT) systems.

Autonomous Vehicles Planning and Control, By Yong Bai & Liang Zhao, ISBN: 9781394355044. Master the future of marine exploration and technology with Autonomous Vehicles Planning and Control, which provides a comprehensive, interdisciplinary guide to the principles, control, and real-world applications of autonomous marine vehicles.

Autonomous Flying Ad-Hoc Networks, Edited by Taskeen Zaidi, Adarsh Kumar, Saurav Mallik, and Keshav Kaushik, ISBN: 9781394231058. Learn to leverage the power of Autonomous Flying Ad-Hoc Networks (FANETs) for everything from urban surveillance to disaster relief, and stay ahead in the rapidly evolving world of drone technology and AI.

AI for Autonomous Vehicles: The Future of Driverless Technology, Edited by Sathiyaraj Rajendran, Munish Sabharwal, Yu-Chen Hu, Rajesh Kumar Dhanaraj, and Balamurugan Balusamy, ISBN: 9781119847465. This title presents the research prospects and security challenges present in the exciting new field of autonomous vehicles by comprehensively introducing advanced concepts of artificial intelligence (AI) in driverless technology.

AUTONOMOUS VEHICLES VOLUME 1: Using Machine Intelligence, Edited by Romil Rawat, A. Mary Sowjanya, Syed Imran Patel, Varshali Jaiswal, Imran Khan, and Allam Balaram, ISBN: 9781119871958. Addressing the current challenges, approaches and applications relating to autonomous vehicles, this groundbreaking new volume presents the research and techniques in this growing area, using Internet of Things, Machine Learning, Deep Learning, and Artificial Intelligence.

AUTONOMOUS VEHICLES VOLUME 2: Smart Vehicles for Communication, Edited by Romil Rawat, Purvee Bhardwaj, Upinder Kaur, Shrikant Telang, Mukesh Chouhan, and K. Sakthidasan Sankaran, ISBN: 9781394152254. The companion to *Autonomous Vehicles Volume 1: Using Machine Intelligence,* this second volume in the two-volume set covers intelligent techniques utilized for designing, controlling and managing vehicular systems based on advanced algorithms of computing like machine learning, artificial Intelligence, data analytics, and Internet of Things with prediction approaches to avoid accidental damages, security threats, and theft.

ROBOTIC PROCESS AUTOMATION, Edited by Romil Rawat, Rajesh Kumar Chakrawarti, Sanjaya Kumar Sarangi, Rahul Choudhary, Anand Singh Gadwal, and Vivek Bhardwaj, ISBN: 9781394166183. Presenting the latest technologies and practices in this ever-changing field, this groundbreaking new volume covers the theoretical challenges and practical solutions for using robotics across a variety of industries, encompassing many disciplines, including mathematics, computer science, electrical engineering, information technology, mechatronics, electronics, bioengineering, and command and software engineering.

FACTORIES OF THE FUTURE: Technological Advances in the Manufacturing Industry, Edited by Chandan Deep Singh and Harleen Kaur, ISBN: 9781119864943. The book provides insight into various technologies adopted and to be adopted in the future by industries and measures the impact of these technologies on manufacturing performance and their sustainability.

AI AND IOT-BASED INTELLIGENT AUTOMATION IN ROBOTICS, Edited by Ashutosh Kumar Dubey, Abhishek Kumar, S. Rakesh Kumar, N. Gayathri, Prasenjit Das, ISBN: 9781119711209. The 24 chapters in this book provide a deep overview of robotics and the application of AI and IoT in robotics across several industries such as healthcare, defense. education, etc.

Printed and bound by CPI Group (UK) Ltd, Croydon, CR0 4YY

17/09/2026

14964282-0004